Engineering materials volume 2

R. L. Timings

Copublished in the United States with
John Wiley & Sons, Inc., New York

Longman Scientific & Technical
Longman Group UK Limited,
Longman House, Burnt Mill, Harlow,
Essex CM20 2JE, England
and Associated Companies throughout the world.

First published 1991
Second impression 1993

British Library Cataloguing in Publication Data
Timings, R. L. (Roger Leslie) *1927–*
Engineering materials.
Vol. 2
1. Materials science
I. Title
620.11

Set in Compugraphic Times 10/11 pt

Printed in Malaysia by CL

Engineering materials volume 2

Contents

Preface

Engineering materials: volume 2 has been written to satisfy the requirements of the Business and Technician Education Council (BTEC) standard units U84/269, U84/270, U84/271, U84/273, U84/274, U84/275, U84/276, U86/324, U86/325, U86/326 and U86/327. These are dealt with to 'H' level standard and, therefore, this text is eminently suitable for students studying for an engineering qualification at the HNC/HND level.

This text follows on naturally from *Engineering materials: volume 1*, with which it is cross-referenced to avoid wasteful repetition. Of particular interest are Chapters 7 and 8 which deal with the electrical and electronic properties of materials, and the manufacture of semiconductor materials and devices.

The broad coverage of *Engineering materials: volumes 1 and 2* ensures that they not only satisfy the requirements of Technician Engineers, but also provide an excellent technical background for undergraduates studying for a degree in Mechanical Engineering, Electrical/Electronic Engineering, Manufacturing Engineering or Combined Engineering.

The author wishes to thank Mr Donald Conway B.Sc.(hons) of Lucas Industries plc., and Mr Gordon Telford of Wacker Chemicals Ltd., for their advice and assistance in preparing Chapter 8, Mr Tony May for reading the manuscript and providing many helpful comments and suggestions and, finally, the many companies, institutions and organisations who have also offered help and advice, some of whose names can be found in the list of acknowledgements.

R. L. Timings
18 February 1990

Acknowledgements

We are indebted to the following for permission to reproduce copyright material:

Butterworths for our Table 9.3 from *Adhesives Handbook* by J. Shields; Inco Europe Limited for our Tables 1.3, 1.4 and 1.5 from *18 per cent Nickel Maraging Steels © 1976*; The National Design Council for our Table 9.4 from *Adhesive Bonding* by J. Shields; and Wacker Chemitronic, a division of Wacker Chemie GMBH, Munich for our Figs 8.1, 8.2, 8.3, 8.4, 8.5, 8.6, 8.8 and 8.10, and for our Tables 8.1, 8.2 and 8.3.

We are also grateful to Edward Arnold for allowing us to reproduce the following copyright material from *The Properties of Engineering Materials* by Raymond Higgins:

our Figs 3.4, 7.6, 7.7, 7.8, 10.2, 10.4, 10.5, 10.11 and 10.12.

1 Alloy steels

1.1 The need for alloying

The limitations of plain carbon steels can be summarised as follows.

(*a*) A high critical cooling rate which leads to cracking when quench hardening.
(*b*) Poor hardenability and a corresponding low value of ruling section.
(*c*) Compared with alloy steels, carbon steels can only attain relatively low values of tensile strength even after quench hardening and tempering unless such properties as ductility and toughness are reduced in value to unacceptable levels.

Alloying elements are added to plain carbon steels to overcome these limitations and, in some instances, to improve the corrosion and heat resistance as well. However, alloy steels are more expensive and more difficult to process than plain carbon steels and should only be used where their special properties can be fully exploited.

1.2 Alloying elements

Steels containing iron and carbon with traces of phosphorus, silicon, and not more than 1.5 per cent manganese are referred to as plain carbon steels. The composition, properties and simple heat treatment of such steels have been fully discussed in volume 1.

Alloy steels are carbon steels, normally containing less than 1.0 per cent carbon, to which other metals and some non-metals (alloying elements) have been added in sufficient quantities to alter the properties of the steels to a significant extent. The more important alloying elements will now be considered.

Aluminium

The presence of up to 1 per cent aluminium in alloy steels enables them to be given a hard, wear-resistant skin by the process *nitriding*.

Chromium

The presence of small amounts of chromium stabilises the formation of hard carbides, and improves the susceptibility of steels to heat treatment. Unfortunately, the presence of chromium also promotes grain growth, therefore chromium is rarely used as an alloying element on its own (Section 1.3). The presence of large amounts of chromium improves the corrosion resistance and heat resistance of steels (stainless steels).

Cobalt

The presence of cobalt induces sluggishness into the heat-treatment transformations and improves the ability of tool steels to operate at high temperatures without softening. It is an important alloying element in super high-speed steels.

Copper

The presence of up to 0.5 per cent copper improves the corrosion resistance of alloy steels.

Lead

The presence of up to 0.2 per cent lead improves the machinability of steels, but unfortunately it also reduces the strength of the steel to which it is added.

Manganese

This element is always present in steel as it combines with residual sulphur from the smelting process and reduces the brittleness caused by the presence of iron sulphide. It also stabilises the γ phase (austenite) and helps to promote the formation of stable carbides. In large quantities (up to 12.5 per cent), manganese improves the wear resistance of steels by causing them to form a hard skin spontaneously when subjected to abrasion.

Molybdenum

The presence of molybdenum in alloy steels raises their high-temperature creep strength, stabilises their carbides, improves the ability of cutting tools to retain their hardness at high temperature, and reduces the susceptibility of nickel-chromium steels to *temper-brittleness* and *weld-decay*. Molybdenum is also present in chromium steel alloys to reduce grain growth (see chromium).

Nickel

The presence of nickel in alloy steels results in increased strength by grain refinement. It also improves the corrosion resistance of a steel.

Unfortunately, nickel is a powerful graphitiser and reduces the stability of any carbides present. Nickel and chromium are often used together in alloy steels where they complement each other's properties (Section 1.3).

Phosphorus

This is a residual element from the smelting process. It causes weakness in the steel and is considered an undesirable impurity. Normally considerable care is taken to keep its presence below 0.05 per cent. However, where maximum strength and toughness is not required, increased quantities of phosphorus can improve machinability and also the fluidity of steel castings.

Silicon

The presence of up to 0.3 per cent silicon improves the fluidity of casting steels without the reduction in mechanical properties associated with phosphorus. Up to 1 per cent silicon improves the heat resistance of steels. Unfortunately silicon, like nickel, is a powerful graphitiser and is never added in large amounts to high carbon steels.

Sulphur

This, like phosphorus, is also a residual element carried over from the smelting process. It is also considered an undesirable impurity since the presence of iron sulphide reduces the strength and toughness of steels. Fortunately sulphur has a greater affinity for manganese than it has for iron, and the presence of manganese sulphide does not impair the mechanical properties of steels. However, sulphur is sometimes alloyed with low-carbon steels to improve their machinability where a reduction in component strength can be tolerated.

Titanium

Up to 1.0 per cent titanium in stainless and maraging steels helps to reduce weld-decay and temper-brittleness. Niobium has the same effect and steels containing either of these alloying elements are said to be 'proofed'.

Tungsten

The presence of tungsten in alloy steels promotes the formation of very hard carbides, and induces sluggishness into the heat treatment transformations. This enables the steels to retain their hardness at high temperature. Tungsten is mainly found in high-speed steels which are used for cutting tools and in high-duty die steels which have to operate at high temperatures.

Vanadium

Vanadium is not used on its own, but is used to enhance the benefits of other alloying elements. The effects of this element in alloy steels are many and various.

(*a*) It promotes the formation of carbides.
(*b*) It stabilises martensite and thus improves hardenability.
(*c*) It reduces grain growth.
(*d*) It enhances the 'hot hardness' of tool steels and die-steels.
(*e*) It improves the fatigue resistance of steels.

It has been mentioned above that some alloying elements, such as chromium, promote the formation of carbides whilst others, such as nickel promote the formation of free graphite. Therefore alloying elements can be classified as: carbide formers, graphitisers, austenite stabilisers and ferrite stabilisers.

Carbide formers

Some alloying elements form very stable carbides which are harder than iron carbide. The formation of such carbides increases the overall hardness of the steel and makes it suitable for tooling purposes. The carbide promoting elements are: chromium, manganese, niobium, molybdenum, titanium, tungsten and vanadium.

Graphitisers

Not all alloying elements tend to combine with carbon when in the presence of iron. Far from forming carbides, such alloying elements as nickel, aluminium and silicon cause instability in any carbide present so that carbon may be precipitated out as free graphite. If any of these elements are required in appreciable amounts, either carbide forming alloying elements must also be present or the carbon content of the steel must be kept very low. Therefore it is not possible to have a high-carbon, high-nickel alloy steel.

Austenite stabilisers

Reference to Fig. 1.1 in Section 1.3 shows that some alloying elements such as cobalt, copper, nickel and manganese raise the A_4 temperature whilst at the same time depressing the A_3 temperature. Therefore, when these elements are added to carbon steels they stabilise the γ phase (austenite) by increasing the temperature over which this phase remains stable. This is because most of these alloying elements do not tend to form carbides and the carbon remains in solid solution in the austenite. When the amounts of the alloying elements present stabilise the austenite to the extent that it is present at room temperature (e.g. austenitic stainless steel), the steel loses its ferromagnetic properties.

Ferrite stabilisers

Other alloying elements such as aluminium, chromium, molybdenum, tungsten, silicon and vanadium have the opposite effect to those described above and stabilise ferrite. This is achieved by raising the A_3 temperature and depressing the A_4 temperature to form what is referred to as the 'γ-loop' (Section 1.3). Since these alloying elements have body-centred

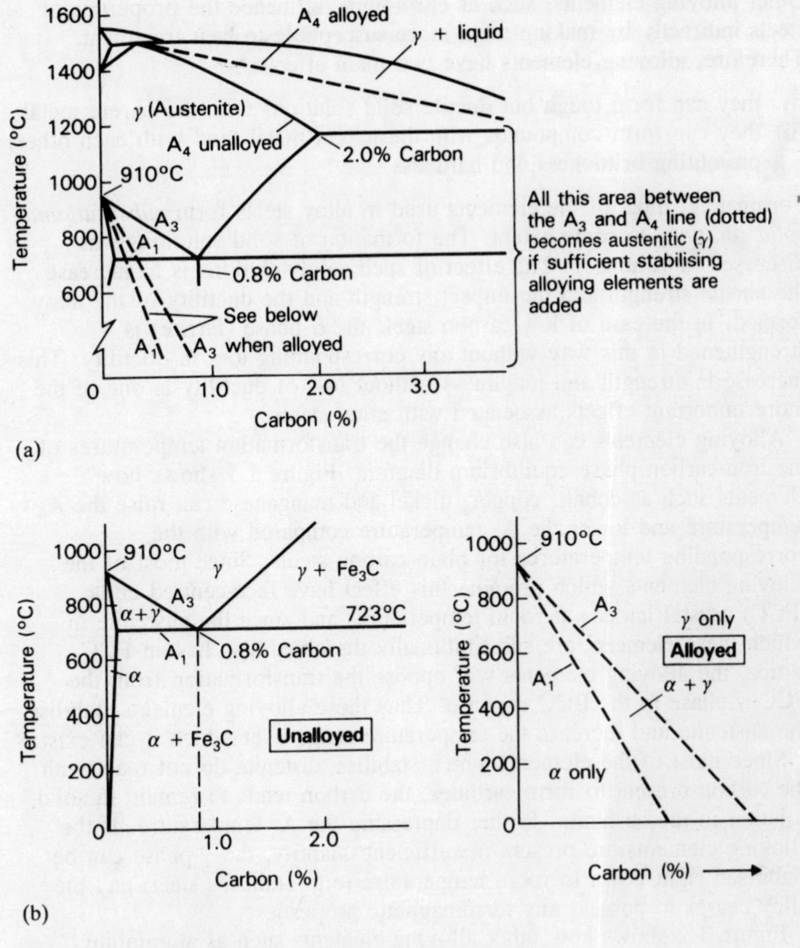

Fig. 1.1 Effect of alloying elements on ferrous metals

cubic (BCC) crystals at room temperature they tend to stabilise the ferrite which also has BCC crystals.

All these effects will be considered more fully in the next section of this chapter.

1.3 The effects of alloying elements

Alloying elements can influence the properties of a steel directly or indirectly. For example, nickel is stronger than iron so its presence increases the strength of the steel directly. Its presence also refines the grain of the steel and this further increases the strength of the steel.

Other alloying elements, such as chromium, influence the properties of steels indirectly by making them more susceptible to heat treatment. Therefore, alloying elements have two main effects:

(i) they can form tough but ductile solid solutions with the parent metal;
(ii) they can form compounds with the parent metal (and with each other) promoting brittleness and hardness.

Fortunately, most of the elements used in alloy steels form *substitutional solid solutions* to some extent. The formation of solid solutions was discussed in volume 1. The effect of such solid solutions is to increase the tensile strength and the impact strength and the ductility of the alloy formed. In the case of low-carbon steels the α-phase (ferrite) is strengthened in this way without any corresponding loss in ductility. This increase in strength and toughness without loss of ductility is one of the more important effects associated with alloy steels.

Alloying elements can also change the transformation temperatures of the iron-carbon phase equilibrium diagram. Figure 1.1 shows how elements such as cobalt, copper, nickel and manganese can raise the A_4 temperature and lower the A_3 temperature compared with the corresponding temperatures for plain carbon steels. Since most of the alloying elements which promote this effect have face-centred cubic (FCC) crystal lattices at room temperature, and since the austenite in which these elements are substitutionally dissolved also has an FCC lattice, the alloying elements will oppose the transformation from the FCC γ-phase to the BCC α-phase. Thus these alloying elements stabilise the austenite and increase the temperature range over which it can exist.

Since most of the elements which stabilise austenite do not react with the carbon present to form carbides, the carbon tends to remain in solid solution in the austenite, further depressing the A_3 temperature. If the alloying elements are present in sufficient quantity, the γ-phase can be stabilised right down to room temperature (e.g. stainless steel) and the alloy ceases to possess any ferromagnetic properties.

Figure 1.2 shows how other alloying elements such as aluminium, chromium, molybdenum, silicon, tungsten and vanadium can have the opposite effect, that is, they raise the A_3 temperature and lower the A_4 temperature to form what is described as the γ-loop. Since this group of alloying elements has BCC crystal lattices at room temperature, as does ferrite, this common lattice structure coupled with the raising of the A_3 temperature has the effect of stabilising and promoting the formation of the α-phase (ferrite). Chromium, molybdenum and tungsten, in particular, form stable carbides and the precipitation of carbon as metallic carbides still further promotes the transformation from austenite to ferrite.

The hardness of alloy steels depends (as in plain carbon steels) on the formation of hard metallic carbides. It has been stated above that the alloying elements chromium, molybdenum and tungsten form very stable carbides which are harder than iron carbide. Hence these alloying elements are widely found in tool and die steels.

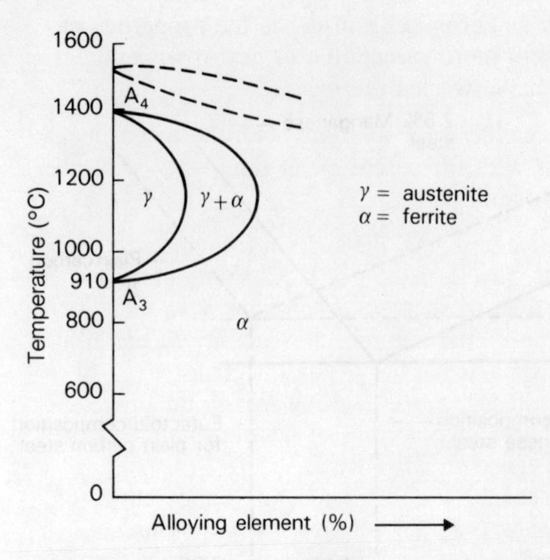

Fig. 1.2 Effect on ferrous metals of alloying elements which stabilise ferrite

Another important effect of the carbide stabilising alloying elements, and in particular tungsten, is the slowing down of the transformation rates, that is the alloy does not have to be cooled as quickly as plain carbon steel to produce a given structure. This enables alloy tool and die steels to be hardened by oil quenching, or even by air-blast quenching, with a correspnding reduction in the possibility of cracking and distortion. This induced sluggishness in the transformation rates also increases the temperature at which the steel may be used without loss of hardness. A steel which can be used at high operating temperatures without its 'temper being drawn' is referred to as having good 'hot hardness' or good 'red hardness'.

Although chromium improves the susceptibility of steel to heat treatment it has one major disadvantage: it promotes grain growth. Therefore, it is most important to heat-treat chrome steels at the lowest possible temperature for the minimum possible time if grain growth and the brittleness associated with grain growth is to be avoided. This is also a problem when attempting to weld such steels.

Fortunately nickel and chromium tend to be complementary to each other when used as alloying elements. Nickel promotes fine grain but tends to unstabilise (graphitise) the carbides, whilst chromium promotes stable carbides but tends to cause grain growth. By careful control of the amount of nickel and chromium present in the alloy, it is possible to produce a steel which has both a fine grain and stable carbides.

Generally, alloy steels contain less carbon than plain carbon steels. It has already been stated that alloy steels rarely contain more than 1 per cent carbon, and the reason for this is that the addition of any significant amount of an alloying element to a plain carbon steel reduces the

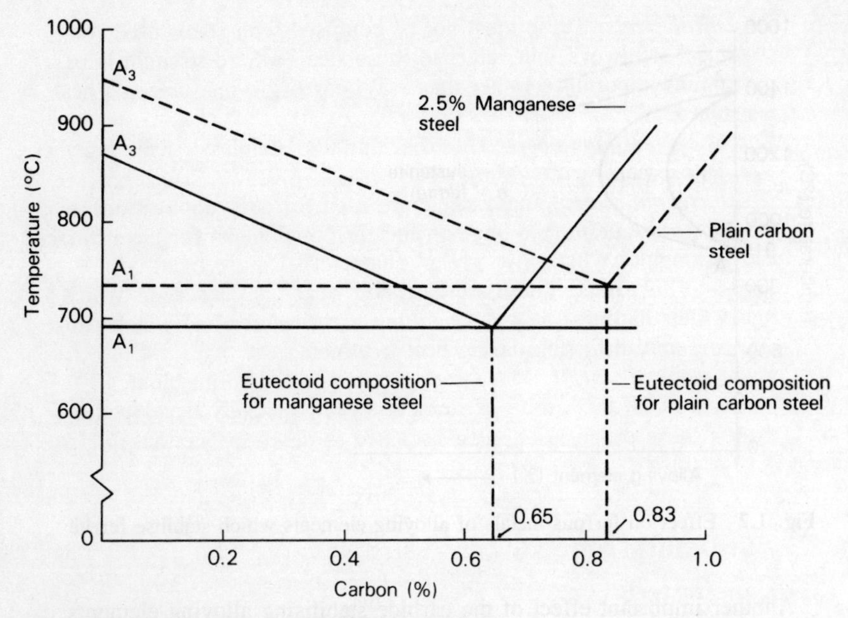

Fig. 1.3 Displacement of eutectoid composition

eutectoid composition below the normal 0.83 per cent carbon. This effect is shown in Fig. 1.3. In this example, the addition of 2.5 per cent manganese reduces the eutectoid composition at which the steel is wholly pearlitic from 0.83 per cent carbon to 0.65 per cent carbon. This is also accompanied by a reduction in the A_1 and A_3 temperatures.

Finally, the addition of alloying elements to a plain carbon steel can improve its corrosion resistance. Although not a particularly reactive metal, iron corrodes (rusts) readily in the presence of oxygen and moisture because the hydroxide film which forms on its surface is porous and offers no protection. Once corrosion commences it tends to become spotaneous, the iron hydroxide (rust) coating reacting with the iron in the steel underneath it. Alloying elements such as aluminium, silicon, copper and chromium cause the formation of corrosion-resistant homogeneous films to form on the surface of the steel if they are added in sufficient quantity (see Chapter 6).

1.4 The classification of steels

The most convenient way to classify the wide range of alloy steels available to the engineer is to group them according to application, and then form sub-groups according to their principal alloying elements. There are five main groups.

(*a*) *Structural steels*. These must not be confused with steels used for structural steelwork, but refers to those steels where strength is of paramount importance rather than, say, corrosion resistance or heat resistance.
(*b*) *Corrosion-resistant steels*. These include the 'stainless' steels together with less expensive low alloy steels.
(*c*) *Heat-resistant steels*. These steels are used for such applications as the valves in automobile engines and for components for gas-turbines and jet engines where low 'creep' characteristics are required.
(*d*) *Tool and die steels*. These are hard and wear-resistant steels which retain their hardness at high operating temperatures and which resist cracking and distorting during heat treatment.
(*e*) *Ferro-magnetic steels*. Most steels show some ferro-magnetic properties, but this group of steels has been specially developed to exploit these properties and is discussed in detail in Sections 7.11 to 7.13 inclusive.

1.5 Structural steels

Manganese steels

All steels contain small amounts of manganese to deoxidise the molten steel during manufacture. A small excess of up to 0.35 per cent is generally provided beyond the amount necessary to ensure proper deoxidation, to combine with any residual sulphur present, and to prevent it forming iron sulphide. Manganese also improves the rolling and forging qualities of steels. However, the true manganese steels contain larger amounts of the alloying element and fall into two groups:

(*a*) those containing from 11 to 14 per cent manganese;
(*b*) those containing from 1 to 2 per cent manganese.

The steels in group (*a*) also contain from 1.0 to 1.3 per cent carbon and before heat treatment they are very hard and lacking in ductility. These steels have to be heat-treated by quenching from 1050°C, after which they become tough with a reasonable ductility and they are sufficiently soft to be machineable. Steels in this group are rapidly hardened by cold-working and form a very hard skin when subjected to abrasion. Therefore they have to be machined with very sharp cutting tools since, if any rubbing occurs during machining, the hard skin which forms can only be removed by grinding before machining can recommence. This resistance to abrasion makes the steel most suitable for such applications as railway points, dredger buckets and stone-crusher jaws.

The steels in group (*b*) contain only 0.25 to 0.55 per cent carbon and although they exhibit similar properties to group (*a*), they do so to a much lesser extent. However, they are much less costly and low-carbon, low-alloy manganese steels are now widely used for automobile components in place of the more costly nickel-chrome steels. For example, a steel containing 0.2 to 0.45 per cent carbon, 1.2 per cent manganese, and 0.8 per cent silicon is sufficiently ductile and machineable for structural purposes, yet it will have a tensile strength of 560 to 910 MPa depending upon heat-treatment.

Nickel steels

Plain nickel steels are widely used and fall into four main groups:

(*a*) structural steels containing up to 6 per cent nickel;
(*b*) corrosion-resistant steels containing up to 20 to 30 per cent nickel;
(*c*) low-expansion steels containing 30 to 40 per cent nickel; and
(*d*) high-permeability magnetic steels containing 50 per cent or more nickel.

In this section only group (*a*) will be considered, and the remaining alloys will be considered in later sections as appropriate.

Group (*a*) steels are used for components of machines and structures which are highly stressed and typical alloys contain 0.1 to 0.55 per cent carbon, 0.3 to 0.8 per cent manganese, and 0.4 to 6.0 per cent nickel.

The steels in this group containing low percentages of carbon are used for case-hardening. Since case-hardening requires the steel to be heated into the γ-phase for prolonged periods, grain growth is usually excessive. However the presence of nickel reduces this grain growth and, after heat-treatment, promotes a very tough, fine-grained core which greatly enhances the mechanical properties of the component.

As stated earlier in this chapter, nickel tends to promote graphitisation of the carbides in the steel and the manganese content has to be increased beyond that normally required for deoxidation in order to counteract this effect.

Since nickel helps to prevent excessive grain growth at high temperatures, it enables fine-grain steels to be produced more easily. Nickel also lowers the critical temperatures slightly and thus makes heat treatment less severe, reducing the chance of cracking and distortion. Figure 1.4 compares the properties of a 1 per cent nickel steel with a 3.5 per cent nickel steel after quenching in oil from 850°C and tempering at various temperatures up to 700°C.

Nickel-chromium steels

These are probably the most widely used of all alloy steels. The compositions most widely used are: carbon 0.1 to 0.55 per cent; nickel 1.0 to 4.75 per cent; chromium 0.45 to 1.75 per cent; manganese 0.3 to 0.8 per cent. The low-carbon steels (less than 0.25 per cent carbon) are used for case-hardening, where the presence of chromium promotes a hard and wear-resistant case, whilst the nickel preserves a tough, fine-grained core.

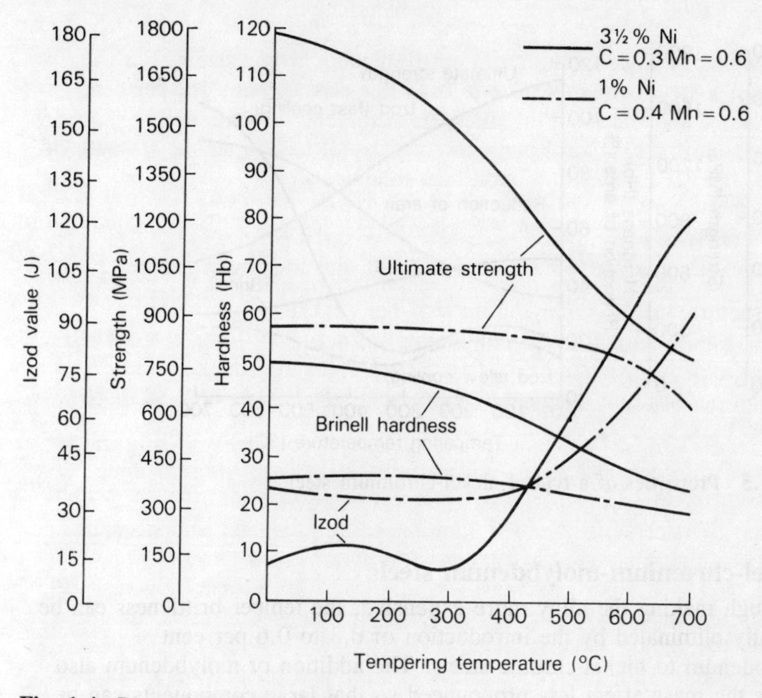

Fig. 1.4 Properties of nickel steels

Alloys containing the higher carbon content (up to 0.55 per cent) are used where high-duty mechanical properties are required. Figure 1.5 shows the properties of a typical steel having a composition of: carbon 0.3 per cent; nickel 3.4 per cent and chromium 0.75 per cent after quenching in oil from 850°C and tempering at various temperatures up to 700°C. The dip in the Izod curve which occurs when the tempering temperature lies between 250°C and 450°C should be noted. It is typical of these steels, and this range of temperatures must not be used for tempering and must be avoided in service. These alloys must be cooled rapidly from the tempering temperature or their impact strength will be low. This marked reduction in impact strength when tempering is slow or when it lies between 250°C and 450°C is called *temper brittleness* and is indicated by the broken line in Fig. 1.5. When the nickel content exceeds 1 per cent and the chromium content exceeds 4 per cent, the alloy may be hardened by quenching in an air blast from a temperature just above the A_3 line. Such alloys are referred to as *air hardening* steels and they possess outstanding properties for small, heavily loaded machine parts and gears.

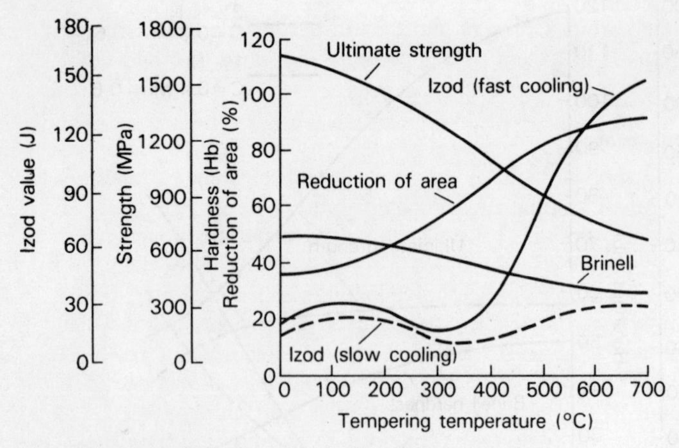

Fig. 1.5 Properties of a typical nickel-chromium steel

Nickel-chromium-molybdenum steels

Although making the alloy more expensive, the temper brittleness can be virtually eliminated by the introduction of 0.3 to 0.6 per cent molybdenum to nickel-chrome steels. The addition of molybdenum also makes the mass effect less pronounced so that large components can be more easily treated. This property of molybdenum is very valuable since it enables large forgings to be cooled slowly without loss of strength and without the setting up of internal stresses which more rapid cooling would cause.

The addition of molybdenum to a nickel-chromium steel also enables an increased amount of manganese to be used without any reduction in impact strength, whilst the increase in manganese content allows the nickel content to be reduced, thus making the steel cheaper. For example, a steel having a composition of: carbon 0.35 per cent; manganese 1.6 per cent; nickel 2.0 per cent and molybdenum 0.6 per cent will have properties almost identical to a more costly alloy whose composition is: carbon 0.35 per cent; manganese 0.6 per cent; nickel 3.0 per cent and chromium 0.8 per cent.

Nickel-chromium-vanadium steels

Vanadium is never used as the sole alloying element in steels but is used in conjunction with nickel or chromium or both. It is a very effective deoxidising element and thus, by eliminating or greatly reducing such impurities as the iron oxide content entrapped in the steel, it improves the mechanical properties generally and the resistance to fatigue in particular. Vanadium also intensifies the effect of the other alloying elements present and enables somewhat smaller quantities of these elements to be used without alteration of the mechanical properties of a steel. Its direct effect is to stabilise the carbides present and thus to harden the steel. Hence it is not normally used in excess of 0.2 per cent except in tool and die steels.

Molybdenum steels

Steels containing molybdenum as the sole alloying element (in addition to carbon and the small amount of manganese present in all steels) are also used for structural purposes. Like silicon steels, to be mentioned later, molybdenum steels give greater strength for a given ductility than is obtainable in plain-carbon steels. Molybdenum steels containing: 0.2 to 0.7 per cent molybdenum; 0.3 to 1.0 per cent manganese; 0.1 to 0.35 per cent silicon and 0.15 to 0.7 per cent carbon, are used for rolled steel sections, forgings and castings. The tensile strengths for these steels range from about 378 MPa to 462 MPa, whilst their elongation percentage can be kept as high as 20 per cent. The grain size in these steels can be controlled accurately and easily.

Chromium-molybdenum steels

In these alloys the chromium content may range from as low as 0.4 per cent to as high as 10 per cent, and the molybdenum content from 0.2 per cent to 1.5 per cent. Those steels containing low percentages of chromium and molybdenum have mechanical properties similar to other low alloy structural steels. However, the steels with high percentages of chromium and molybdenum may be used as rolled steel sections, forgings and castings with tensile strengths up to 1120 MPa whilst still maintaining elongation values of 18 to 22 per cent. The high chromium content in these latter steels makes them highly corrosion resistant.

Chromium steels

Steels in which chromium is the principal alloying element can have tensile strengths up to 980 MPa, elongations of 10 to 15 per cent and relatively high impact strengths. Such steels may have 0.2 to 1.6 per cent chromium and up to 1.0 per cent manganese. Great care is required in their heat treatment as chromium in a steel promotes grain growth. However the high manganese content helps to offset this effect.

Steels containing 5 to 10 per cent chromium and 1 to 4 per cent silicon are heat resistant and widely used for the valves in automobile engines. These alloys are usually referred to as silicon-chromium steels.

Silicon steels

Small amounts of silicon are frequently used in many alloy steels, but the amount does not usually exceed 0.8 per cent. If the silicon content is increased to 1.25 per cent (together with carbon 0.5 to 0.65 per cent and manganese 0.6 to 0.9 per cent) a steel with a high resistance to fatigue is produced. Such steels are frequently used for springs when quenched in oil from 857°C to 900°C and tempered between 475°C and 525°C.

Table 1.1 Typical structural (low-alloy) steels

BS970 spec	Type of steel	Composition (%) C	Mn	Ni	Cr	Mo	Condition	R_e (MPa)	R_m (MPa)	I (J)	A %	Applications
150M28	Low manganese	0.28	1.50	—	—	—	Normalised	355	590	—	20	A cheap medium-duty alloy for automobile components
530M40	Nickel-manganese	0.40	0.90	1.00	—	—	Quench hardened from 850°C in oil, tempered at 600°C	495	695	91	25	Crankshafts, connecting rods, axles and general components in the automobile and machine tool industries
608M38	Manganese-molybdenum	0.38	1.50	—	—	0.50	Quench hardened from 850°C in oil, tempered at 600°C	1000	1130	70	19	A lower-cost substitute for nickel-chrome-molybdenum steels for highly stressed components

653M31	Nickel-chromium	0.31	0.60	3.00	1.00	—	Quench hardened from 820–840°C in oil, tempered at 600°C	820	930	105	23	Highly stressed components such as: differential shafts, half-shafts, stub axles, connecting rods, pinion shafts, high-tensile studs and bolts
817M40	Nickel-chromium-molybdenum	0.40	0.55	1.50	1.20	0.30	Oil quenched from 840°C and tempered at 600°C		2010	27	14	Highly stressed gears for the automobile and machine tool industries
								990	1080	70	22	Highly stressed components where resistance to shock and fatigue is important
945M38	Manganese-nickel-chromium-molybdenum	0.38	1.40	0.75	0.50	0.20	Oil quenched from 840°C, tempered at 600°C	960	1040	85	21	A cheaper alloy than 817M40, but still suitable for highly stressed components where resistance to shock and fatigue is important

R_e = Yield stress, R_m = Tensile strength, I = Izod number, A = elongation %, Mn = manganese, C = carbon, Ni = nickel, Cr = chromium, Mo = molybdenum

Steels containing 0.5 to 1.0 per cent silicon and 0.7 to 0.95 per cent manganese are now being used increasingly for structural purposes. These steels, like the molybdenum steels mentioned above, give greater strength for a given ductility than can be achieved in a plain-carbon steel, and at lower cost than for molybdenum steels. A selection of structural steels is given in Table 1.1, together with their properties and some typical uses.

1.6 Corrosion-resistant steels

These are alloy steels containing large amounts of nickel and chromium which promote the formation of homogeneous, corrosion-resistant oxide films on the surface of the metal (see Chapter 6).

Nickel steel

Alloys containing 0.4 to 0.5 per cent carbon and 20 to 30 per cent nickel are austenitic in structure even when cooled slowly. Besides being extremely tough, they are highly resistant to corrosion by sea-water, steam and hot gases. They also have low coefficients of thermal expansion and are used for steam turbine blades and for internal-combustion engine valves. Since they are austenitic, these alloys are also non-magnetic and use is often made of this property. To stabilise the carbides present the alloys always contain 1.4 per cent manganese and up to 0.5 per cent chromium. The only heat treatment required is cooling in air from 800°C to render the alloy machineable.

Alloys containing from 30 to 40 per cent nickel are also available but they are very costly and are only used where a negligible coefficient of thermal expansion is required.

Chromium-molybdenum steels

These have already been introduced as high-strength structural steels. Alloys containing up to 10 per cent chromium and 1.5 per cent molybdenum combine very high strength and toughness with very high corrosion resistance, particularly to acids.

Stainless steels

There are several distinctive types of stainless steels and they all contain fairly large amounts of chromium and, sometimes, nickel. The chromium content ranges from 4 to 22 per cent and the nickel content from 0 to 26 per cent. Stainless steels can be categorised into four main groups.

(*a*) *Martensitic* stainless steels containing 10 to 14 per cent chromium.
(*b*) *Ferritic* stainless steels containing 14 to 18 per cent or 23 to 30 per cent chromium.
(*c*) *Austenitic* stainless steels containing 15 to 20 per cent chromium and 7 to 10 per cent nickel.
(*d*) *High austenitic* stainless steels containing 22 to 26 per cent chromium and 12 to 14 per cent nickel.

Probably the most important group of stainless steels are the austenitic alloys (group (*c*)). Within this group lies the 18/8 stainless steel alloy which is the most widely used of all. Stainless steels in the 18/8 group are austenitic in structure even when cooled slowly and they are not responsive to heat treatment, except that they can be annealed from 1100°C. After annealing they are very tough but ductile. They are hardened by cold-working, resulting in increased tensile strength and decreased ductility. Austenitic stainless steels retain these properties at fairly high temperatures, whilst maintaining their toughness and impact resistance at very low temperatures. This range of operating temperatures is somewhat exceptional.

18/8 stainless steels have excellent corrosion-resisting properties, but they are liable to intercrystalline corrosion after being heated to between 600°C and 900°C. This must be borne in mind when these steels are welded since the parent metal will be at this temperature either side of the weld zone. Under these circumstances, this form of intercrystalline corrosion is referred to as *weld-decay*. This susceptibility can be reduced by quenching the steel from 1050°C after welding or by the addition of such alloying elements as molybdenum or titanium (1.6 per cent) or both. Stainless steels which include these additional alloying elements are said to have been 'proofed' against weld-decay.

Stainless steels can be manipulated by all the usual engineering processes, and they machine quite well provided the cutting tools used are kept sharp. For maximum ductility the carbon content should be kept low, whilst for cutlery and general engineering the carbon content is increased to 0.3 to 0.55 per cent and the nickel content is eliminated to ensure the promotion of hard, stable carbides. Table 1.2 lists some corrosion resistant steels, together with their properties and some typical applications.

Ferritic stainless steels are susceptible to *sigma phase* formation if subjected to prolonged heating at 475°C. The sigma phase, which is formed from ferrite in the presence of chromium, promotes hardness and brittleness in the steel. This leads to cracking and fatigue failure and is referred to as *temper brittleness*. Ferritic steels, containing a high chromium content, must be cooled rapidly from 650°C after welding to prevent the onset of *weld decay*. The sigma phase can also occur in austenitic stainless steels following prolonged heating at 800°C or slow cooling from this temperature.

1.7 Heat-resistant steels

These alloy steels, examples of which are also included in Table 1.2, are designed to resist corrosion and oxidation by reactive gases at high temperatures, whilst retaining their strength at such temperatures. They must also be creep resistant and free from any tendency to temper brittleness, carbide precipitation, and the formation of the sigma phase.

Table 1.2 Corrosion and heat-resistant steels: derived from BS 970: Pt 3

Type of steel	Composition						Mechanical properties				Heat treatment	Applications
	C	Mn	Cr	Ni	Ti	Si	R_m	R_e	A	H_B		
403S17 Ferritic	0.04	0.45	14.0	0.50	—	0.80	510	340	31	—	Condition soft. Cannot be hardened except by cold-work	Soft and ductile; can be used for fabrications, pressings, drawn components, spun components. Domestic utensils
420S45 Martensitic	0.30	1.0	13.0	1.0	—	0.80	1470 1670	— —	— —	450 534	Quench from 950°–1000°C Temper 400–450°C Temper 150–180°C	Corrosion-resistant springs for food processing and chemical plant. Corrosion-resistant cutlery and edge tools
302S25 Austenitic	0.1	1.0	18.0	8.50	—	0.80	61 896	278 803	50 30	170 —	Condition soft solution treatment from 1050°C	18/8 stainless steel widely used for fabrications and domestic and decorative purposes

321S20 Austenitic weld-decay resistant	0.1	0.80	18.0	8.50	1.60	0.80	649 803	278 402	45 30	180 225	Condition soft solution treatment from 1050°C Can only be work-hardened	18/8 which can be safely fabricated by welding, used for brewing, food-processing, and chemical plant
401S45 Valve steel	0.4	0.50	8.0	0.5	—	3.0	—	—	—	225 (min)	Quench harden in oil from 1030–1060°C Temper at 750–850°C	Heat-resistant steel of relatively low cost General-purpose steel
349S54 Valve steel	0.5	10.0	22.0	4.50	—	0.25	—	—	—	321 (min)	Soften by solution treatment at 1160–1190°C Harden by precipitation treatment at 750–850°C for 6 to 15 hours	A high-quality, high-cost valve steel suitable for hostile environments, e.g. furnace and chemical-plant components

C = carbon, Mn = manganese, Cr = chromium, Ni = nickel, Ti = titanium, Si = silicon, R_m = Tensile stress (MPa), R_e = Yield stress (MPa), A = elongation %, H_B = Brinell hardness

Such steels contain up to 30 per cent chromium together with up to 3.5 per cent silicon. Nickel is also present to limit the grain growth at sustained high temperatures which would otherwise result from the high chromium content.

To maintain their strength at high temperatures, heat-resisting alloys are 'stiffened' by the addition of one or more of the following elements in small quantities: aluminium, carbon, molybdenum, titanium and tungsten.

1.8 Maraging steels

These are high-strength, high-alloy steels containing approximately 18 per cent nickel. Some typical compositions are shown in Table 1.3. The high nickel content has a considerable graphitising effect so that the carbon content must not exceed 0.3 per cent. It was found that balanced additions of cobalt and molybdenum to iron-nickel martensite gave a combined age-hardening effect far greater than when these alloying elements were used separately. The ability to age harden the martensite

Table 1.3 Composition ranges — weight per cent — of the 18 per cent Ni-Co-Mo maraging steels[1]

	Wrought				**Cast**
Grade	18Ni1400	18Ni1700	18Ni1900	18Ni2400	17Ni1600
Nominal 0.2% proof stress:					
N/mm^2 (MPa)	1400	1700	1900	2400	1600
$tonf/in^2$	90	110	125	155	105
$10^3 lbf/in^2$	200	250	280	350	230
kgf/mm^2	140	175	195	245	165
hbar	140	170	190	240	160
Ni	17–19	17–19	18–19	17–18	16–17.5
Co	8.0–9.0	7.0–8.5	8.0–9.5	12–13	9.5–11.0
Mo	3.0–3.5	4.6–5.1	4.6–5.2	3.5–4.0	4.4–4.8
Ti	0.15–0.25	0.3–0.5	0.5–0.8	1.6–2.0	0.15–0.45
Al	0.05–0.15	0.05–0.15	0.05–0.15	0.1–0.2	0.02–0.10
C max.	0.03	0.03	0.03	0.01	0.03
Si max.	0.12	0.12	0.12	0.10	0.10
Mn max.	0.12	0.12	0.12	0.10	0.10
Si + Mn max.	0.20	0.20	0.20	0.20	0.20
S max.	0.010	0.010	0.010	0.005	0.010
P max.	0.010	0.010	0.010	0.005	0.010
Ca added	0.05	0.05	0.05	none	none
B added	0.003	0.003	0.003	none	none
Zr added	0.02	0.02	0.02	none	none
Fe	Balance	Balance	Balance	Balance	Balance

[1]The composition ranges given are those originally developed by Inco which broadly cover current commercial practice. Slight changes in these ranges have been made in some national and international specifications.

Table 1.4 Advantages of nickel maraging steels

Excellent mechanical properties	Good processing and fabrication characteristics	Simple heat treatment
1. High strength and high strength-to-weight ratio. 2. High notched strength. 3. Maintains high strength up to at least 350°C. 4. High impact toughness and plane strain fracture toughness.	1. Wrought grades are amenable to hot and cold deformation by most techniques. Work-hardening rates are low. 2. Excellent weldability, either in the annealed or aged conditions. Pre-heat not required. 3. Good machinability. 4. Good castability.	1. No quenching required. Softened and solution treated by air cooling from 820–900°C. 2. Hardened and strengthened by ageing at 450–500°C. 3. No decarburization effects. 4. Dimensional changes during age hardening are very small — possible to finish machine before hardening. 5. Can be surface hardened by nitriding.

gave this group of steels its name — *maraging steels*. Furthermore, it was found that the iron-nickel-cobalt-molybdenum matrix was amenable to supplemental age-hardening by small additions of aluminium and titanium. Because of the high cost of the alloying elements in maraging steels they are only used where their special properties can be fully exploited. Table 1.4 lists the advantages of nickel maraging steels over conventional alloy steels, whilst Table 1.5 lists some typical applications. The heat treatment of maraging steels by solution and precipitation process is described in Section 2.10. The welding of maraging steels is discussed in Section 5.15.

1.9 Tool and die steels

Plain carbon steels with a carbon content between 0.7 per cent and 1.5 per cent make excellent cutting tools for low-strength materials, such as wood, where the keen edge attainable with such steels is a distinct advantage for hand tools. Such steels are no longer adequate as cutting tool materials for metal, plastic or wood machining under modern production conditions. Quench hardened high-carbon steels are very brittle and have to be tempered to improve their toughness.

Table 1.5 Typical applications of nickel maraging steels

Aerospace	Tooling and machinery	Structural engineering and ordnance
Aircraft forgings (e.g. undercarriage parts, wing fittings).	Punches and die bolsters for cold forging.	Lightweight portable military bridges.
Solid-propellant missile cases.	Extrusion press rams and mandrels.	Ordnance components.
Jet-engine starter impellers.	Aluminium die-casting and extrusion dies.	Fasteners.
Aircraft arrestor hooks.	Cold reducing mandrels in tube production.	
Torque-transmission shafts.	Zinc-base alloy die-casting dies.	
Aircraft ejector release units.	Machine components: gears index plates lead screws	

Unfortunately, tempering also reduces their hardness and wear resistance. Further, it is the ease with which the temper of plain carbon steel can be drawn which renders such steels unsuitable for the high-speed machining of modern high-duty alloys because of the high temperatures generated in the cutting zone.

The addition of such alloying elements as chromium, cobalt, manganese, molybdenum, tungsten and vanadium makes tool and die steels harder, more wear resistant, more shock resistant, less liable to shrink and warp and better able to operate at high temperatures. For example, a correctly hardened high-speed steel can retain its hardness and continue cutting at temperatures approaching dull-red heat.

Most of the alloying elements used in tool and die steels are *refractory metals* with very high melting points. They also form very hard stable carbides and, having body-centred cubic crystal lattices, they limit the range of temperatures over which austenite can exist, thus stabilising the ferrite and the hard tetragonal martensite.

The heat treatment of alloy steels will be dealt with in detail later in Chapter 2. However, at this point, it is sufficient to say that the low-alloy tool and die steels are quench hardened in oil from temperatures only slightly above those used for plain carbon steels of equivalent carbon content. They are tempered at similar temperatures to plain carbon steels. The more heavily alloyed steels have much slower transformation rates and can be air-hardened (air-blast quenched) for thinner sections or oil quenched for heavier sections. In order that full advantage can be taken of the ability of the high-alloy tool and die steels to retain their hardness at high operating temperatures, the maximum amounts of tungsten and molybdenum must be present in solid solution in the austenite before quenching. For this reason, very high hardening temperatures are required (as much as 1300°C in some instances) and great care has to be taken to avoid grain growth. Despite the large amount of carbide stabilising alloying elements present, a substantial amount of austenite still remains after the initial quench hardening. To transform the retained austenite into martensite a *secondary hardening* treatment is required. This involves quenching the already hard steel from 550°C. With some steels it is necessary to repeat the secondary hardening treatment two or even three times before the transformation of austenite into martensite is complete. This secondary hardening process must not be confused with tempering. Tempering increases toughness at the expense of hardness, whilst secondary hardening increases both hardness and toughness. Once the martensite has been formed the steel can be used at temperatures up to 700°C before tempering and softening sets in. This is because of the sluggishness of the transformations in such heavily alloyed steels. Cutting tool materials will be considered further in Chapter 4.

2 The heat treatment of steels

2.1 Non-equilibrium transformations

The heat treatment of plain carbon steels was introduced in volume 1 by relating the processes of full annealing, sub-critical annealing, normalising and quench hardening to the iron-carbon phase equilibrium diagram as shown in Fig. 2.1. These transformations can only occur if heating and cooling proceeds sufficiently slowly so that all the diffusion processes associated with the transformations are completed, that is, for all the transformations to achieve equilibrium. However, most heat treatment processes involve heating and cooling the steel more rapidly so that *equilibrium is not achieved* and this leads to the formation of microstructures which cannot be forecast from the iron-carbon phase equilibrium diagram.

Reference to Fig. 4.8 in *Engineering materials: volume 1* shows that, depending upon whether the steel is being heated or cooled, stable austenite exists at temperatures above the Ac_1/Ar_1 line, and that the steel is fully austenitic above the $Ac_3-Ac_{cm}/Ar_3-Ar_{cm}$ lines. Further, carbon dissolves interstitially in γ (gamma) iron to form the solid solution known as austenite and this enables the carbon to diffuse quickly throughout the iron so that coring is negligible. However, despite the ease and rapidity with which diffusion of the carbon in γ iron occurs under equilibrium conditions, if the austenite is cooled very quickly from above its upper critical temperature to a much lower temperature (say 200°C) there is insufficient time for total diffusion to occur. This results in a rapid transformation from the face-centred cubic structure of γ iron to the body-centred cubic structure of α iron *before* the carbon can diffuse out of solution and form iron carbide, thus causing supersaturation of the body-centred structure. This, in turn, causes such distortion of the crystal

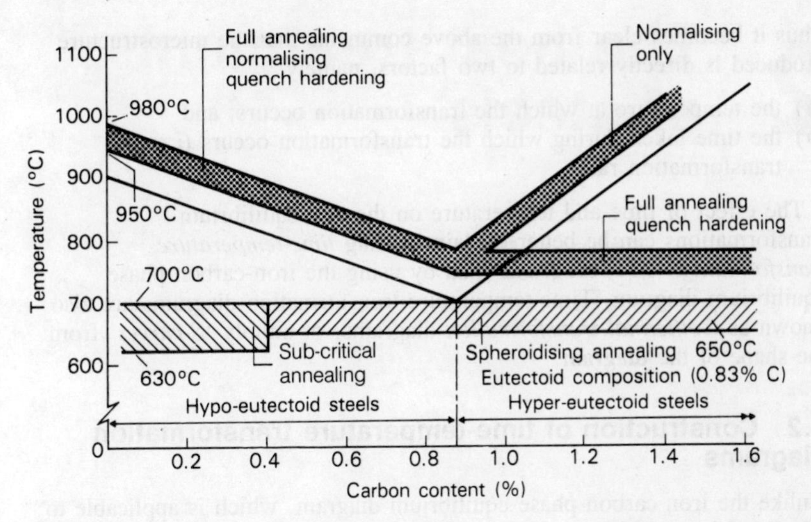

Fig. 2.1 Heat treatment temperatures for plain carbon steels related to the iron-carbon phase equilibrium diagram

lattice structure that slip (see Section 10.2) becomes virtually impossible and the steel exhibits the property of hardness. In fact, such sudden quenching produces a solid solution of carbon in body-centred cubic crystals which is over one thousand times supersaturated and is thus very unstable. Under the microscope a polished and etched specimen of plain, high-carbon steel, quench hardened from above the Ac_{cm} line, shows a structure of acicular (needle-like) crystals called *martensite* after the metallurgist, Martens, who first identified it.

If the quenching is less severe so that the transformations can proceed isothermally (constant temperature), then some iron carbide can form and precipitate out.

(*a*) If the temperature of the quenching bath permits the transformations to occur at 250°C the particles formed will be those of *lower bainite*. These are very fine and can only be seen under a microscope of very high magnification. Lower bainite is similar in appearance to martensite but rather less hard and appreciably tougher.
(*b*) If the temperature of the quenching bath permits the transformations to occur nearer 550°C, the particles formed are called *upper bainite*. These particles are less fine and more 'feathery' in appearance than lower bainite. They are also softer and very much tougher.
(*c*) If the temperature of the quenching bath permits the transformations to occur above 550°C but below the A_1 line (723°C), then *pearlite* will be formed with a further increase in toughness and loss of hardness.

Thus it becomes clear from the above comments that the microstructure produced is directly related to two factors, namely:

(*a*) the temperature at which the transformation occurs; and
(*b*) the time taken during which the transformation occurs (i.e. the transformation rate).

The effect of time and temperature on the non-equilibrium transformations can be better explained using *time-temperature transformation (TTT) diagrams* than by using the iron-carbon phase equilibrium diagram. Time-temperature transformation diagrams are also known as *isothermal transformation* diagrams, or simply '*S-curves*' from the shape of the diagram.

2.2 Construction of time-temperature transformation diagrams

Unlike the iron-carbon phase equilibrium diagram, which is applicable to all plain carbon steels, the time-temperature transformation curves refer to only one steel of a particular composition at a time. Thus, if a comparison is to be made between several different steels, each steel will have to be represented by its own particular time-temperature transformation diagram. These diagrams are equally applicable to both plain carbon and alloy steels. However, they are essential for predicting the outcome of the heat treatment of alloy steels, as such steels cannot be referred to the iron-carbon phase equilibrium diagram. To produce a time-temperature transformation diagram for a plain carbon steel, as shown in Fig. 2.2, the following procedure may be adopted.

(*a*) A large number of specimens approximately 12 mm diameter by 1.5 mm thick are produced from a sample of the steel being examined.
(*b*) These specimens are suspended in a salt bath at a temperature just above that required to ensure the steel is fully austenitic. This is the 'austenising bath' in Fig. 2.3.
(*c*) A suitable number of specimens are transferred to a second salt bath. This is referred to as the 'incubation' bath in Fig. 2.3. For a plain carbon steel, this furnace is held at a pre-determined temperature below 723°C. For example, a temperature of 250°C would be suitable if a structure of lower bainite is required. In this bath the transformation of the austenite to carbide takes place dependent upon the time the specimens remain in the bath.
(*d*) The specimens are then removed from the incubation bath one by one at predetermined times and quenched in water as shown in Fig. 2.3. This final quench halts any transformations which were taking place in the incubation bath, and converts any residual austenite into martensite.
(*e*) The quenched specimens are polished, etched and examined

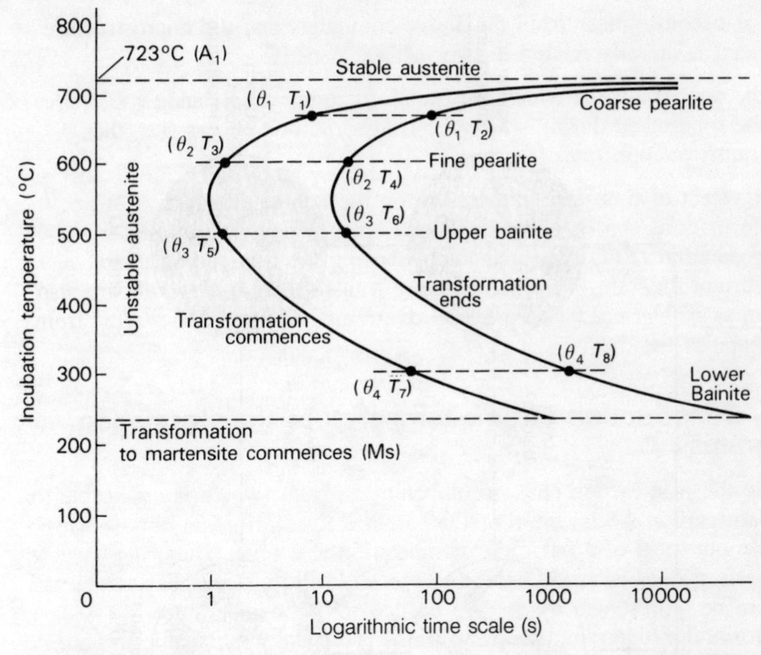

Fig. 2.2 Typical time-temperature transformation (TTT) curves for a plain carbon steel

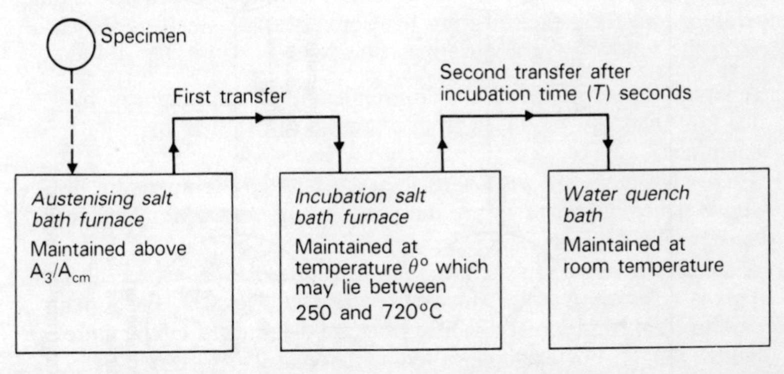

Fig. 2.3 Heat treatment sequence for producing time/temperature transformation diagrams

microscopically to assess the extent to which the transformations have occurred during incubation. Figure 2.4 shows the results of this sequence of treatments. The greater the time interval, the greater will be the degree of transformation for any given incubation temperature.

(*f*) This sequence of events is repeated for a range of temperatures from 250°C to just below 723°C (θ_1, θ_2, θ_3, ... etc.) and sets of results

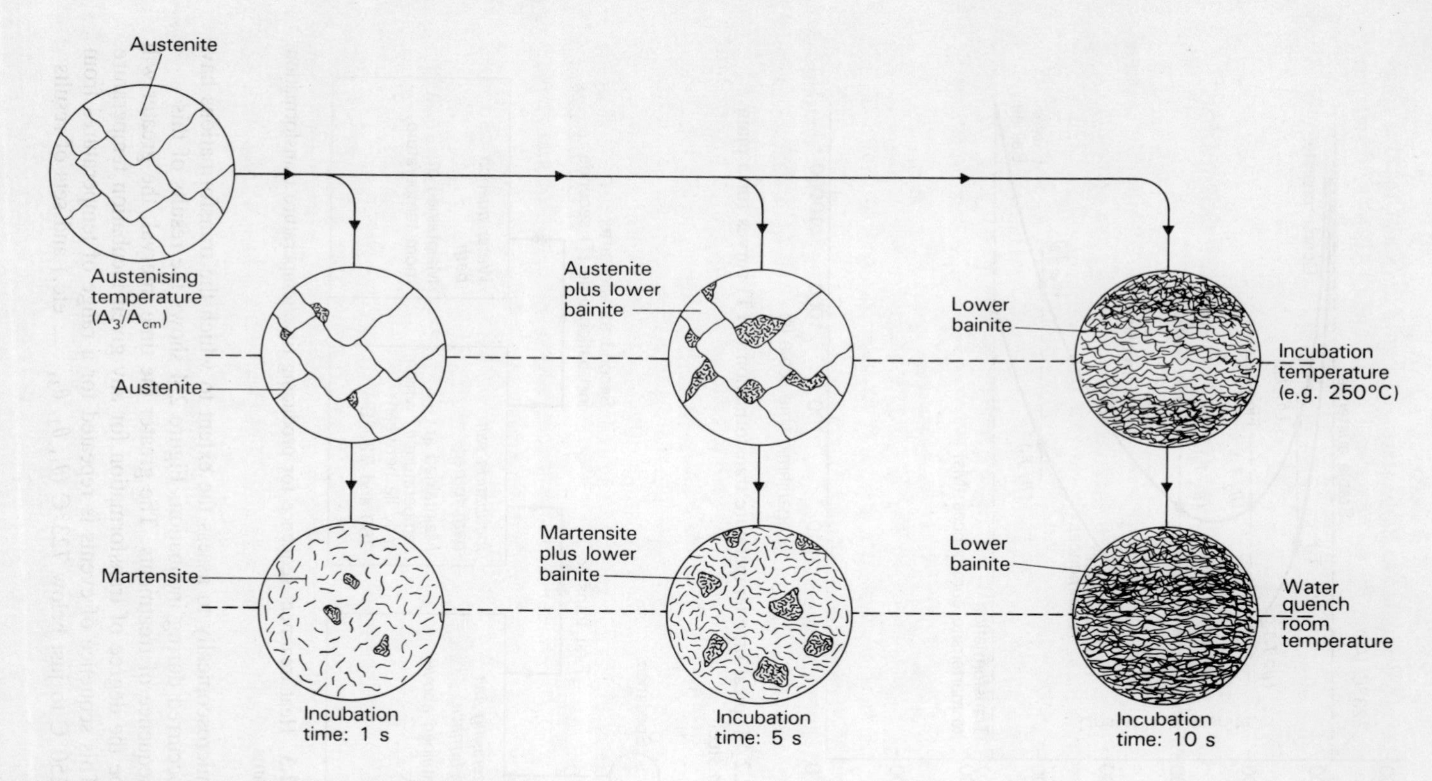

Fig. 2.4 Typical incubation transformations

are obtained similar to those shown in Fig. 2.2 except that the time intervals (T_1, T_2, T_3, ... etc.) will be different. All these results are then plotted on a common set of axes and a time-temperature transformation diagram is produced as shown in Fig. 2.2.

2.3 Interpretation of time-temperature transformation diagrams

Figure 2.5 shows the time-temperature transformation diagram for a plain carbon steel of eutectoid composition. Since the steel is of eutectoid composition its structure will be entirely stable austenite above 723°C. However, below this temperature the austenite will become increasingly unstable and the two curved lines indicate the times taken for the transformations to begin and end as previously explained.

Just below 723°C there is considerable inertia in the transformation process as there is little instability in the austenite. However, as the temperature falls, the instability in the austenite increases so that at 550°C the time taken before the transformation commences is at a minimum, as is the time for the completion of the transformation. The transformations between 723°C and 550°C are initiated by precipitation of iron carbide, and the transformation products range from coarse pearlite just below 723°C to upper bainite (fine pearlite) at about 555°C.

Although the austenite becomes increasingly unstable as the temperature falls below 550°C, the lower temperature causes the diffusion of carbon in the iron to become increasingly more sluggish. This latter factor has more influence than the increasing instability of the austenite below 550°C and the time taken for the transformations to commence and finish increases again. This is shown in Fig. 2.5. Whereas the transformations above 550°C are initiated by the precipitation of iron carbide and result in a structure of dark, feathery upper bainite, below 550°C the transformations are initiated by the precipitation of ferrite and result in a structure of acicular lower bainite at 220°C.

If the temperature of the austenised steel is lowered sufficiently rapidly by severe quenching so that the transformation into bainite is avoided, the austenite is transformed directly into martensite. The amount of martensite present is indicated by the 'M' lines. In Fig. 2.5 this would mean cooling the steel in less time than that indicated by the point 'A' on the diagram (less than one second). At M_s the martensite only just commences to form and very little will be present: whilst at M_{90} the transformation into martensite is 90 per cent complete. Total transformation is not achieved unless the steel is quenched to M_f (−50°C) in less than one second (time 'A'). Thus some retained austenite is always present when the steel is quenched to room temperature.

Figure 2.6 shows the time-temperature transformation diagram for a 0.4 per cent (hypo-eutectoid) plain carbon steel. Reference to the iron

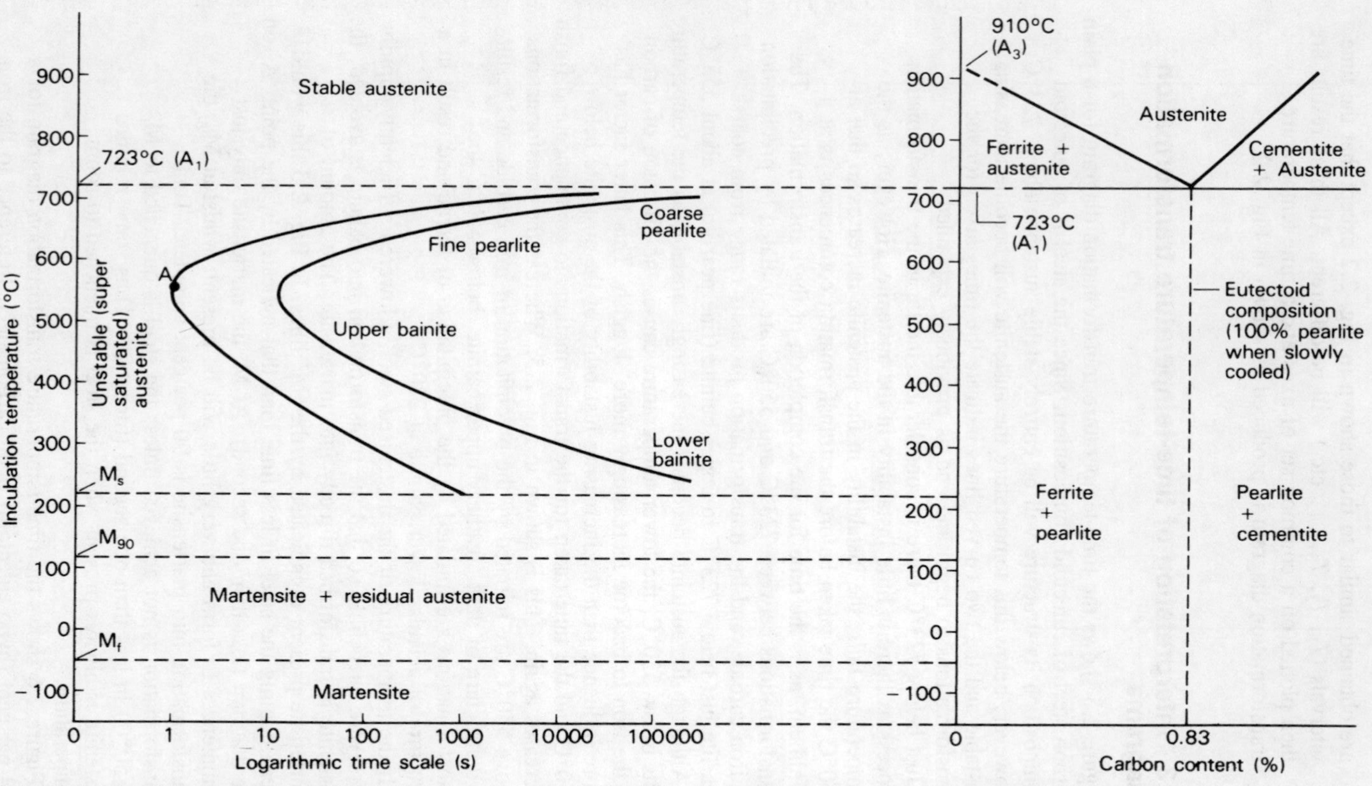

Fig. 2.5 Time/temperature transformation curves for a plain carbon steel of eutectoid composition (0.83 per cent C)

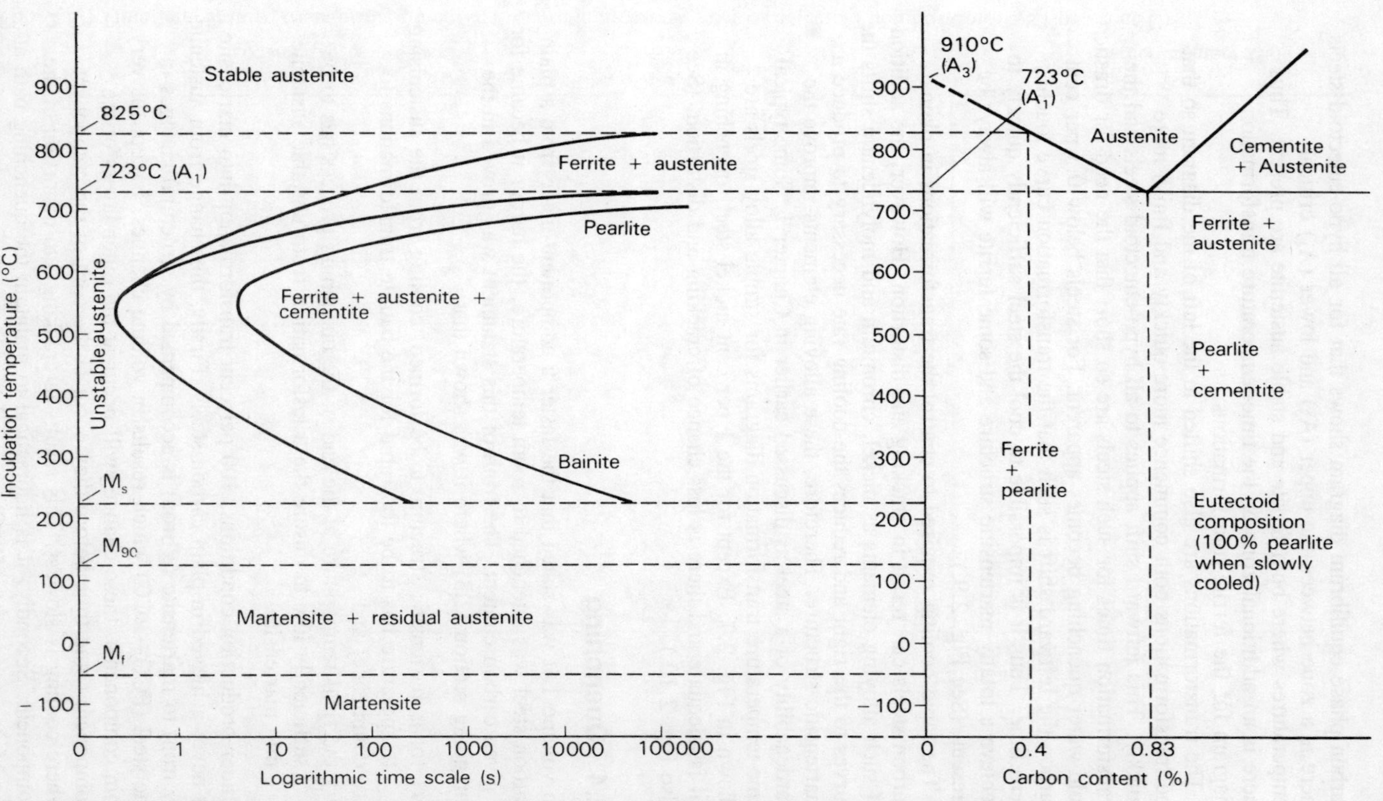

Fig. 2.6 Time/temperature transformation curves for a plain carbon steel of 0.4 per cent carbon content

carbon phase equilibrum diagram shows that for all hypo-eutectoid steels there is a zone between the upper (A_3) and lower (A_1) critical temperatures where both ferrite and stable austenite are present. Thus there is an additional zone on the time-temperature transformation diagram for the ferrite transformations.

The transformations are also shifted to the left of the diagram so that the transformations both commence more quickly and finish more quickly. This leftward shift applies to all hypo-eutectoid steels and the transformation times for such steels are so short that the need for drastic cold-water quenching becomes apparent. For steels below 0.3 per cent carbon the leftward shift is such that the transformation curve touches zero time. Thus it is impossible to cool the steel sufficiently quickly to achieve a totally martensitic structure and some ferrite will always be present. (See Fig. 2.9.)

The rapid cooling required to control the transformations in plain carbon steels can result in cracking and distortion. However, the addition of such alloying elements as nickel, chromium and molybdenum shifts the curves to the right and reduces the cooling rate necessary to produce a martensitic structure. Therefore, these alloying elements improve the hardenability of a steel as discussed earlier in Chapter 1. Some typical time-temperature transformation diagrams for simple alloy steels are shown in Fig. 2.7. Except for the 1 per cent nickel steel, quenching in oil is adequate and there is less chance of cracking and distortion. (See also Fig. 2.10.)

2.4 Quenching

In volume 1 it was stated that the faster a component made from a plain carbon steel was quenched to room temperature, the harder it became for a given carbon content. The truth of this statement was proved in the foregoing section (2.3) where it was shown that:

(*a*) for a martensitic structure to be formed, cooling from the austenising temperature has to be too rapid for the bainite transformations to commence; and
(*b*) the final temperature at the end of the quenching process has to be sufficiently low to ensure the transformation from unstable austenite into martensite.

Under production conditions, 100 per cent transformation into martensite is never achieved in plain carbon steels. Firstly, the change from austenite (γ iron) to martensite (α iron) is accompanied by structural changes in the steel (FCC$\rightarrow$BCC) which results in volume changes. Except for very thin components, these changes will occur at the outer layers of the component some time before they occur at the core of the component where cooling is slowest. This results in cracking and distortion of the component. Secondly, it is impractical to maintain the quenching bath at the M_f temperature ($-50°C$) since all liquid quenching media are frozen at this temperature.

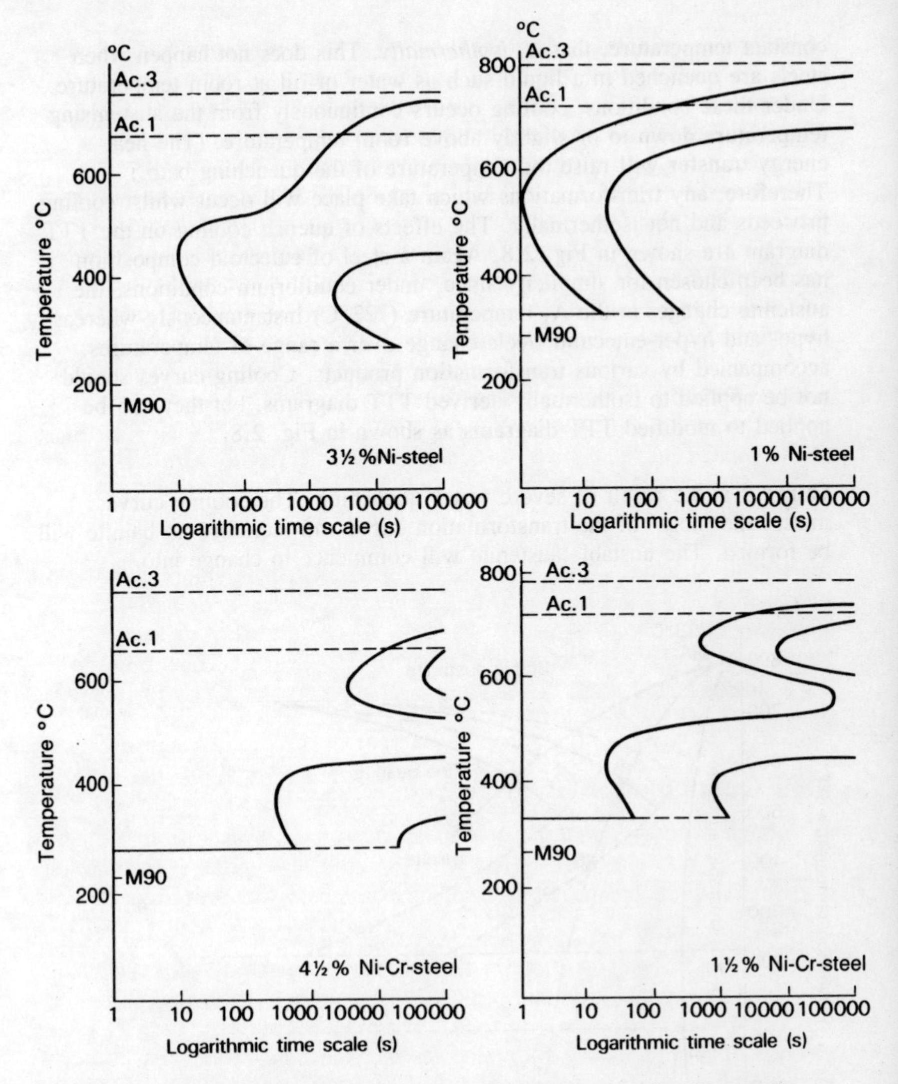

Fig. 2.7 Some typical time/temperature transformation curves for alloy steels

Thus quenching becomes a compromise between minimising the chance of cracking and/or distorting the work and achieving the maximum practical transformation of austenite into martensite. To this end, the process of quench hardening will now be considered in greater detail.

The time-temperature transformation diagrams discussed in Section 2.2 were derived by quenching the austenised steel in an incubation bath whose temperature lay between 723°C and 250°C and holding the specimen at the incubation bath temperature whilst transformation of the metal structure took place. The transformations therefore took place at a constant temperature, that is, *isothermally*. This does not happen when steels are quenched in a liquid such as water or oil at room temperature. Under these conditions, cooling occurs continuously from the austenising temperature down to or slightly above room temperature. (The heat energy transfer will raise the temperature of the quenching bath.) Therefore, any transformations which take place will occur whilst cooling proceeds and not isothermally. The effects of quench cooling on the TTT diagram are shown in Fig. 2.8. Again a steel of eutectoid composition has been chosen for simplicity since, under equilibrium conditions, the austenite changes at the A_1 temperature (723°C) instantaneously whereas hypo- and hyper-eutectoid steels change over a range of temperatures, accompanied by various transformation products. Cooling curves should not be applied to isothermally derived TTT diagrams, but they can be applied to modified TTT diagrams as shown in Fig. 2.8.

Curve 1 is the result of severe water quenching. The cooling curve misses the 'nose' of the transformation curve and therefore no bainite will be formed. The unstable austenite will commence to change into

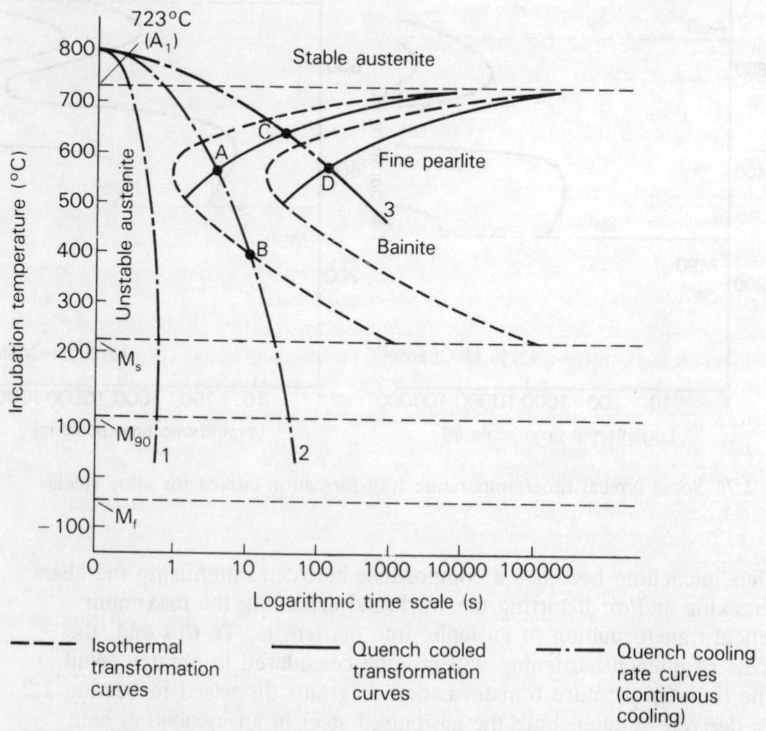

Fig. 2.8 Continuous cooling curves for a plain carbon steel of eutectoid composition (0.83 per cent C)

martensite at the M_s temperature and the percentage change will increase until the metal reaches the temperature of the quenching bath. Since this will normally lie between 0°C and 100°C, some 90 per cent martensite will be present together with some residual austenite. To achieve this condition quenching has been more rapid than the *critical cooling rate*. The critical cooling rate is defined by the cooling curve which just grazes the nose of the appropriate transformation curve for a given steel.

Curve 2 is the result of a less rapid quenching and does not achieve the critical cooling rate. Thus some transformations into bainite will occur between A and B in Fig. 2.8, and the remaining austenite will commence to transform at the M_s temperature and this latter transformation will cease when the temperature of the quenching bath is reached. Thus the final composition of the steel will contain a mixture of martensite together with some softer bainite and even softer residual austenite. Hence the hardness of curve 1 will not be achieved. A thick component could well achieve the conditions of curve 1 at its surface, whilst only achieving those of curve 2 at its core where cooling proceeds more slowly and this effect will be discussed more fully in Section 2.5.

Curve 3 is the result of slow cooling as, for instance, when normalising a component. There will be time for complete transformation of the unstable austenite into fine pearlite together with some upper bainite between C and D in Fig. 2.8, depending upon the section thickness and rate of cooling. Although the temperature must pass through the M_s point, no martensite will be formed since there has been sufficient time for all the unstable austenite to transform completely into pearlite and bainite and there is no residual austenite to transform into martensite.

Figure 2.9 shows the curve for a low carbon steel. It can be seen that the nose of the cooling curve cuts the zero time point of the diagram so that no critical cooling curve can exist. Thus there can be little direct transformation from unstable austenite into martensite, as most of the austenite transforms directly into pearlite and bainite between A and B and only very limited hardening can occur. No attempt is made to quench harden such steels in practice.

It has already been stated that the effect of adding alloying elements such as nickel, chromium and molybdenum can shift the curve of the TTT diagram to the right (Section 2.3). Figure 2.10 shows how a 4.5 per cent nickel-chrome steel can be quenched in oil and still achieve a direct transformation from unstable austenite directly into martensite. Although the cooling curve is less steep than for a water quench, it is still to the left of the 'nose' of the curve and, therefore, exceeds the critical cooling rate. Obviously, there is much less chance of cracking and distortion occurring when steels can be satisfactorily hardened at a slower cooling rate.

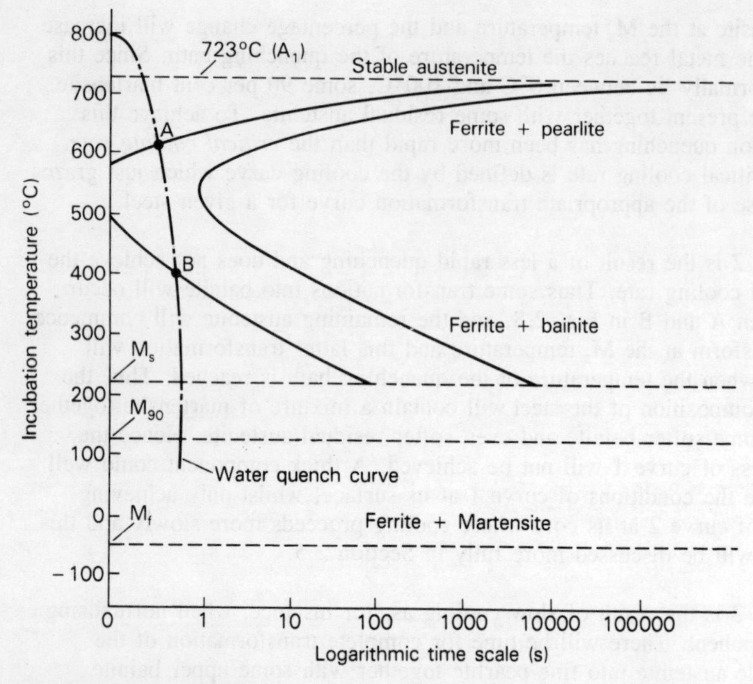

Fig. 2.9 Effect of water-quenching a 0.3 per cent plain carbon steel

2.5 Hardenability

It has already been explained that the hardness of a plain carbon steel depends upon its carbon content and the rate of cooling from the hardening temperature for a given steel. When a thick component is quenched from its hardening temperature it will take longer for the inner core of the workpiece to cool than for the surface layers which are in contact with the quenching medium (water or oil). This leads to a variation in hardness across the section of the material component as shown in Fig. 2.11(*a*). This variation in hardness is referred to as *mass effect*.

Since plain carbon steels have a high critical cooling rate it follows, therefore, that large sections cannot be fully hardened throughout and this is shown in Fig. 2.11(*b*). Thus plain carbon steels have *poor hardenability*. However, a 3 per cent nickel steel containing only 0.3 per cent carbon has a lower critical cooling rate and it will harden uniformly across comparatively thick sections. Such an alloy steel is said to have *good hardenability*.

Hardness and hardenability should not be confused. It has already been stated that a 1.0 per cent carbon steel has poor hardenability compared with a 3 per cent nickel steel containing only 0.3 per cent carbon. Nevertheless, because of its higher carbon content, the 1.0 per cent plain carbon steel will have a very much higher surface hardness.

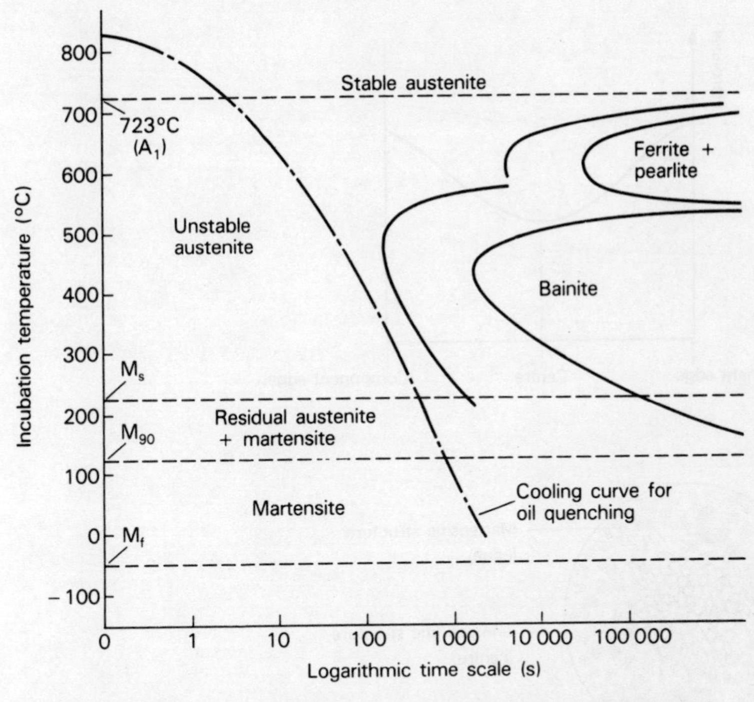

Fig. 2.10 Effect of oil-quenching a 4.5 per cent nickel-chromium steel

Lack of uniformity of structure and hardness in steels with a poor hardenability can seriously affect their mechanical properties. For this reason it is necessary to specify the maximum diameter of bar (*ruling section*) for which the stated mechanical properties can be achieved under normal heat-treatment conditions. Examples of how the ruling section can affect the mechanical properties of a carbon steel and an alloy steel are shown in Table 2.1. One of the main reasons for adding alloying elements such as nickel and chromium to steel is to reduce the mass effect and to increase the ruling section for which the required properties can be achieved. Figure 2.12 shows the effect of ruling section on the transformations in a typical plain carbon steel. The diameter of steel (1) is less than the ruling section and both surface layers and core transform to martensite. The diameter of steel (2) is greater than the ruling section thus, although its surface layers will still be martensitic, its core will be pearlitic.

2.6 The Jominy (end quench) test

This test is used to determine the *hardenability* of steels. It involves heating a specimen to just above its upper critical temperature so that it is fully austenitic, and then quenching it by spraying a jet of water against

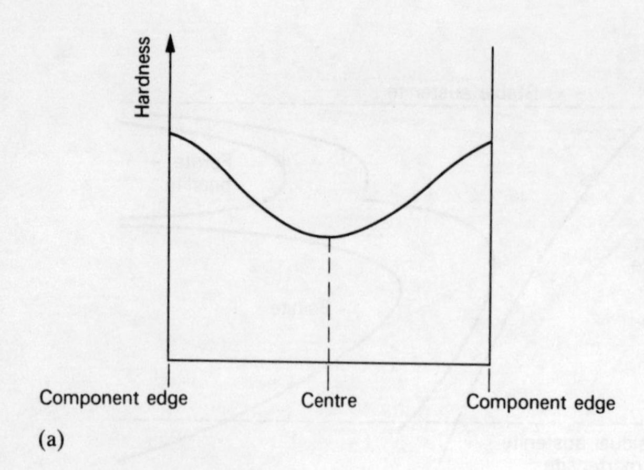

(a)

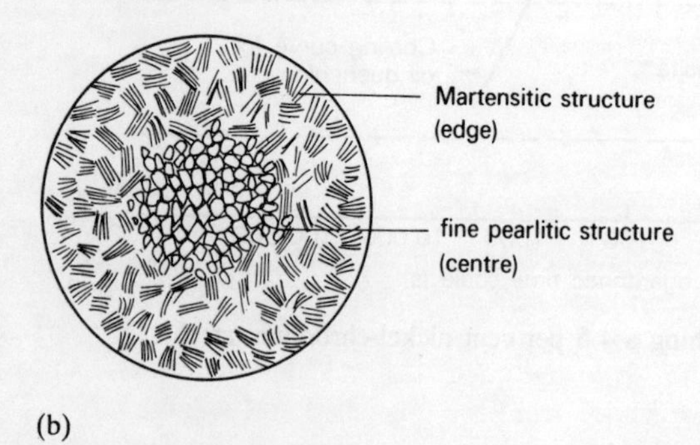

(b)

Fig. 2.11 Mass effect (hardenability) (*a*) cross-section hardenability of a plain carbon steel (*b*) effect of hardenability on structure

Table 2.1 Effect of ruling section on mechanical properties

BS970 Spec	Condition	Limiting ruling section (mm)	Tensile strength (MPa)	Minimum elongation (%)
070M55 (carbon steel)	Hardened and tempered	19	850 → 1000	12
		63	770 → 930	14
		100	700 → 850	14
835M30 (alloy steel)	Hardened and tempered	63	1080 → 1240	11
		100	1000 → 1160	12
		150	930 → 1080	12
		250	850 → 1000	13

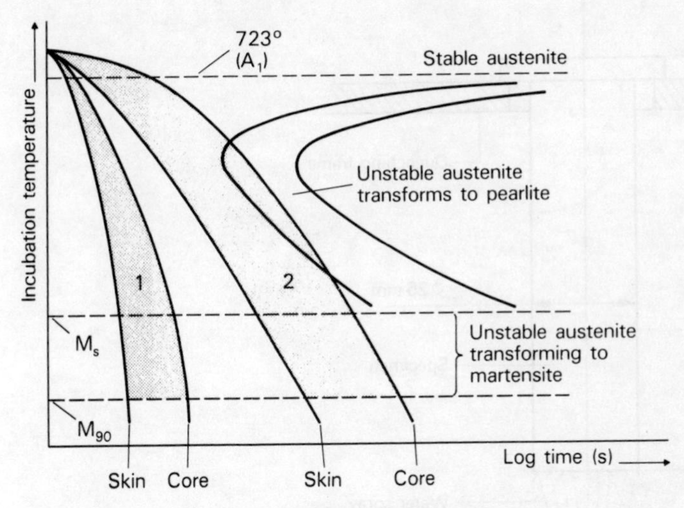

Fig. 2.12 Effect of ruling sections on transformations

its lower end as shown in Fig. 2.13. This figure also shows details of the specimen and the test. The specimen cools very rapidly at the quenched end and progressively less rapidly towards the opposite (shouldered) end. When cold, a flat is ground along the side of the specimen and its hardness is tested every 3 mm from the quenched end. The hardness readings are plotted against distance from the quenched end to give hardenability curves as shown in Fig. 2.14.

It can be seen that the hardness is more uniform for the alloy steel than it is for the plain carbon steel. Thus the alloy steel has better hardenability. This is purely an empirical test and there is no mathematical relationship between the results of the Jominy end test and the ruling section for any particular steel.

2.7 Tempering

Quench hardened plain carbon steels and low-alloy steels are brittle and hardening stresses are present. In such a condition they are of little practical use and they have to be reheated (*tempered*) to relieve the stresses and to reduce the brittleness. Tempering causes the transformation of martensite into the less brittle structure now to be described. Unfortunately, any increase in toughness as a result of tempering is accompanied by some decrease in hardness.

Tempering always tends to transform unstable martensite back to the stable pearlite of the equilibrium transformations. This is because tempering causes the dissolved carbon atoms to precipitate out as iron carbide particles. These particles increase in size as the tempering

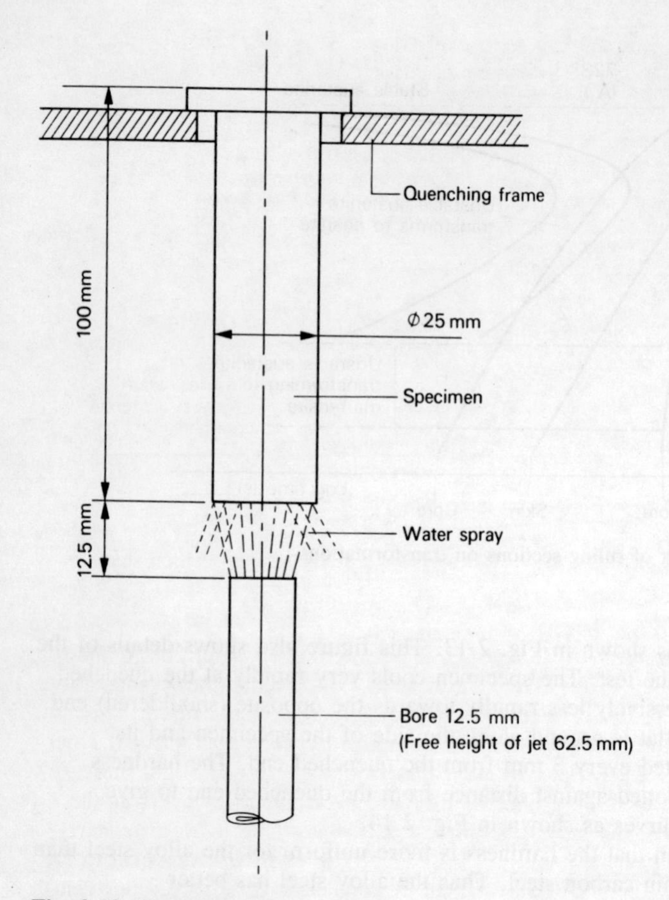

Fig. 2.13 Jominy end test

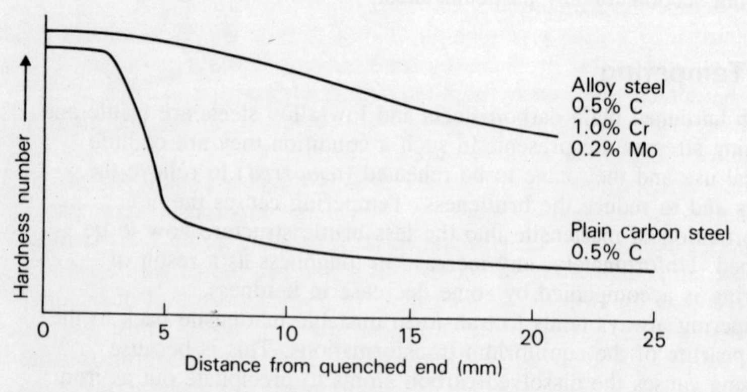

Fig. 2.14 Typical hardenability curves

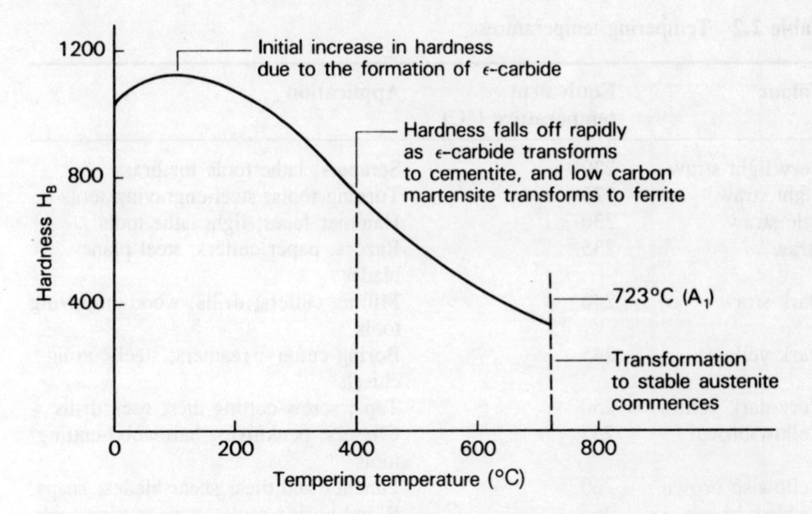

Fig. 2.15 Effect of tempering on hardness for a 0.83 per cent carbon steel

temperature increases. At temperatures between 100°C and 200°C the iron carbide which forms is not the normal Fe_3C (cementite) composition, but ε-carbide (epsilon-carbide) which is different in composition. This leaves the remaining martensite with a reduced carbon content of 0.3 per cent. From Fig. 2.15 it can be seen that at first there is a slight increase in hardness due to the presence of the hard ε-carbide. However, as the temperature rises, the hardness falls off as the unstable martensite starts to transform to pearlite. At 400°C the ε-carbide starts to transform to the more usual Fe_3C composition, and the residual low carbon (0.3 per cent) martensite starts to transform into ferrite accompanied by a reduction in hardness with a corresponding increase in toughness and ductility.

As the tempering temperature continues to rise towards the A_1 temperature (723°C) the precipitation of iron carbide (Fe_3C) results in a structure similar to that produced by spheroidising annealing. In fact, quench hardening followed by high-temperature tempering is often used in place of spheroidising annealing since it is a faster process and gives a more uniform dispersion of the iron carbide.

Plain carbon and low-alloy steels are usually tempered below 300°C where hardness and wear resistance is of primary importance. Examples of tempering temperatures for such steels are listed in Table 2.2. At these tempering temperatures the structure is of a fine pearlite called *troostite*. To differentiate the troostite of tempering from the troostite of quenching, the former is called *secondary troostite* (or just 'troostite'), whilst the latter is called *primary troostite* (or 'bainite'). However, the term troostite is falling into disuse and nowadays it is more usual to use the terms 'tempered martensite' and 'bainite'.

Table 2.2 Tempering temperatures

Colour*	Equivalent temperature (°C)	Application
Very light straw	220	Scrapers; lathe tools for brass
Light straw	225	Turning tools; steel-engraving tools
Pale straw	230	Hammer faces; light lathe tools
Straw	235	Razors; paper cutters; steel plane blades
Dark straw	240	Milling cutters; drills; wood-engraving tools
Dark yellow	245	Boring cutters; reamers; steel-cutting chisels
Very dark yellow	250	Taps; screw-cutting dies; rock drills
Yellow-brown	255	Chasers; penknives; hardwood-cutting tools
Yellowish brown	260	Punches and dies; shear blades; snaps
Reddish brown	265	Wood-boring tools; stone-cutting tools
Brown-purple	270	Twist drills
Light purple	275	Axes; hot setts; surgical instruments
Full purple	280	Cold chisels and setts
Dark purple	285	Cold chisels for cast iron
Very dark purple	290	Cold chisels for iron; needles
Full blue	295	Circular and band saws for metals; screwdrivers
Dark blue	300	Spiral springs; wood saws

*Appearance of the oxide film that forms on a polished surface of the material as it is heated.

2.8 Martempering

This is a process for hardening alloy steels without the risk of distortion and cracking which is present when quench hardening in water or oil. The process consists of heating the steel to its austenising temperature and then quenching it in a salt bath furnace maintained at just above the M_s temperature for the steel. The steel is maintained at this lower temperature until the structure is uniformly heated throughout and the transformation to martensite is complete. At this point the work is removed from the low temperature salt bath furnace and it is allowed to cool naturally to room temperature. The transformations are shown in Fig. 2.16 where it can be seen that both the surface of the metal and its core pass through the M_s to M_{90} range at the same time. This uniform transformation of both the surface and the core of the material, coupled with a relatively mild quench into a salt bath furnace rather than into oil or water at room temperature, results in the risk of distortion and cracking due to internal stresses being reduced to a minimum. A uniform martensitic structure will have been achieved throughout the work and this has to be tempered as for any other hardened steel.

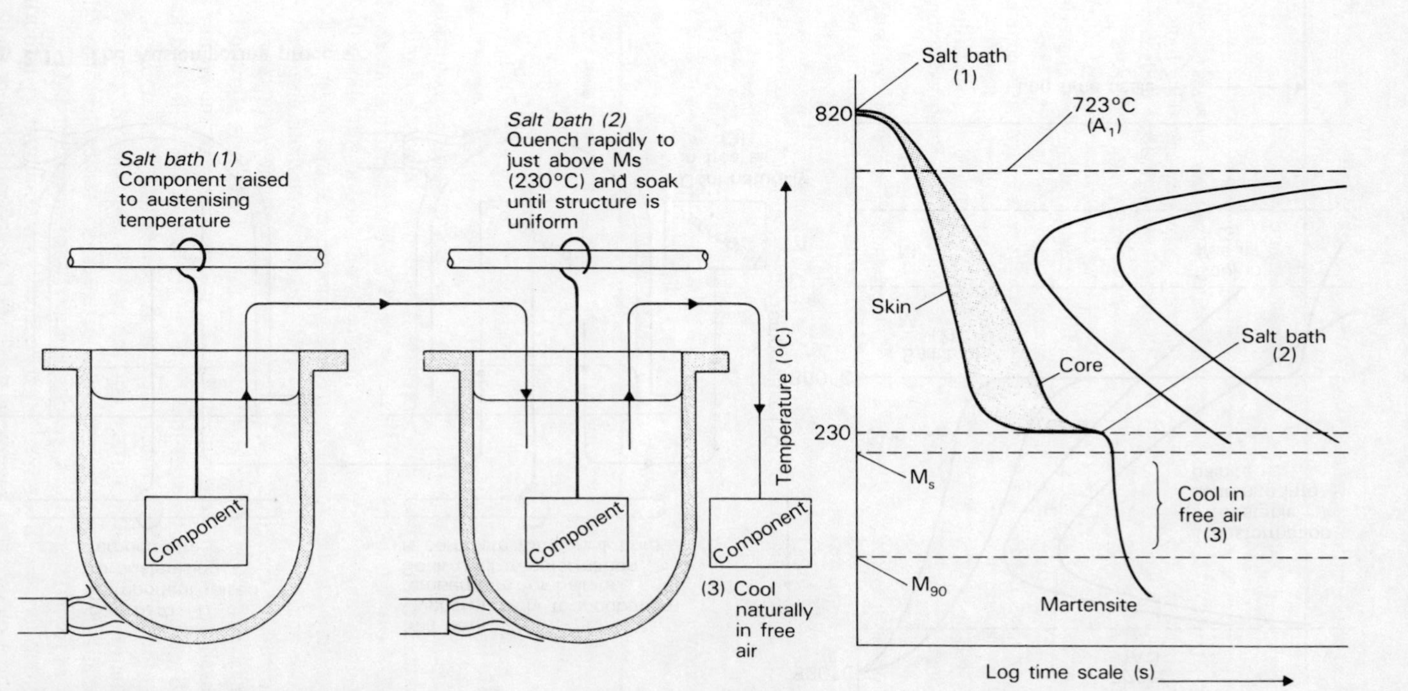

Fig. 2.16 The Martempering process

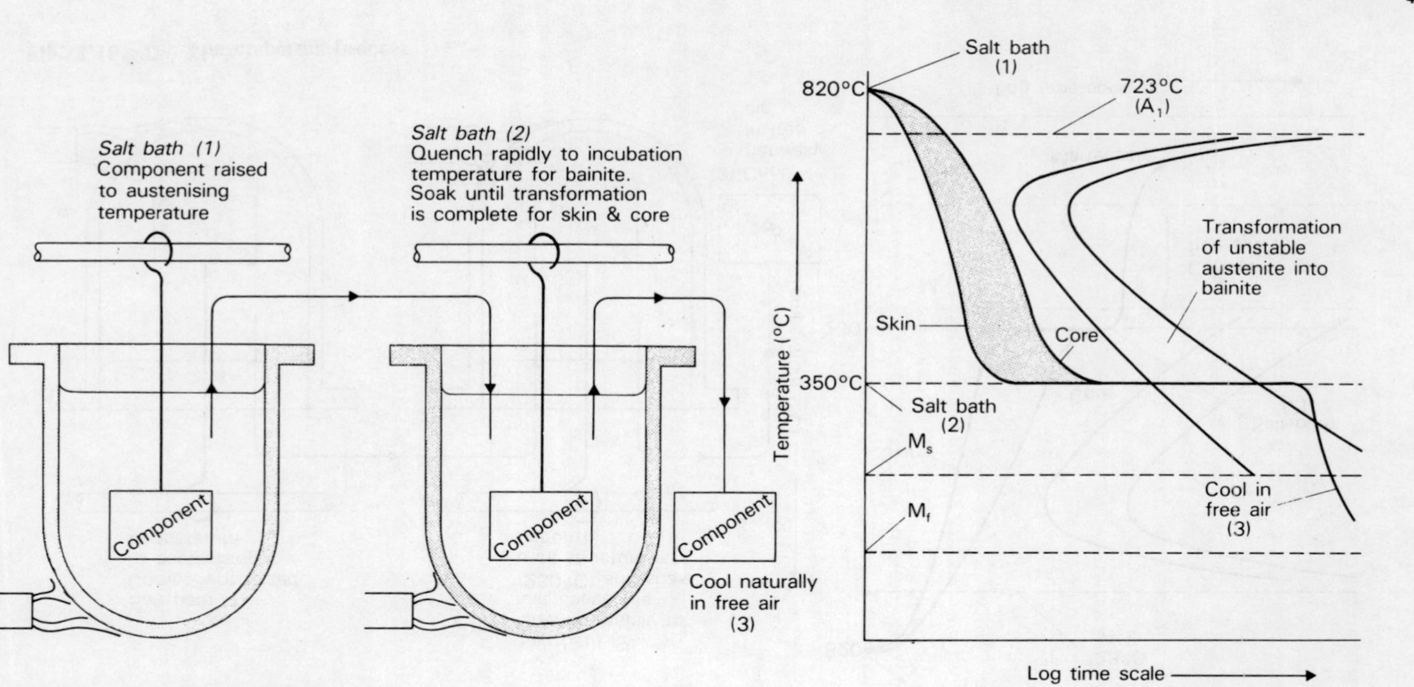

Fig. 2.17 The Austempering process

2.9 Austempering

This process is identical to that used in producing specimens when constructing time-temperature transformation diagrams. Reference to Fig. 2.17 shows that the work is first raised to above its austenising (A_3) temperature and then quenched in a salt bath at the required incubating temperature above the M_s temperature. The work remains in the incubating salt bath until both the surface and the core of the work have transformed into lower bainite. Once the transformations are complete the work can cool down to room temperature naturally. The relatively high temperature of the quenching bath and the slow subsequent cooling reduces the possibility of cracking and distortion due to internal stresses. Austempering is a true isothermal process.

The mechanical properties of the lower bainite structure produced by austempering are similar to those for tempered martensite. Whilst austempering is widely used for the heat treatment of alloy steel components, it can only be used on plain carbon steels of high carbon content and small cross-sections (less than 10 mm thick). This is due to the fact that it is difficult to cool a plain carbon steel quickly enough in the incubation salt bath to avoid the 'nose' of the transformation curve. The lower critical cooling rates (better hardenability) of alloy steels overcome this problem.

2.10 Solution and precipitation treatment of maraging steels

Such steels are annealed by *solution treatment* to absorb the precipitated compounds, in a similar manner to the aluminium-copper alloys, at a temperature of 820°C for a time dependent upon the section thickness. This is usually about 15 to 30 minutes for thin sections and up to 1 hour per 25 mm for thick sections. Air cooling provides an adequate quenching rate because of the high alloy content, and the structure will become martensitic. This is *not* the hard, brittle tetragonal martensite associated with high carbon steels and alloy tool steels, but a BCC martensite which is softer and tougher and capable of being machined and flow-formed.

After processing, the alloy can be precipitation age-hardened, again in a similar manner to aluminium-copper alloys, by heating the alloy to 480°C for 3 hours. The effects of over-ageing are slight even after 200 hours. The exception is the 18Ni2400 grade which needs to be aged for 12 hours, although this time can be reduced by ageing at 540°C. Age-hardening precipitates intermetallic compounds such as $TiNi_3$ and it is these compounds which give maraging steels their characteristic high strength and toughness.

3 Ceramics and composites

3.1 Introduction to ceramics

Ceramics are inorganic, non-metallic materials which are processed and may be used at high temperatures. They consist mainly of silicon chemically combined with non-metallic elements such as oxygen, carbon and nitrogen. Metallic compounds also frequently present. Their internal structure and bonding accounts for their unique properties. Ceramics are used in engineering for a wide range of products including cutting tool tips, abrasives, piezo-electric transducers, insulators, magnets, refractories, and fibres for reinforcement and optical data transmission.

(*a*) *Crystalline ceramics*. These are widely used for cutting tools and abrasives. They may be single-phase materials such as aluminium oxide (corundum), or mixtures of such compounds. Some of the carbides and nitrides also belong to this group.

(*b*) *Amorphous ceramics*. Although exhibiting the characteristics of solids in as much that they are of definite and permanent shape at ambient temperatures, this group of ceramic materials is not crystalline and the molecules are not arranged in regular geometric patterns. They are usually regarded as super-cooled liquids. This group of ceramic materials includes the 'glasses' as used for such applications as glazing, mirrors, optical lenses, reinforcement fibres for GRP products and optical fibres for data transmission.

(*c*) *Bonded ceramics*. This group includes the 'clay' products. These are complex materials containing both crystalline and amorphous consitituents in which individual crystals are bonded together by a glassy matrix after 'firing'. The uses of ceramic products from this group include electrical insulators and refractories for furnace linings.

Ceramic insulators are suitable for use out of doors (electricity grid system) as their hard glaze renders them weather resistant. They can also be used at high temperatures (sparking-plug bodies).

3.2 The basic structure of ceramics

The crystalline group of ceramics may be subdivided into:

(*a*) those with simple crystal structures;
(*b*) those with complex crystal structures.

Examples of those ceramic materials with simple crystal structures are magnesium oxide and silicon carbide. Magnesium oxide is widely used in refractory furnace linings for steel making, and for refractory electrical insulation in mineral-insulated, copper-sheathed cables. Silicon carbide is the 'green grit' abrasive used for sharpening carbide tipped cutting tools.

The element silicon is present in most ceramic materials in the form of complex silicates. (Note that *silica* is silicon oxide, SiO_2, whilst *silicates* are compounds of metal ions with silicon and oxygen — e.g. $MgSiO_3$.) Silicon and carbon have many similar chemical properties and, just as carbon compounds can be built up into long polymer chains to produce 'plastic' materials, silicon-oxygen groups can also be built up into long chain, sheet or three-dimensional framework structures.

3.3 Chain structures of ceramics

Figure 3.1 shows a single SiO_4 group tetrahedron which forms a 'building block' for many different ceramic structures. For example, Figure 3.2 shows a typical double chain structure based on SiO_4 group tetrahedra cross linked by ionically bonded metal ions. Unlike organic polymer materials (plastics) where adjacent chain molecules are only held together by relatively weak Van der Waals forces, the stronger ionic bonds in ceramic materials make them harder and stronger than organic polymers. However the ionic bonds cross-linking adjacent silicate chains are weaker than the covalent bonds linking the silicate groups. Thus ceramic materials tend to fracture along the ionic bonds parallel to the silicate chain. Note that single chains based on SiO_4 groups may also be formed but these are less important.

The differences in behaviour between metallic crystals and ionically bonded ceramic crystals can be explained by reference to Fig. 3.3. In a metallic crystal the positively charged metal ions are prevented from repelling each other by the negatively charged electron 'cloud' surrounding each metal ion. The mutual attraction of the positive metal ions and the negatively charged electron 'clouds' results in a state of equilibrium which holds the metal ions in position in the crystal, providing no external disturbing force is applied. It can be seen in Fig. 3.3 (*a*) that when such an external disturbing force (F) is applied to a plane of metal ions they tend to move along a slip plane, so that the ions

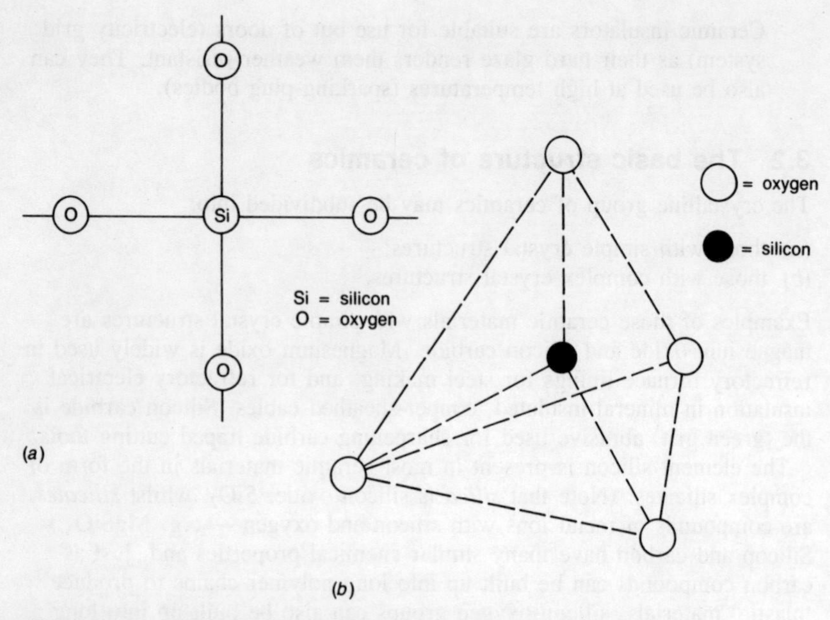

Fig. 3.1 The SiO_4 group (*a*) silicon–oxygen bond (*b*) three-dimensional representation of the SiO_4 group tetrahedron

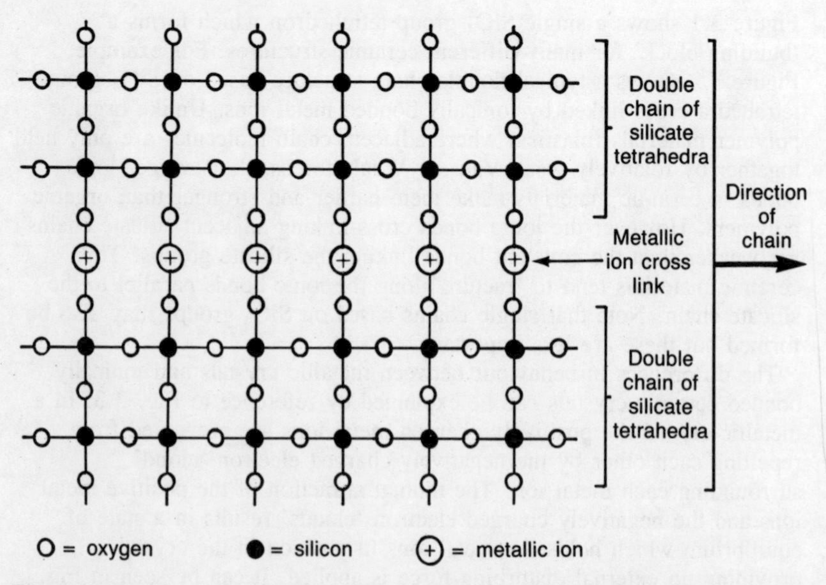

Fig. 3.2 Double chain structure of silicate tetrahedra with metallic ion cross-linking

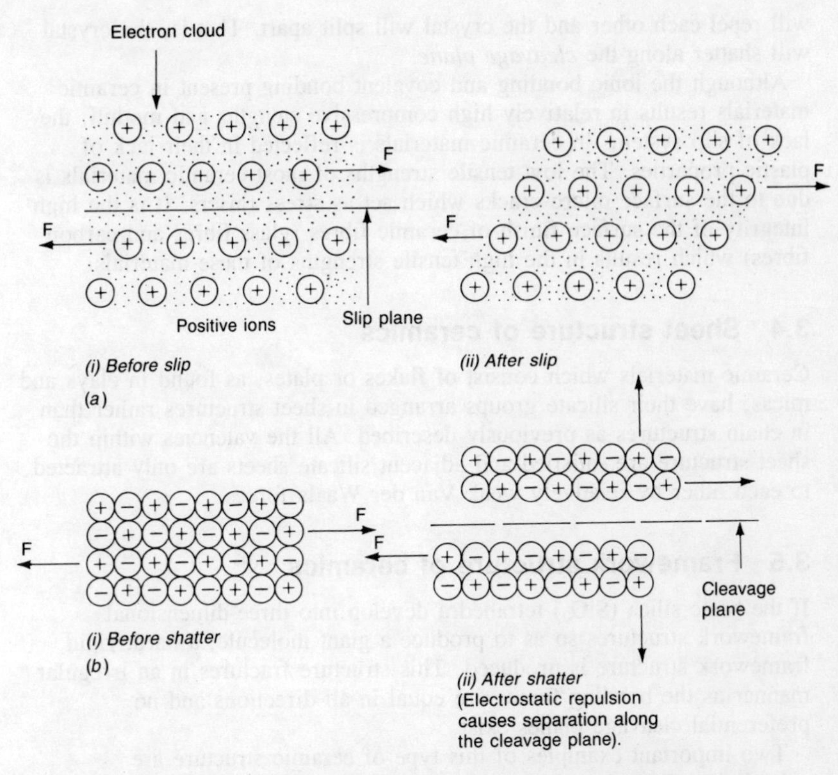

Fig. 3.3 Deformation of crystals (*a*) metallic crystal (*b*) ionic crystal

become nearer to those of an adjacent plane. This results in an increase in the electrostatic repulsion forces between adjacent planes of metal ions and they try to move back to their original postions (spring-back of elastic materials). The repulsion forces reach a maximum at the elastic limit for the material but, if overcome by the external disturbing force, they move to the next incremental position, and equilibrium between the positive metal ions and the negative electron 'cloud' is once more achieved and *slip* will have occurred. The plane along which slip occurs by dislocation is called the *slip plane* and the theory of slip by dislocation was introduced in volume 1, and will be developed further in Section 10.2.

In an ionically bonded ceramic crystal as shown in Fig. 3.3 (*b*), it can be seen that positively charged metal ions and the negatively charged silicate ions are arranged so that each ion is surrounded by ions of opposite charge and this equilibrium system holds the ions rigidly in place. However if an external disturbing force of sufficient magnitude is applied the equilibrium position of the ions will be disturbed. This will result in ions with like charges being brought closer together so that they will repel each other and the crystal will split apart. That is, the crystal will shatter along the *cleavage plane*.

Although the ionic bonding and covalent bonding present in ceramic materials results in relatively high compressive stengths and moduli, the lack of slip systems in ceramic materials is reflected in their lack of plastic properties. The low tensile strengths of most ceramic materials is due to the surface micro-cracks which act as stress raisers. It is the high integrity of the surface finish of ceramic fibres (glass fibres and carbon fibres) which results in the high tensile strengths of these materials.

3.4 Sheet structure of ceramics

Ceramic materials which consist of flakes or plates, as found in clays and micas, have their silicate groups arranged in sheet structures rather than in chain structures as previously described. All the valencies within the sheet structure are satisfied and adjacent silicate sheets are only attracted to each other by relatively weak Van der Waals forces.

3.5 Framework structure of ceramics

If the basic silica (SiO_2) tetrahedra develop into three-dimensional framework structures so as to produce a giant molecule, a hard, rigid framework structure is produced. This structure fractures in an irregular manner as the bonding forces are equal in all directions and no preferential cleavage planes exist.

Two important examples of this type of ceramic structure are *cristobalite* which is used in refractory bricks for high temperature furnace linings and *quartz* which occurs naturally in sandstones and in beach sands. Quartz is used as an abrasive (sand paper) and, industrially made quartz crystals are used for resonant frequency applications (e.g. in electronic oscillator circuits). Quartz is extremely resistant to solution in water, hence its long life in beach sands. Unfortunately quartz is unsuitable for high-temperature applications since the Si–O–Si bonds change their angles at 573°C. Since silica is a rigid compound, this change of angles causes an abrupt change of volume resulting in cracking, unless the rate of heating or cooling is very slow.

3.6 Refractory products (clays)

These are ceramic materials whose silicate tetrahedra form sheet structures as described earlier. The flaky particles of dry clay are transformed into a plastic mass (dough) by the addition of water. The water molecules are attracted to the surface layers of the 'sheets' by polarisation forces and provide boundary layer lubrication between adjacent sheets. Clays are frequently 'blended' by mixing two or more ingredients to control the composition and plasticity. This is particularly important when producing high-grade articles such as porcelain insulators, chemical ware and whiteware. The blend consists of finely ground clay mixed with such non-plastic materials as ground quartz, flints and fluxes. The ground quartz and flints are non-absorbent and reduce the amount of water required to make the blend plastic. This reduces shrinkage, distortion and cracking during drying. Fluxes are added to reduce the fusion point of vitreous (glass) content of the blend and thus reduces the firing temperature required for complete vitrification. Clay products may be shaped by hand or by machine moulding.

Drying is used to remove water from the moulded mass before firing. Excessive drying shrinkage can lead to distortion and even cracking; therefore drying ovens should operate at between 85°C and 96°C and the atmosphere of the oven should have a high humidity. This prevents the surface of the mould clay product drying out before the core of the product and ensures uniform drying throughout the mass of the product.

Firing vitrifies the moulded and dried clay product so as to give it strength, rigidity and durability. Firing temperatures depend upon the composition of the clay and the physical characteristics of the product. The process of firing takes place in a series of stages as follows.

110°C–260°C. The final shrinkage water and pore water (hygroscopic water) is driven off.

430°C–650°C. The clay minerals break down into silica and alumina and any chemically combined water molecules are driven off as water vapour. At this point the clay loses its ability to absorb water and can never again become plastic dough. However there is, as yet, very little change in the strength and porosity of the moulding.

800°C–900°C. At this temperature an oxidising atmosphere should be maintained in the furnace to burn off residual organic material and to oxidise any iron pyrites which may be present as an impurity. All gas-forming reactions should also be completed at this stage whilst the moulding is still porous and such gases as are produced can easily escape.

900°–1000°C. At these temperatures vitrification and firing shrinkage begins to take place. Vitrification is the gradual fusion of some of the compounds present to form a liquid which fills up the pores in and between the particles forming the moulding. Full vitrification is reached at about 1400°. Further heating is unnecessary and could lead to the softening and collapse of the moulding. When cooled to ambient temperatures this vitreous liquid solidifies to provide a matrix which cements together the inert particles. This decreases the porosity and increases the strength of the product.

3.7 Refractory products (common)

These are heat-resistant products such as firebricks and fireclays which are used for furnace linings. Refractoriness is defined as the ability of a material to withstand high temperatures without appreciable deformation under service conditions. It is generally assessed by the softening or melting point of the material and, since refractory materials in service are subjected to loads of varying intensity, it is important to assess their refractoriness under similar loads. Some refractories, such as high alumina (aluminium oxide) firebricks and fireclays, soften gradually over a wide range of temperatures and collapse under load well below their true fusion temperatures. Others, such as silica (silicon oxide), soften over a narrow range of temperatures and continue to exhibit good load-bearing properties close to their fusion temperatures. However, the melting temperature of silica is very much lower than that of alumina and this must be taken into account when determining their relative usefulness for load-bearing refractories.

The majority of commercial refractories used in high-temperature processes and environments are represented by the 'common' refractories because of their availability and low price. Such refractories consist of crystalline and some amorphous constituents cemented together by a vitreous (glassy) matrix. One of the most widely used range of 'common' refractories are based upon alumina and silica compositions and vary from almost pure silica through a wide range of alumina silicates, to almost pure alumina.

The silica-alumina phase equilibrium diagram is shown in Fig. 3.4. Note the *Cristobalite* and *Tridymite* are allotropes of silica, and *Mullite* is the name given to the composition formed when excess alumina reacts with residual silica ($3Al_2O_3.2SiO_2$). This composition corresponds, by

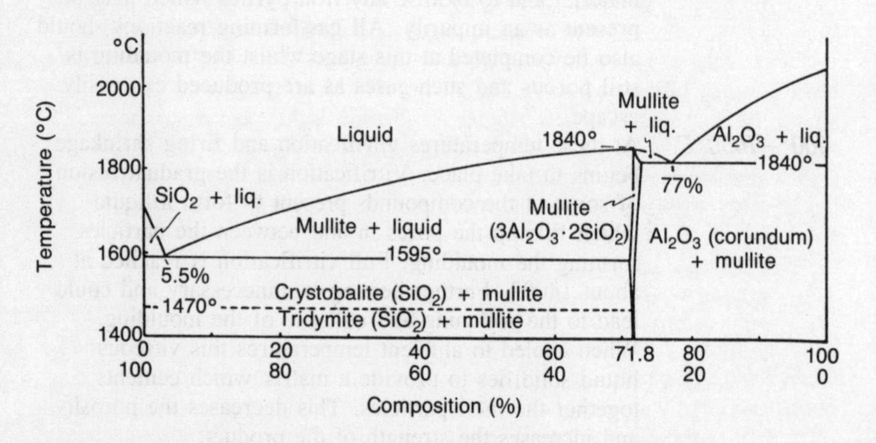

Fig. 3.4 Silica-alumina phase equilibrium diagram

weight, to 71.8 per cent alumina and 28.2 per cent silica. *Corundum* is the name given to crystalline alpha alumina which forms when excess mullite dissociates at 1840°C.

A detailed explanation of the silica-alumina phase equilibrium diagram is beyond the scope of this book, but a number of practical points should be noted.

(*a*) Figure 3.4 shows that refractories containing between 4 per cent and 8 per cent alumina should be avoided since they melt completely at or a little above the eutectic temperature of 1595°C.

(*b*) Refractoriness increases with an increasing alumina content and the load-bearing properties of the material also increases. This may appear to contradict an earlier statement that alumina has a wide softening range and tends to collapse before it melts. However, it must be remembered that silica has a relatively low melting point compared with alumina and that refractoriness depends upon the amount of liquid present and the viscosity of that liquid.

(*c*) Silica-alumina refractories soften over a wide range of temperatures. Softening commences at the temperature at which the silica starts to melt and liquid starts to form, and extends to the temperature at which the refractory is totally molten. As the alumina content increases above the eutectoid point of 5.5 per cent alumina, the quantity of liquid produced at 1595°C will decrease and the refractory will become stronger at that temperature.

(*d*) For severe conditions where the temperature exceeds 1800°C the alumina content must exceed 71.8 per cent to yield a solid phase containing only corundum and mullite in a vitreous matrix.

Apart from silica and alumina, other 'common' refractories are magnesite, forsterite, dolomite, silicon carbide and zircon. Refractories can be classified chemically as acid, basic or neutral. Magnesite is a basic refractory found in the linings of iron and steel making plant where high phosphorous content ores are used. The phosphorous content in the metal is reduced to a safe level because the phosphorous reacts with the basic furnace lining to form a basic slag. This slag provides a cheap source of high phosphorous content fertilizers. Although the furnace lining is gradually eaten away this process allows low cost, low grade ores to be used. Silica brick is an acid refractory, and silica-aluminium brick is a neutral refractory providing the alumina content is high enough.

3.8 Refractory products (high grade)

The 'common' refractories discussed so far depend upon a ceramic bond when fired. The term ceramic bond, as used in the ceramic industry, is not a true chemical bond but a cementing or binding together of refractory particles in a vitreous (glassy) matrix. This ceramic bond is responsible for the cold strength of ceramic materials but its presence reduces the refractoriness of the material at high temperatures. Thus for 'high grade' refractories, where strength is required at high temperatures, the vitreous matrix should be reduced to the minimum necessary to provide adequate strength at room temperature.

When the temperature during firing is sufficiently high and the duration of the firing process sufficiently long, the glassy matrix may be gradually replaced by crystals. This results from the dissolution of some of the compounds and a change in the crystal structure of mullite. This results in an interlocking crystal structure which greatly improves the bonding, and, therefore, the strength of the material under load at high temperatures (refractoriness).

Phase equilibrium diagrams for refractories, such as the example shown in Fig. 3.4, indicate that generally higher refractoriness can be obtained by using pure, high melting point oxides since no eutectoid compositions will be present. The presence of even small quantities of impurities not only lowers the melting point considerably but also reduces the refractoriness under load to a greater extent than is indicated by the phase equilibrium diagram. Thus, 'high grade' refractory materials consist of very pure basic ingredients and the elimination of the ceramic bond by special processing techniques. Some individual 'high grade' refractories will now be considered in greater detail.

3.9 Oxides

Pure oxide refractories consist essentially of such oxides as alumina, beryllia, magnesia, thoria, and zirconia. This is not an exhaustive list but represents the more readily available refractory oxides. Products from these materials are made by slip-casting, pressing, and extrusion. Firing takes place at about 1800°C at which temperature the fine oxide particles sinter rapidly and solid surface reactions occur between the individual particles. This results in a *crystalline bond* which produces a coherent mass and the fired article so produced is said to be a *self-bonded* refractory. Since the ceramic bond is eliminated there is no glassy matrix with its low melting point to reduce the refractoriness of the product. The crystalline bond is composed of crystals of the same materials as the particles being bonded and this results in self-bonded products having a refractoriness approaching that of the particle material itself.

3.10 Borides

Borides exhibit the properties of very high hardness coupled with good resistance to chemical attack and, depending upon the boride, melting points ranging from 1800°C to 2500°C. Although borides are more resistant to oxidation at high temperatures than carbides, they do start to oxidise at about 1400°C. The more commonly available industrial borides are those of chromium, molybdenum, titanium, tungsten, and zirconium.

3.11 Nitrides

Although nitrides have high melting points, they have a low resistance to oxidation and chemical attack generally. However, 'Borazon', which is a synthetic boron nitride produced under high temperatures and pressures, has a hardness approaching that of diamond and can withstand temperatures up to 1925°C without appreciable oxidation.

3.12 Carbides

Although carbides have very high melting points, they lack resistance to oxidation at high temperatures. The more important refractory materials are the carbides of boron, silicon, titanium and zirconium. Carbides for cutting tools will be considered in Chapter 4. Silicon carbide (carborundum) refractories are the oldest and the most widely used, their properties depending upon the type and quantity of bond.

The most widely used binding agent is refractory clay which forms a ceramic bond on firing. However, as with all ceramic bonds, the refractory produced starts to soften at the relatively low temperatures of 1200°C to 1500°C. Other binding materials used are silicon nitride and silicon. When silicon nitride is used as the binding material the refractory produced has high strength and a high resistance to thermal shock. Silicon has only limited application as a binding material since its melting point is only 1426°C.

Self-bonded silicon carbides are formed by mixing them with temporary organic binding agents, after which they are pressed to shape and fired at 1700°C. The temporary binder is burnt off and a crystalline bond between the silicon carbide particles develops. This results in a product of high refractoriness, high strength, high density, high abrasion resistance, and high resistance to chemical attack. All silicon carbide refractories have high thermal conductivity and low coefficients of thermal expansion but, unfortunately, they tend to oxidise slowly to silica in the temperature range 900°C to 1300°C.

Boron carbide is very hard and abrasion resistant. Components are produced by hot pressing or by bonding with sodium silicate and firing. Although the melting point of boron carbide is 2450°C, its maximum usable temperature is restricted to about 980°C as boron carbide oxidises rapidly at higher temperatures and it also reacts with hot or molten ferrous metals.

Cerium, molybdenum, niobium, tantalum, tungsten and zirconium carbides can be used at temperatures above 2000°C in neutral or reducing atmospheres, whilst niobium, titanium and vanadium carbides can be used above 2500°C in an atmosphere of nitrogen. Hafnium carbide has the highest melting point of any known substance at 2900°C.

3.13 General properties of crystalline ceramics

The ceramic materials considered so far have been crystalline or largely crystalline with some amorphous content. All these materials share the following general properties to a greater or lesser degree.

Refractoriness

The refractoriness of the various groups of ceramics has already been dealt with in some detail. All ceramic materials have a greater refractoriness than most metals. The melting temperatures of some typical high purity ceramic materials are listed in Table 3.1.

Table 3.1 Melting points of some refractory ceramics

Refractory material	Melting point °C
Halfnium carbide	3900
Tantalum carbide	3890
Thorium oxide	3315
Magnesium oxide	2800
Zirconium oxide	2600
Beryllium oxide	2550
Aluminium oxide	2050
Silicon nitride	1900

Strength

Ceramic materials have high compressive strengths compared with their tensile strengths and the ability to retain this strength at high temperatures is one of their most important properties. Titanium diboride, for example, has a compressive strength of 250 MPa at 2000 °C which makes it one of the strongest materials known at such a high temperature. The lack of tensile strength in ceramic materials is due to the presence of micro-cracks. These act as potential stress raisers and the lack of ductility in ceramic materials prevents any stress concentrations at the micro-cracks from being relieved by the onset of plastic flow. The reason why ceramic materials lack ductility and suffer from sudden cleavage was explained in Section 3.3.

Hardness

Ceramic materials are harder than any pure or alloyed metallic materials even after heat treatment. This hardness makes ceramics useful as abrasives and as cutting tool tips, and they will be referred to again in this context in Chapter 4. The Knoop hardness number for some hard ceramic materials is listed in Table 3.2.

Table 3.2 Hardness of some ceramic materials (room temperature)

Ceramic material	Knoop hardness number
Cubic boron nitride*	7000
Boron carbide	2900
Silicon carbide	2600
Aluminium oxide	2000
Beryllium oxide†	1220

* compare with DIAMOND, Knoop hardness 7000
† compare with quench hardened, high carbon STEEL, Knoop hardness 700

Electrical properties

Ceramic materials have been used for electrical insulation purposes for a long time. Glazed porcelain insulators are used for such purposes as supporting high- and medium-voltage overhead electric cables and also telephone and telegraph cables. The hard glaze prevents the insulators from 'weathering'. Softer plastic insulators would be unsuitable for such purposes since they would quickly 'weather' and become roughened. They would then become covered in dirt from atmospheric pollution, and this surface layer of dirt would provide a conducting path over the insulator leading to a 'flash-over'. Any dirt deposited on hard glazed procelain is quickly washed away by rain as it cannot adhere to the smooth surface. Unglazed ceramics are used for formers for wire-wound resistors and for heating elements.

Magnesium oxide powder is used in mineral insulated, copper sheathed cables and in sheathed heating elements. This material has the advantage that it is able to retain its insulating properties at very high temperatures which would destroy more conventional insulating materials. Ceramic materials are also used for low-loss high-frequency insulators, ferromagnets, and semiconductor devices.

3.14 Shaping methods

Traditional shaping methods for ceramics (some of which have already been mentioned) consist of hand and machine moulding, powder pressing, the extrusion and rolling of a plastic, clay-water mixture, and 'slip-casting'.

Slip-casting

This process consists of filling a porous mould with a liquid suspension of the powdered clay (slip). On standing, the powder forms a deposit on the walls of the mould. After sufficient time has elapsed for the deposit to have acquired the specified thickness, the surplus slip is poured off leaving the 'casting' behind in the mould. This technique is used for the production of thin-walled products. The casting is dried to give it sufficient strength to withstand handling, after which its is removed from the mould and fired.

Isostatic pressing

Mass produced components such as sparking plug bodies can be made by the compaction of ceramic powder in dies, followed by *sintering*. Whereas 'firing' converts the amorphous ceramic particles in the mix into a glassy vitreous 'bond' or matrix, sintering is the heating of the compact of high purity crystalline ceramic material, in a controlled atmosphere to prevent oxidation, to a temperature which is sufficient to cause diffusion and recrystallisation across the particle boundaries causing them to bond together.

Simple pressing in metal dies leads to non-uniform density distribution in the compact resulting in the distortion and cracking of the final product. This can be overcome by the use of *isostatic pressing*. In this technique the powder to be compacted is placed in a strong rubber bag mould and subjected to external fluid pressure, as shown in Fig. 3.5. This ensures uniform compaction in all directions. Unfortunately this technique does not allow the forming of complex shapes of high

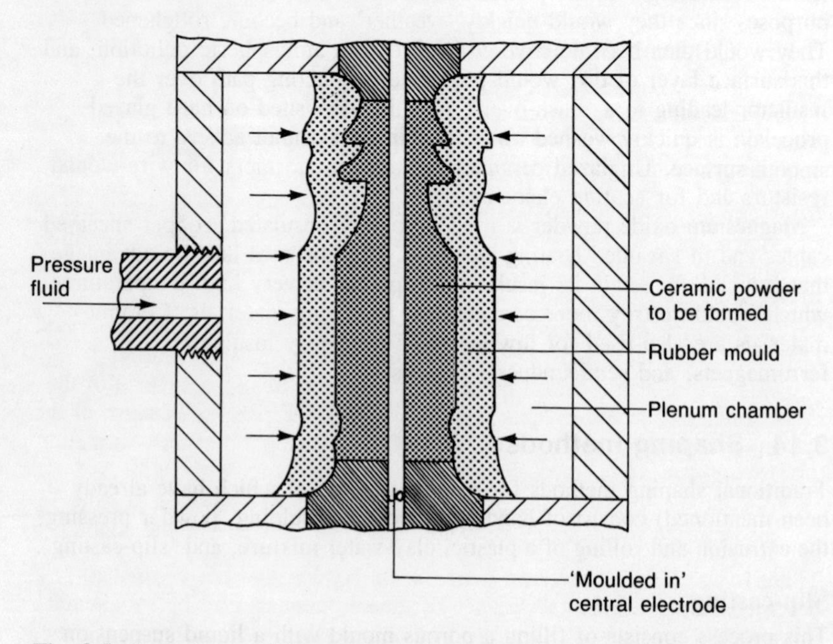

Fig. 3.5 Isostatic mould for sparking-plug body

accuracy. However, machining of a compact, which has been formed by isostatic pressing, prior to sintering can be used to improve the accuracy of the finished product.

Hot-pressing

Unfortunately ordinary sintering processes do not increase the density of many materials. This applies particularly to materials which dissociate at the high temperatures involved in sintering. For example silicon carbide dissociates to silicon and carbon at temperatures exceeding 2300°C. However, many of these materials can be densified at lower temperatures by the simultaneous application of pressure and heat. Hot pressing results in densities approaching the theoretical maximum and also results in a finer grain structure. Unfortunately this process is not only expensive but it can only be applied to simple shapes such as equiaxial cylinders. Uniform densification becomes difficult when the length/area ratio becomes large.

Gas pressure bonding

The Battelle process of gas pressure bonding is, in fact, hot isostatic pressing. The rubber bag is replaced by a flexible, thin-walled, metal pressure vessel and the liquid is replaced by hot pressurised gas. Alumina can be densified to 99.3 per cent of its theoretical maximum using gas at a temperature of 1290°C and a pressure of 69 MPa.

3.15 Ceramic coatings

Flame spraying

There are two basic variants of the flame-spraying technique for depositing ceramic coatings.

(*a*) Ceramic powder is melted as it is passed through an oxy-acetylene flame, an oxy-hydrogen flame, or a plasma-arc. The molten ceramic material is sprayed onto the surface of the component in a similar way to metal spraying.

(*b*) A prefabricated ceramic rod is used in place of the powder described above. The heat source both melts and 'atomizes' the ceramic material and the molten particles are sprayed onto the surface of the component. Although this is a more expensive process because of the cost of prefabricating the ceramic rods, it ensures that only molten particles are sprayed onto the component surface. In the process previously described there is a danger that some unmolten powder is also carried over in the spray, reducing the integrity of the coating.

There is no chemical reaction between the coating and the component being coated, therefore the surface of the component must be roughened to provide a mechanical 'key'. In fact better adhesion is achieved if the surface to be protected is sprayed with metal prior to spraying with a ceramic material. Sprayed ceramic coatings are not impermeable and, therefore, only offer limited protection against oxidation or chemical corrosion of the substrate. Although the substrate is cold in most spraying processes, higher densities and lower permeabilities can be achieved if the substrate is heated before and during the spraying process.

Chemical vapour deposition (CVD) consists of heating the substrate and passing a vapour or mixture of vapours over it. Reactions take place at the surface with the deposition of a dense impermeable coating. For example, passing methyl trichlorosilane (CH_3SiCl_3) over heated graphite results in the deposition of a layer of silicon carbide on the surface of the graphite component. The coefficients of thermal expansion of the coating and the substrate must be closely matched to prevent cracking or peeling of the coatings. Deposits of this type (e.g. titanium carbide on steel) are useful for increasing the resistance of the component to abrasive wear and to chemical attack.

Thin-walled components have been manufactured experimentally by building up the thickness of the deposit and then removing the substrate. For example, thin-walled impermeable silicon carbide tubes have been manufactured by depositing the silicon carbide on a graphite mandrel and then burning away the graphite.

3.16 Finishing processes

Many ceramic products used in engineering applications require some finishing process. Since fired ceramics are very hard they can only be shaped or finished by grinding using diamond-impregnated grinding wheels or machining using diamond-tipped tools. Therefore it is economically important to ensure that the fired component only requires the minimum of machining. Although finishing improves the dimensional accuracy of a fired component, the surface finish produced by grinding or machining tends to be unsatisfactory for engineering applications due to the porosity of the unglazed material. Pore-free ceramics are now available which can be finished to high accuracies and very low values of surface roughness. However such materials and processes tend to be costly.

3.17 Joining ceramics to metals

To obtain a satisfactory joint between ceramic materials and metals, it is essential to ensure that the materials being joined have closely matching coefficients of thermal expansion. Also, if there are any residual stresses in the assembly, these must be such that the ceramic component is maintained in compression.

Metals and ceramic materials can be joined using intermediate layers of glass as a bond. Alternatively, the ceramic component in the assembly can be metallised with silver or nickel and soft-soldered to the metal component. Some of the strongest joints produced by the latter process are used in the manufacture of high-power thermionic valves used for such purposes as power amplifiers in transmitters. For this application, high-temperature metallisation of the ceramic component is carried out using a molybdenum-manganese alloy or titanium. The layer so formed is built up by nickel plating and then brazed to the metal component of the assembly.

3.18 Glass

Unlike the ceramic materials considered so far, glass can be considered as a product of the fusion of inorganic materials which has cooled to a rigid condition without crystallisation. That is, it behaves as a super-cooled liquid. At high temperatures, when molten, glasses form normal liquids. Their atoms are free to move and they respond readily to shear stresses. In fact molten glass can be used as a lubricant for some metal-forming processes where extreme pressures are involved. As super-cooling takes place without crystallisation, a sudden change in the coefficient of thermal contraction occurs, as shown in Fig. 3.6, due to

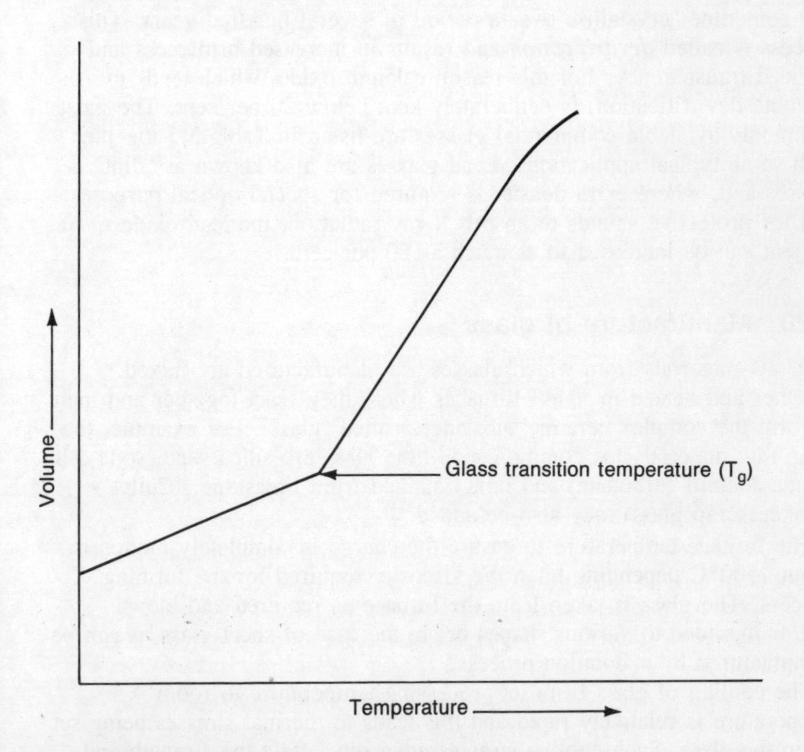

Fig. 3.6 Glass transition temperature (T_g) for silicon glass

the rearrangement and more efficient packing of the atoms. this is the glass transition (T_g) or *fictive* temperature for glass. Below this temperature no further rearrangement of the atoms takes place and any further contraction is due only to reduced thermal agitation of the atoms. Other physical properties which change at the glass transition temperature are: density, viscosity, refractive index, and electrical resistivity. Although an excellent insulator at ambient temperatures, glass becomes a conductor of electricity at red heat.

3.19 Types of glass

Although silica is, theoretically, an excellent glass-making material, its high melting temperature makes it uneconomical to use alone and basic metal oxides are added to lower the fusion temperature and viscosity of the melt. A eutectic mixture of 75 per cent silica with 25 per cent sodium oxide reduces the fusion temperature to 793°C and results in the formation of sodium disilicate ($Na_2O \cdot 2SiO_2$). Unfortunately such a glass would be water soluble and unsuitable for all practical purposes. The addition of calcium oxide to the mixture produces soda-lime glass which is insoluble in water and widely used for window glass and bottles. Glass will sometimes crystallise over a period of several hundred years. This process is called *devitrification* and results in increased brittleness and reduced transparency. For this reason calcium oxide, which tends to promote devitrification, is deliberately kept below 20 per cent. The most commonly available commercial glasses are listed in Table 3.3 together with some typical applications. Lead glasses are also known as 'flint' glasses and, where extra density is required for special optical purposes and for protective shields to absorb X-ray radiation, the lead oxide content can be increased to as much as 80 per cent.

3.20 Manufacture of glass

The raw materials from which glasses are manufactured are mixed together and heated in 'tank' furnaces where they react together and melt to form the complex ceramic substances called 'glass'. For example, the main raw materials for common-soda lime glass are silica sand, soda ash (crude sodium carbonate) and lime obtained from limestone. 'Cullet' (broken scrap glass) may also be added.

The furnace temperature to ensure the charge is completely molten is about 1500°C depending upon the viscosity required for the forming process. The glass is taken from the furnace as required and blown and/or moulded to various shapes or, in the case of sheet glass, it can be manufactured by a flotation process.

The cooling of glass from its processing temperature to room temperature is relatively rapid and this leads to thermal stresses being set up in the glass. Such cooling stresses adversely affect the strength and other physical properties of the glass, but they can be relieved by suitable

Table 3.3 Types of glass

Type of glass		Composition								Properties and Uses
		SiO_2	Na_2O	CaO	MgO	K_2O	PbO	Al_2O_3	B_2O_3	
Soda-lime glass		70—75	12—18	5—14	0—4	0—1	—	0.5—2.5	—	Window glass, bottles and general usage.
Leaded glasses	(a)	53—68	5—10	0—6	—	1—10	15—40	0—2	—	High electrical resistance — lamp and valve envelopes.
	(b)	40	2.5—5	—	—	2.5—5	45—50	0—5	—	High refractive index and dispersive powers. Used for lenses, prisms, and other optical devices. Used for cut 'crystal' glass for tableware.
Borosilicate glass		73—82	3—10	0—1	—	0.4—1	0—10	2—3	5—20	Low thermal expansion and resistant to chemical attack. Heat-resistant cooking and tableware (Pyrex) and laboratory apparatus. Will perform a seal with some metals.
Aluminosilicate glass		70	2—4	—	—	2—4	—	3	20	High softening temperature $T_g = 800°C$
High-silica glass (vitreous silica)		96	—	—	—	—	—	—	3	Made by removing the alkalies from borosilicate glass after melting and shaping. Very low thermal expansion; very high T_g, can be used continuously at 800°C.

heat-treatment processes. The glass is annealed for sufficient time for the glass transition range of 100°C to 200°C, after which it can be cooled relatively rapidly to room temperature.

Alternatively the glass may be tempered. This consists of heating the glass up to its annealing temperature and cooling the surface rapidly by an air blast. This results in the surface of the glass becoming rigid whilst the interior is still plastic. On cooling down to room temperature contraction occurs and the surface of the glass is left in compression whilst the interior is in tension. Since glass usually fails at its surface due to induced tensile forces, the fact that the surface is prestressed in compression results in a considerable increase in strength. Tempered glass has a strength and impact resistance three to five times greater than annealed glass whilst retaining the same appearance, clarity, hardness and coefficient of thermal expansion as the original glass. Once tempered, the glass cannot be cut, machined or ground as this would upset the stress system leading to disintegration of the glass. All forming and cutting must occur before tempering. 'Toughened' glass windscreens are made from tempered glass and shatter into small but relatively harmless granules when broken.

3.21 Mechanical properties of glasses

The mechanical properties of glasses have little in common with crystalline materials such as the metals. Normal tests for tensile and impact strengths show glasses to be brittle at room temperature. Glasses are elastic to the point of failure and show no previous yield or plastic deformation, whereas the most brittle metals show some plastic deformation. The failure of glass is always due to a tensile stress component in its loading, even when the applied load is essentially compressive.

It has already been explained that whilst plastic flow occurs in a metal due to 'slip', glass lacks slip planes and only viscous flow can occur. The ease with which viscous flow in glass can occur increases as the temperature rises. However viscous flow in glass at room temperature is negligible, although large areas of glass have been known to increase in thickness at their lower edges over a long period of time running into many years.

3.22 Vitreous silica

Vitreous silica occurs naturally as quartz and, unlike the glasses previously described, it has a tetrahedral crystalline structure similar to the ceramic materials described earlier in this chapter. Vitreous, or crystalline, silica has a very high purity and is manufactured by fusing pure quartz crystals or glass sand in electric arc furnaces or by means of an oxy-hydrogen flame. No fluxes are used and there are no other ingredients present to produce a low temperature eutectic. Thus the fusion temperature is about 1750°C. It is extremely difficult to produce a homogeneous glass free from blowholes, despite the high viscosity of the glass. Vitreous silica is available as a translucent solid when the minimum silica content is 99.6 per cent, and it is available as a transparent solid when the minimum silica content is 99.9 per cent.

Transparent silica (fused quartz) has a high transparency to ultraviolet and infra-red radiations. It is much stronger, more impervious to gases, and more resistant to devitrification than translucent silica. The high fusion point of fused quartz together with its low and regular thermal expansion make it highly resistant to thermal shock and continuous operation at high temperatures, the upper limit being about 1100°C.

Typical applications for transparent vitreous silica are optical instruments requiring a high transparency to a wide range of radiation frequencies, and for high-temperature applications such as the envelopes of tungsten-halogen electric lamps. The cheaper, translucent vitreous silica is used for chemical laboratory equipment, electrical insulating materials, and cooking hobs with flush, easy-clean surfaces.

Polycrystalline glass (Pyroceram) is produced by adding nucleating agents to a conventional or to a special-purpose glass batch whilst the glass is molten. After processing to the desired form, the polycrystalline glass is then heat treated. Sub-microscopic crystallites are formed and, although polycrystalline glass is not ductile, it has much greater strength and hardness than commercial glasses.

3.23 Composite materials

Ceramic materials generally show brittle characteristics when subjected to tensile loading. This is due to crack propagation as shown in Fig. 3.7. The use of composite materials overcomes this problem by preventing any crack from running.

In its simplest form a composite material consists of two independent and dissimilar materials in which one material forms the matrix to bond together the other, reinforcing, material. The matrix and the reinforcement are chosen so that their desirable mechanical properties complement each other, whilst their deficiencies are neutralised. In particular circumstances there may be more than one type of reinforcement present at the same time.

Metallic alloys do not qualify as composite materials since, although there may be hard, strong particles present in a softer matrix in the solid state, these are derived from a single homogeneous liquid. In a true composite material the constituent components are always separate and distinguishable.

The reinforcement in a composite material may be in many different forms: for example, fibres in glass reinforced polyesters (GRP), steel rods and mesh in reinforced concrete, nodules in the form of natural

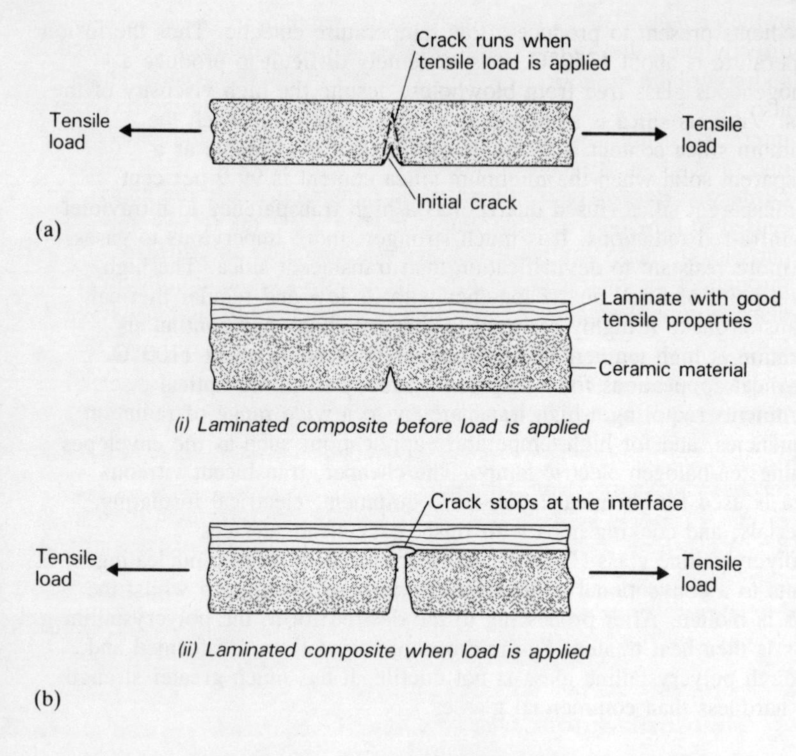

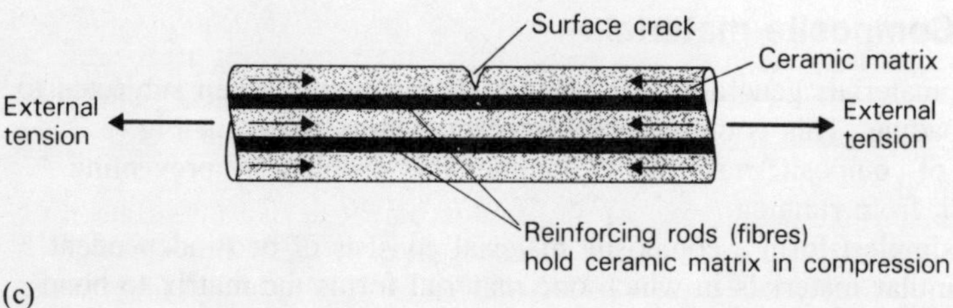

Fig. 3.7 Principles of reinforced composite materials: (*a*) crack propagation in a non-reinforced ceramic material, (*b*) behaviour of a laminated composite when in tension, (*c*) fibre reinforcement

stones in concrete, and particles in cermets and cemented carbides. In fact composite materials can be classified according to the type of reinforcement.

(*a*) *Fibre reinforcement*. This is present in such materials as natural wood, glass reinforced polyester and epoxy resins, and reinforced concrete.
(*b*) *Particle hardening*. For example, where the ductility and toughness of metals are combined with the hardness and strength of ceramics as

in the *cermets*. The particles increase the strength of the composite directly since they suffer elastic strains when the material is stressed and contribute to the load-bearing capacity of the composite. They also increase the strength and hardness of the composite indirectly by interfering with dislocations along the slip planes. (Rather like sliding one sheet of sandpaper over another.)

(*c*) *Dispersion hardening*. Unlike particle hardening where the matrix is caused to flow between the reinforcing particles, the reinforcing particles in dispersion hardening are much smaller and diffuse throughout the matrix in the solid state. For example aluminium dispersion hardened by the diffusion of aluminium oxide particles.

No matter which mechanism of reinforcement is used in a composite material, cohesion between the matrix and the reinforcement is essential and must occur in one or more of the following ways.

(*a*) *Mechanical bonding* between the matrix and the reinforcement by mechanical 'keying' and friction.
(*b*) *Physical bonding* between the matrix and the reinforcement by Van der Waals forces acting between the surface molecules of the various constituents.
(*c*) *Chemical bonding* by chemical reactions at the interfaces of the various constituents. However some of the compounds formed can be weak.

3.24 Reinforced concrete

This is a composite material which combines the *particle reinforcement* of the aggregate with the *fibre reinforcement* of the reinforcing rods or mesh. A hydraulic cement matrix binds the various constituent materials together. The use of parallel fibres and rods for reinforcement was discussed in volume 1. Not only is the reinforcement area fraction important, but the positioning of the reinforcement is also important. For instance, since the tensile strength of concrete is virtually non-existent, the steel reinforcing bars must be introduced as closely as possible to the point of maximum tensile stress. Figure 3.8 shows a simple reinforced concrete beam. It can be seen that the reinforcement is concentrated near the surface on the tension side. Although concrete is very strong in compression, in practice, it is also usual to introduce some reinforcement on the compression side of the beam to guard against reversed stresses. These can occur whilst lifting the beam into position as shown in Fig. 3.9(*b*).

Simple reinforcement, such as has been considered so far, is never used for highly stressed structural members. This is because when a load is applied to a beam it commences to deflect until the stresses in the beam balance the applied load. In a homogeneous beam, such as a steel girder, this deflection is of little importance providing it is kept within reasonable limits. However, in a concrete beam, bending of the

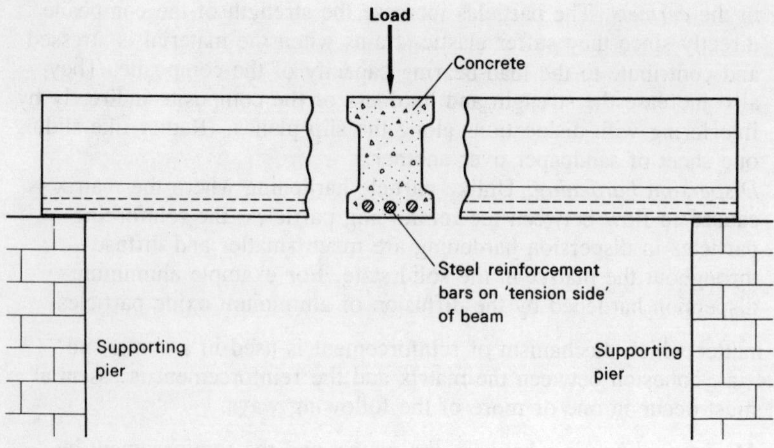

Fig. 3.8 Simple reinforced concrete beam

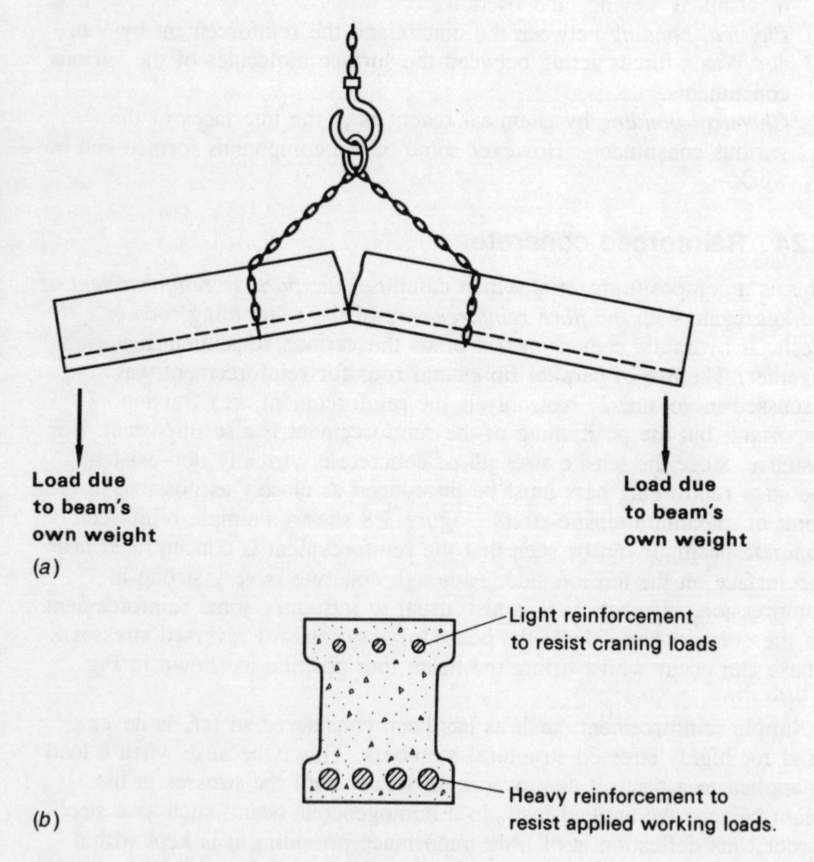

Fig. 3.9 Additional reinforcement to resist handling loads

reinforcement (which is flexible and elastic) can be disastrous as it would permit the concrete matrix to crack as shown in Fig. 3.10.

To overcome this problem the steel reinforcement is stressed in tension by means of hydraulic jacks whilst the concrete is cast into the mould. When the concrete has set, the hydraulic jacks are released and the *prestressed* reinforcement tries to return to its original length and places the concrete in compression. Members made in this manner are referred to as *prestressed concrete*. The design load and prestressing are such that the concrete is always in compression and no cracking occurs. In the event of accidental overloading, any cracks in the concrete are closed immediately the excess load is removed providing the reinforcement is not stressed beyond its elastic limit. Figure 3.11 shows a section through a prestressed concrete beam in the course of manufacture.

The success of simple and prestressed reinforcement depends upon two main factors.

(*a*) Bonding of the concrete to the reinforcement to ensure transmission of the load without slip and to ensure that the beam and reinforcement are subjected to equal strain. This is virtually impossible to achieve in practice.

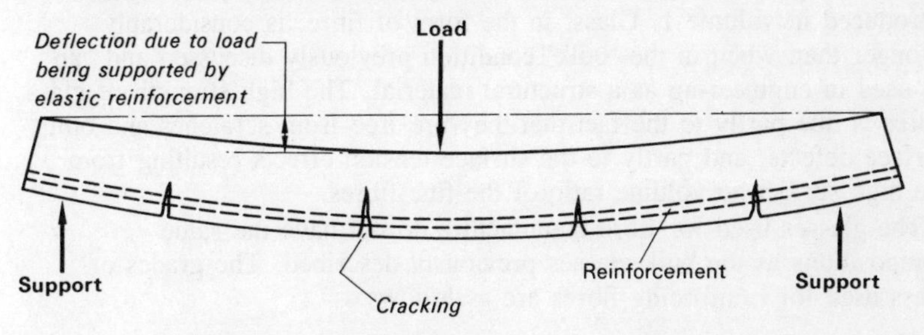

Fig. 3.10 Cracking of a simple reinforced beam in service

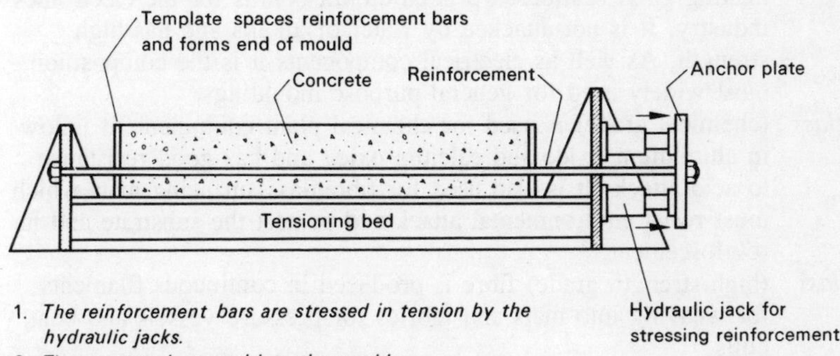

1. *The reinforcement bars are stressed in tension by the hydraulic jacks.*
2. *The concrete is poured into the mould.*
3. *When the concrete is set, the jacks are released. The reinforcement, behaving in an elastic manner, tries to shrink back to its original length and compresses the beam.*

Fig. 3.11 Prestressing concrete beams

(*b*) Compatibility between the concrete and the reinforcement so that the steel is not attacked chemically and corroded. A numer of concrete structures have become unsafe or even collapsed from this cause.

An alternative to prestressing is *post-tensioned concrete*. Here, the reinforcement is applied and tensioned after construction. This allows the degree of tensioning of the reinforcement to be closely controlled at the time of installation and for the tension to be corrected from time to time thereafter. Since the reinforcement passes through precast ducts, it is not in intimate contact with the concrete, and corrosion due to chemical incompatibility is unlikely. Post-tensioning is used for large structures such as bridges. Ducts are cast in the concrete through which are passed high-tensile steel hawsers. The hawsers are stretched hydraulically until the required stress is achieved. They are then locked in position by taper plugs. Figure 3.12 shows various systems of post-tensioning.

3.25 Glass fibres

The use of glass fibres to reinforce a polyester matrix has already been introduced in volume 1. Glass, in the form of fibre, is considerably stronger than when in the 'bulk' condition previously discussed and can be used in engineering as a structural material. The high strength of glass fibres is due partly to the fact that they are free from scratches and other surface defects, and partly to the surface tension effects resulting from the high surface to volume ratio of the fine fibres.

The glasses used for fibre manufacture do not have the same compositions as the bulk glasses previously described. The grades of glass used for reinforcing fibres are as follows.

E-glass (electrical grade) has good insulation properties and is used for making glass reinforced printed circuit boards for the electronics industry. It is not attacked by water or alkalis and has high strength. As well as electrical components it is the composition most widely used for general-purpose mouldings.

C-glass (chemical grade) is used for chemical plant equipment. It is low in aluminium oxide and calcium oxide and has good resistance to acid attack. It is also used for fibreglass surfacing mats which must resist environmental attack and protect the substrate and its reinforcement.

S-glass (high-strength grade) fibre is produced in continuous filaments for weaving into mats and fabrics for pressure vessels and boat hulls.

M-glass (high-modulus grade) is an expensive material reserved for very high strength applications where exploitation of its special properties offsets its high cost.

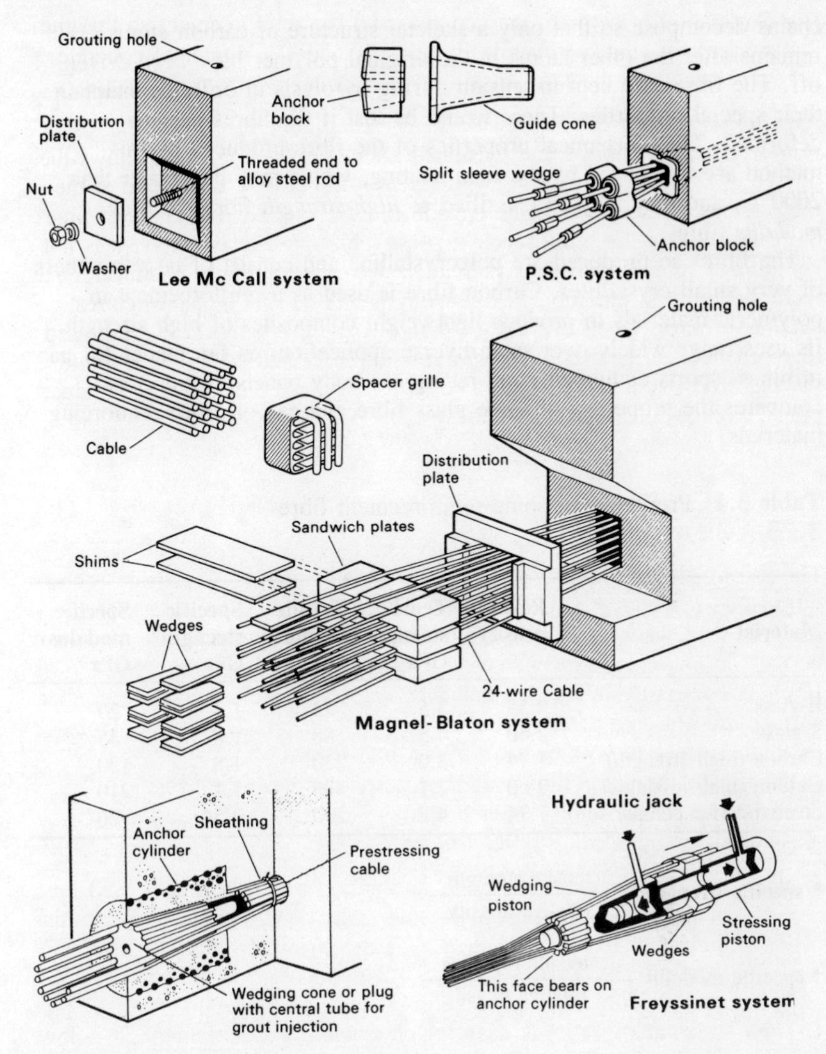

Fig. 3.12 Post-tensioning concrete beams

3.26 Carbon fibres

These have higher elastic modulus and a lower density than glass fibres and therefore, can be used to reinforce composite materials having a higher strength to weight ratio. Carbon fibres are produced by the *pyrolysis* of polyacrylonitrile filaments in an inert atmosphere. Pyrolysis is the decomposition of substances at high temperatures. The polymer chains decompose so that only a skeletal structure of carbon atoms remains after the other atoms in the original polymer have been driven off. The fibres are kept in tension during pyrolysis in order to maintain their special properties. These would be lost if the fibres became deformed. The mechanical properties of the fibre produced by this method are influenced by the final heating, which may be greater than 2000°C, and they may be classified as *high-strength* fibres or *high-modulus* fibres.

The fibres so produced are polycrystalline and consist of large numbers of very small crystallites. Carbon fibre is used as a reinforcement in polymeric materials to produce lightweight composites of high strength. Its uses range widely over such diverse applications as fan blades for gas turbines, sports equipment, and racing-car body panels. Table 3.4 compares the properties of some glass fibre and carbon fibre reinforcing materials.

Table 3.4 Properties of some reinforcement fibres.

Material	Relative density	Tensile strength GPa	Tensile modulus GPa	Specific strength* GPa	Specific modulus† GPa
E-glass	2.55	3.5	74	1.4	29
S-glass	2.50	4.5	88	1.8	35
Carbon (high strength)	1.74	3.0	230	1.8	130
Carbon (high modulus)	2.00	2.1	420	1.1	210
Steelwire (for comparison)	7.74	4.2	200	0.54	26

$$* \text{ specific strength} = \frac{\text{tensile strength}}{\text{relative density}}$$

$$\dagger \text{ specific modulus} = \frac{\text{tensile modulus}}{\text{relative density}}$$

3.27 'Whiskers'

'Whiskers' are single hair-like crystals with a high aspect ratio and very high strengths. The crystal diameter may range from 0.5 micron to 2 micron in diameter by up to 20 mm long. The tensile strength of a carbon whisker can be as high as 21 GPa compared with a carbon fibre which has a tensile strength of 3 GPa. This is due to the relative freedom of dislocations in such crystals. In fact they usually have only a single dislocation running along the longitudinal axis. The properties of some whiskers are listed in Table 3.5. Such whiskers are difficult and costly to manufacture and are only used for special applications. Boron and carbon

Table 3.5 Properties of some reinforcement 'whiskers'

Material	Relative density	Tensile strength GPa	Tensile modulus GPa	Specific strength* GPa	Specific modulus* GPa
Alumina	3.96	21.0	430	5.3	110
Boron carbide	2.52	14.0	490	5.6	190
Carbon (graphite)	1.66	20.0	710	12.0	430

* See Table 3.4

whiskers in a polymeric matrix are the most commonly used, but alumina whiskers are being used to reinforce the metal nickel. This composite retains its strength at high temperatures.

3.28 Particle hardened materials

These materials are commonly known as *cermets* since they are metals hardened and strengthened by particles of ceramic materials uniformly distributed throughout the metallic matrix (similarly to the aggregate in concrete). The principles of particle hardening were introduced earlier in the chapter. Cermets are widely used for cutting tools and will be considered further in Section 4.16. For example, cemented carbides consist of particles of tungsten carbide or mixtures of tungsten and titanium carbides in a matrix of metallic cobalt. The preformed compact of the powdered ingredients is then *sintered* at a temperature above the recrystallisation temperature of the cobalt. Such materials combine the toughness and ductility of the matrix with the hardness and strength at high temperatures of the ceramic particles. Alternatively, cermets may be made by infiltrating the spaces between the solid ceramic particles with molten metal. Whichever method is used, it is essential that a strong bond should exist at the interface of the metal and the ceramic particles. Some typical cermets are listed in Table 3.6.

3.29 Dispersion hardened materials

The use of alumina 'whiskers' to harden and strengthen metallic nickel has already been discussed. Alumina in the form of spheroidal particles can be used to increase the tensile strength of composites rather than the hardness. If aluminium is ground to a fine powder in the presence of oxygen under pressure, aluminium oxide (alumina) is formed on the surfaces of the particles. The grinding process causes some of the surface alumina to disintegrate and become distributed throughout the mass of the aluminium powder. The mixture is then sintered and consists of about 6 per cent alumina particles dispersed through a matrix of aluminium. This material is known as *sintered aluminium powder* (SAP) and is stronger than pure aluminium. Although not as strong as duralumin alloy at room temperature, SAP holds its strength at much higher temperatures, as shown in Fig. 3.13. Alumina particles may be used to dispersion strengthen other materials than aluminium, for example silver and nickel.

Table 3.6 Typical cermets

Type	Ceramic particles	Metal matrix	Applications
Borides	Titanium boride	Cobalt/nickel	Mostly cutting tool tips
	Molybdenum boride	Chromium/nickel	
	Chromium boride	Nickel	
Carbides	Tungsten carbide	Cobalt	Mostly cutting tool tips
	Titanium carbide	Cobalt or tungsten	Tool tips and abrasives
	Molybdenum carbide	Cobalt	
	Silicon carbide	Cobalt	
Oxides	Aluminium oxide	Cobalt/chromium	Disposable tool tips; refractory sintered components, e.g. spark plug bodies, rocket and jet engine parts
	Magnesium oxide	Cobalt/nickel	
	Chromium oxide	Chromium	

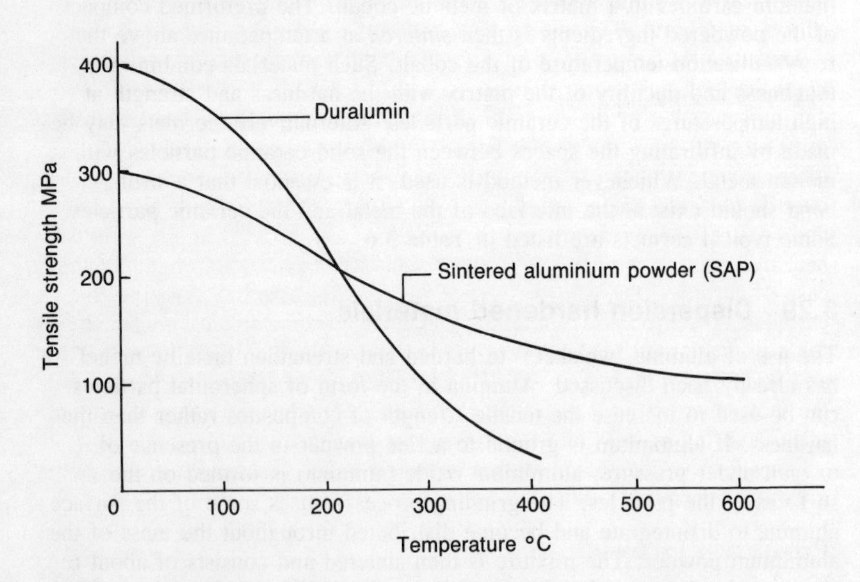

Fig. 3.13 Comparative effects of temperature on duralumin and sintered aluminium powder

4 Tool materials

4.1 Introduction

The choice of a cutting tool material for a particular application is determined principally by technical and economic requirements. The technical requirements are dependent upon the process, the material being processed, the rate of processing, the equipment being used and the condition of that equipment. The economic requirements are related to such factors as initial tool cost, tool life, refurbishment costs versus 'throw-away' and replacement costs, rate of processing, and batch size.

Tool materials must have sufficent strength to resist the forces acting upon the tool, sufficient hardness to resist wear and, in the case of cutting tools, the ability to maintain a sharp cutting edge. Tool materials must also be resistant to softening at the elevated temperatures which often accompany the processing of engineering materials. Such properties may be *intrinsic*, as in carbides and diamonds which are naturally hard and heat resistant, or they may be *conferred*, as in metal alloys where heat treatment is required to increase their hardness and strength. In the case of carbides, the wear resistance may be increased by 'coating' the carbides. The problems relating to the selection of a suitable tool material for a particular process can be complex and require detailed knowledge and wide experience. The following simple example describes some of the factors which need to be taken into account when selecting a suitable manufacturing process and the tooling for that process.

Consider the component shown in Fig. 4.1, the flat can be produced in a variety of ways.

Production by hand

If only one or two components are to be produced, then the flat could be filed on by hand. Engineers' files are usually made from a hardened and

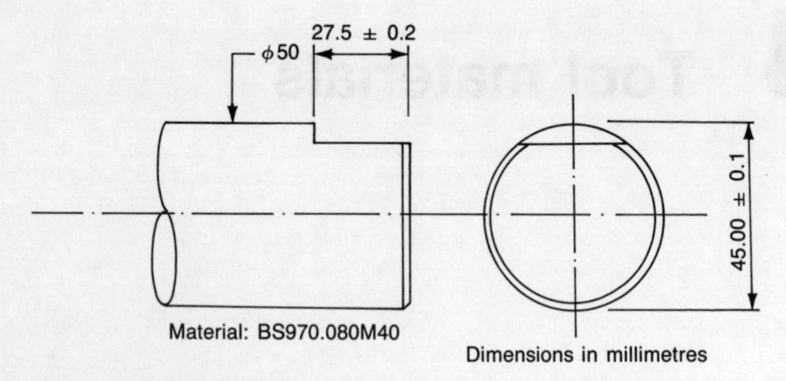

Fig. 4.1 Component with flat

tempered plain carbon steel with a high carbon content (1.2 per cent carbon). For a hand operation this is a very suitable material. It takes a keen edge which reduces the manual effort required and insufficient heat is generated whilst filing to draw the temper of the tool. It also has a relatively low initial cost.

Production by machining

If larger batches of the component are to be made, it would not be economical to produce the flat by hand and a milling process would most likely be used. The choice then lies between the use of *high-speed steel* or *cemented carbide* cutting tools. The economic rate of production (productivity) expected from a machining process precludes the use of high-carbon-steel tools since the heat generated by cutting would quickly soften the cutting edge. For small and medium batch sizes, a conventional side and face cutter made from high-speed steel would be suitable. The cost of the cutter is reasonable, it would have an adequate life between regrinds and it could be resharpened relatively easily. However if the rate of production had to be increased to satisfy the demands of the customer, and the batch size warranted the financial outlay, then a cutter with cemented carbide teeth would be used. This choice assumes that the milling machine is sufficiently powerful to exploit the advantages of cutting with carbides. The choice then has to be made between a milling cutter in which the carbide inserts are brazed permanently in place and a milling cutter in which disposable inserts are clamped in place. The former type of cutter has a lower initial cost but refurbishment by grinding the cutting edges is relatively costly, whilst the latter type will have indexable inserts so that refurbishment merely involves indexing the inserts so as to present a fresh and sharp cutting edge. When all the cutting edges have become dulled, the inserts are rejected and new, standard inserts are clamped quickly and easily in place.

This example shows that the choice of tool material and its application involves some quite complex technical and economic decisions even for a simple operation. The tooling engineer would solve this relatively simple operation by personal experience of previous, similar jobs but, for more complex machining operations, mathematical modelling would have to be resorted to in order to arrive at the correct choice.

One simple technique is the use of a break-even diagram. Figure 4.2 shows such a diagram comparing the batch size and relative costs of using a high-speed steel drill and a solid carbide drill. It can be seen that the initial cost of the carbide drill is greater than that for the high-speed steel drill. However, the unit production costs with the carbide drill is lower since it can be used at higher cutting speeds. Where the lines cross is the *break-even* point. To the left of this point it is more economical to use the high-speed steel drill, but to the right of this point it is more economical to use the carbide drill despite its greater initial cost.

Some of the more common areas of application for tool materials are: cutting tools (machining); press tools (cutting and forming); moulds and dies; drawing and extrusion dies; and electrodes for ECM and EDM machining. The properties required by tool materials for these various applications will now be considered in detail.

4.2 Requirements for cutting tools (machining)

Cutting-tool materials for the machining of metallic and non-metallic materials require the following general properties.

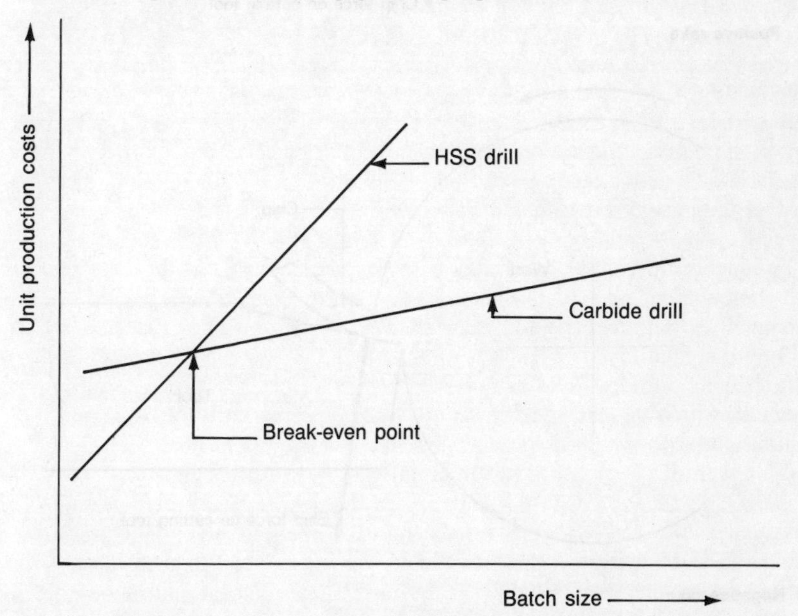

Fig. 4.2 Comparison of HSS and carbide drills

Strength

The tool material will require sufficient strength to resist the cutting forces acting upon the tool. The nature and magnitude of these forces, in turn, depend upon a number of factors. Figure 4.3 shows the difference between positive and negative rake cutting. With positive rake geometry

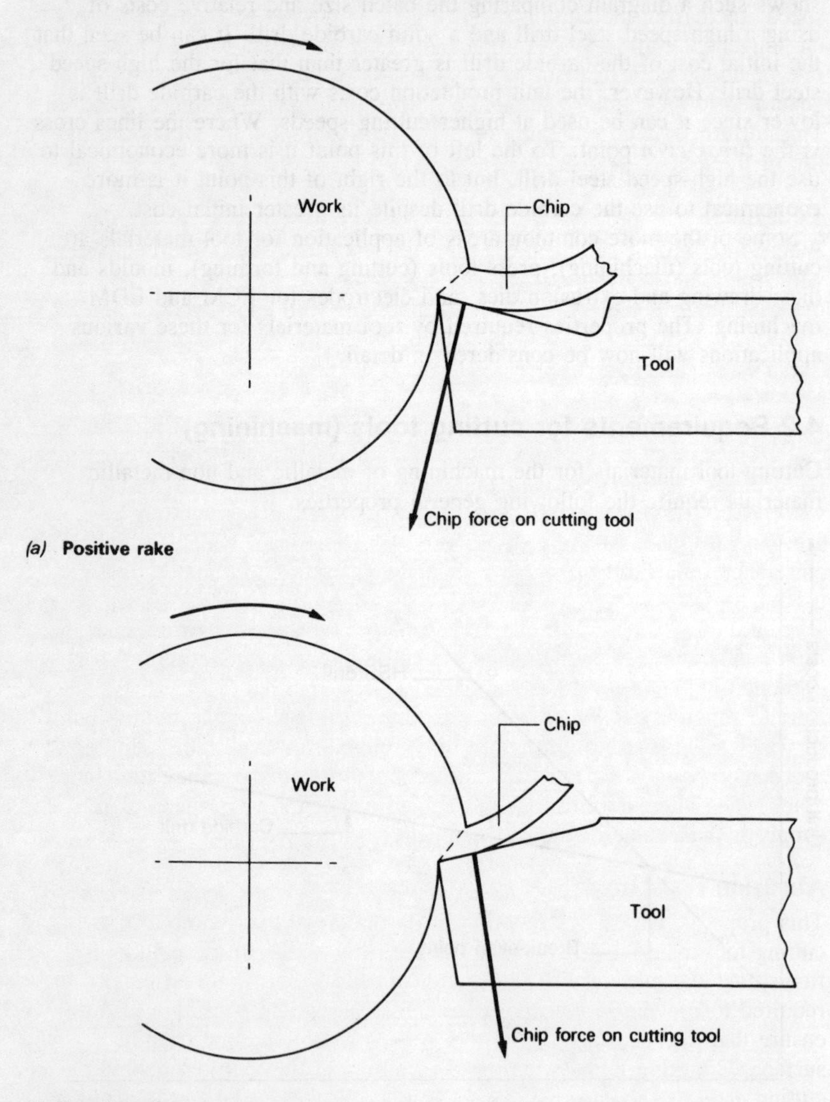

Fig. 4.3 Negative rake cutting

the tip of the tool is in shear, whilst with negative rake geometry the tip of the tool is in compression. This is why brittle tool materials such as tungsten carbide are generally used with a zero or negative rake cutting geometry.

The magnitude of the cutting forces depends upon the properties of the material being cut and the area of cut, but is independent of the cutting speed. Increasing the cutting speed increases the power required but not the tangential cutting force on the tool. This increased power (increased rate of doing work or using energy) results in greater heat energy being generated in the cutting zone and a corresponding rise in temperature. This in turn leads to softening of the cutting edge and a reduction in the life of the tool or even its destruction.

Toughness

Where intermittent cutting takes place or when cast or forged components with rough and uneven surfaces are being rough machined, the tool material requires the property of toughness as well as strength. Unfortunately toughness is usually achieved at the expense of hardness.

Hardness

In order to retain a sharp cutting edge over a reasonable tool-life, the cutting tool must be substantially harder than the material being cut. Unfortunately many very hard materials are also brittle and relatively weak. Thus very hard materials, which will retain a keen cutting edge, are generally used for light, high-speed finishing cuts, whilst for roughing and general machining less hard but tougher cutting tool materials are used.

A cutting tool material must be capable of retaining its hardness at the high temperatures encountered in the cutting zone (see Section 4.9). Although high-carbon steels can achieve very high hardness values, they start to soften at relatively low temperatures (just above the boiling point of water), and, therefore they are only suitable for hand tools. Cutting tool materials must be resistant to thermal shock, that is, they must not crack when alternately heated and cooled as occurs when the coolant supply is inadequate or applied manually.

Abrasion resistance

This property, which is a function of hardness, is also required in a cutting tool material to withstand the scouring action of the chip as it flows over the rake face of the tool. Not only is abrasion resistance required to prevent 'cratering' just behind the cutting edge, but also to ensure that the rake face of the tool retains a smooth, low friction surface. Cratering is the wearing of a hollow in the tool just behind the cutting egde as shown in Fig. 4.4. This weakens the tool resulting in reduced tool life caused by failure of the tool. Roughening of the rake face of the tool increases the friction between the chip and the tool. This increased friction leads to increased heating of the tool and a reduction in

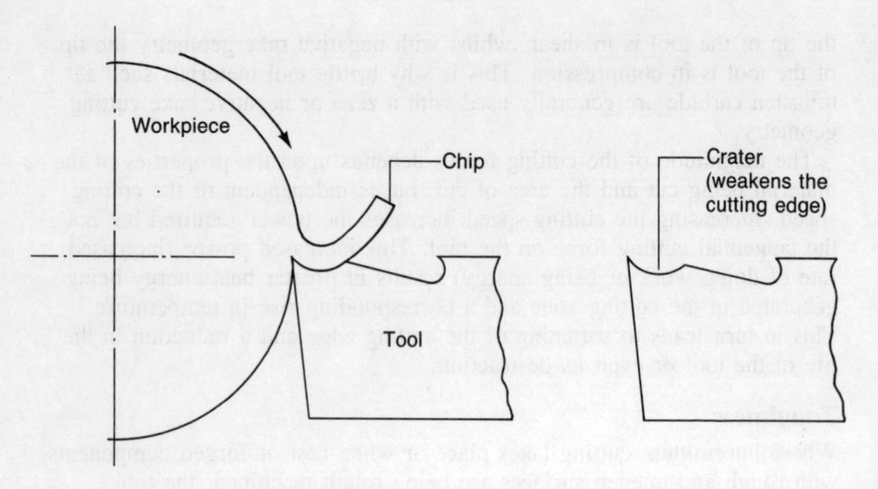

Fig. 4.4 Cratering of the cutting tool

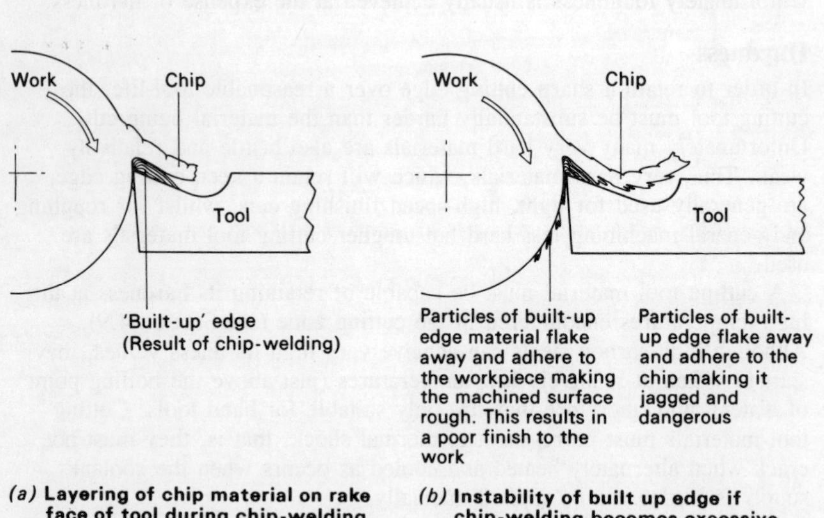

Fig. 4.5 Chip welding (built-up edge)

tool life. It also leads to chip welding and the formation of a 'built-up edge', as shown in Fig. 4.5, which results in a further lowering of the cutting efficiency of the tool and a poor finish on the work. Additional information on metal cutting and the requirements of metal cutting tools can be found in *Manufacturing technology: volumes 1 and 2*.

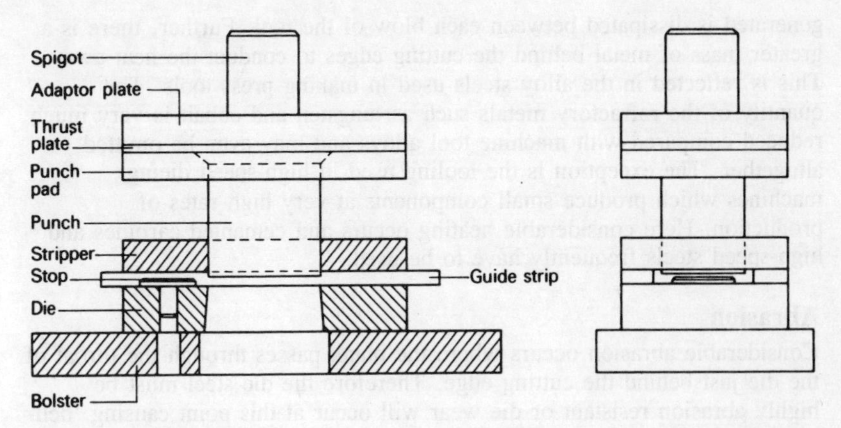

Fig. 4.6 Blanking tool

Compatibility

Cutting tool materials must not react chemically with the materials being cut under normal cutting conditions, either to cause corrosion of the rake face of the tool (which would aggravate mechanical erosion) or for the cutting tool material and the material being cut to have an affinity for each other which could aggravate any tendency towards chip welding.

4.3 Requirements for press tools (cutting)

The requirements for press tools depend upon whether the pressing operation involves cutting (e.g. blanking or piercing) or forming (e.g. bending or cupping). Figure 4.6 shows a section through a blanking tool, and it can be seen that the conditions are different to those for cutting metal on a machine tool.

Strength

It is apparent from Fig. 4.6 that the punch and die are in compression close to their cutting edges. Therefore material with a high compressive strength is required. At the same time the punch and die are subjected to considerable shock loads each time they close on the material being cut. Therefore the punch and die materials also require to be tough and shock resistant.

Hardness

As for any cutting tool, the punch and die must retain sharp cutting edges and have an economical life. Therefore the cutting tool material must again be substantially harder than the material being cut. Unlike machining, the cutting action in press tools is intermittent and any heat generated is dissipated between each blow of the tool. Further, there is a greater mass of metal behind the cutting edges to conduct the heat away. This is reflected in the alloy steels used in making press tools. The quantity of the refractory metals such as tungsten and cobalt is very much reduced compared with machine tool alloys and may even be omitted altogether. The exception is the tooling used in high-speed dieing machines which produce small components at very high rates of production. Here considerable heating occurs and cemented carbides and high-speed steels frequently have to be used.

Abrasion

Considerable abrasion occurs where the blank passes through the throat of the die just behind the cutting edge. Therefore the die steel must be highly abrasion resistant or die wear will occur at this point causing 'bell-mouthing' which will not only reduce the cutting efficiency, resulting in a heavy 'burr' round the edge of the blank, but may also result in the blanks not dropping clear and becoming jammed in the tool. When this occurs the tools and the press can suffer severe damage.

4.4 Requirements for press tools (forming)

Figure 4.7 shows a U-bending tool and Fig. 4.8 shows a cupping tool. It can be seen that the properties required for these tools are somewhat different to those for blanking and piercing tools.

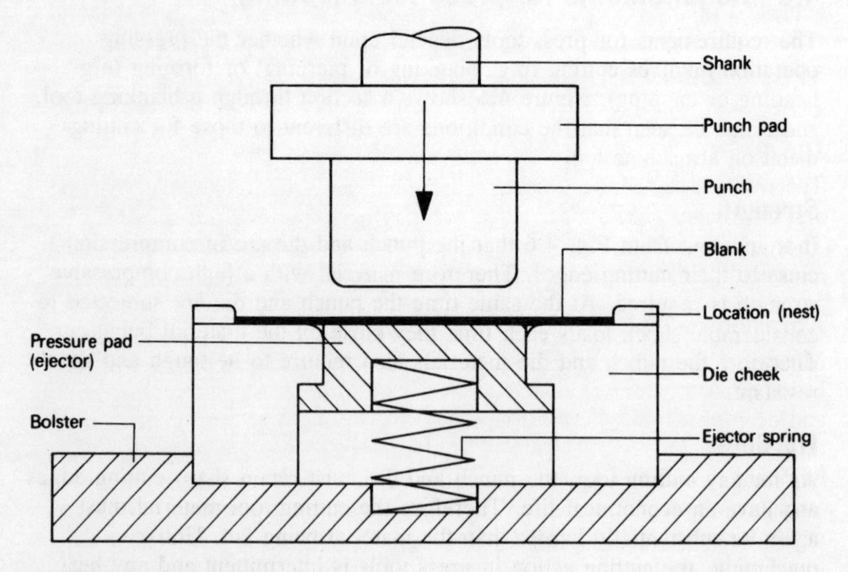

Fig. 4.7 U-bending tool

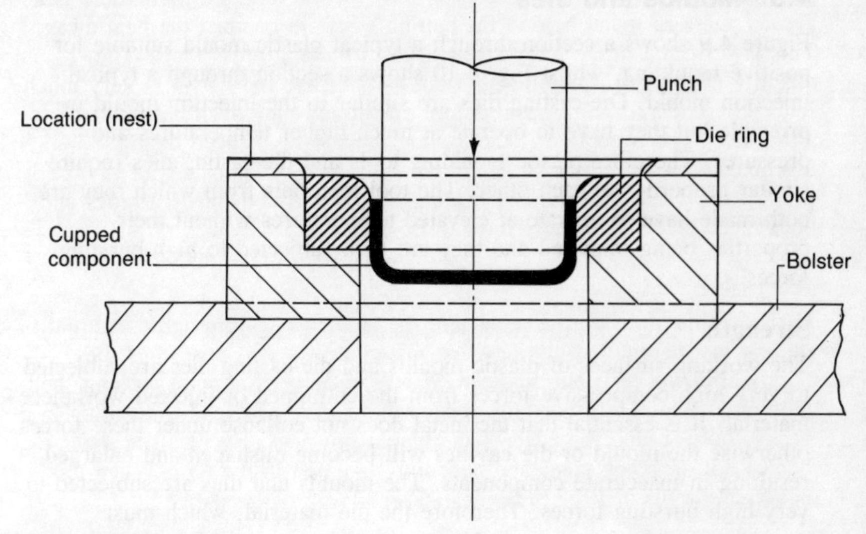

Fig. 4.8 Cupping tool

Strength

In both tools the working surfaces will be in compression whilst forming is taking place. In addition, the die ring of the cupping tool will be subjected to severe hoop stresses. Therefore, in both cases, it is usual to use die cheeks of a hard, wear resistant material set in a yoke of low or medium carbon steel. As for all press tools, forming tools must also have a high resistance to impact loading.

Hardness

The wearing surfaces are usually hardened and ground to reduce friction and to prevent marking the surfaces of the material being formed. Thorough hardening also prevents the die cheeks from collapsing under load and introducing inaccuracies in the formed component.

Abrasion resistance

Most of the wear on forming tools is caused by abrasion as the blank bends or flows to its finished shape. Large forming tools, cupping tools and deep-drawing tools are frequently made from high-grade cast iron as this material has excellent anti-friction and wear-resistant properties. A lubricant should be used to reduce die wear and special lubricants are available which will withstand the high point pressures which often occur in the forming process. These pressures are very much greater than those found in bearings and ordinary mineral lubricating oils are only suitable for the most undemanding forming processes.

4.5 Moulds and dies

Figure 4.9 shows a section through a typical plastic mould suitable for positive moulding, whilst Fig. 4.10 shows a section through a typical injection mould. Die-casting dies are similar to the injection mould in principle but they have to operate at much higher temperatures and pressures. Therefore plastic moulding tools and die-casting dies require similar properties to each other. The tool materials from which they are both made have to operate at elevated temperatures without their properties being impaired and they are both subjected to high bursting forces.

Strength

The working surfaces of plastic moulds and die-casting dies are subjected to very high compressive forces from the entrapped or injected workpiece material. It is essential that the metal does not collapse under these forces otherwise the mould or die cavities will become mishapen and enlarged resulting in inaccurate components. The moulds and dies are subjected to very high bursting forces. Therefore the die material, which must

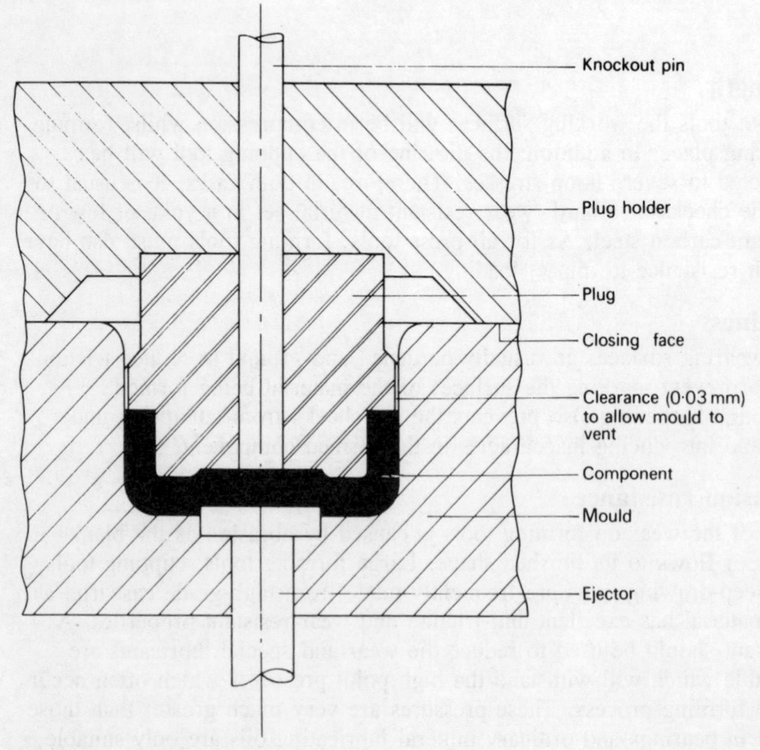

Fig. 4.9 Positive mould

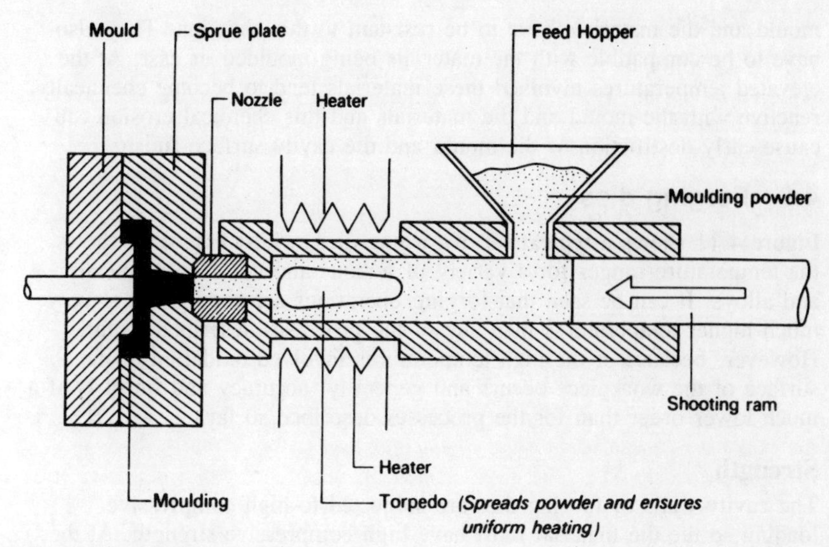

Fig. 4.10 Injection mould

combine high tensile and high compressive strengths, is reinforced by being set in a yoke of medium carbon steel. Die-casting dies are generally subjected to considerably greater forces than plastic moulds.

Hardness

Both plastic mouldings and die castings are expected to have high levels of accuracy and surface finish so that finishing costs are reduced to a minimum. They may also contain a considerable amount of fine detail. Since moulds and dies are very costly to produce, it follows that they must have a long working life and be capable of producing many thousands of components before replacement. Therefore their working surfaces must be extremely hard and wear resistant. Because of the high quality of lustre and finish expected on plastic mouldings, the mould cavity is usually hard chrome plated prior to finishing to size and polishing. Hard chrome plating, which provides a relatively thick, wear resistant surface, must not be confused with decorative chrome plating which is only a few microns thick. At the same time the moulds and dies are constantly heating up and cooling down. This thermal shock must not cause surface cracks in the mould or die which could lead to the surface flaking away.

Abrasion resistance

The moulding powder and granules used in the manufacture of plastic components and the scouring action of the molten metal in die casting both cause abrasive wear on the walls of the mould and die cavities. The mould and die materials have to be resistant to this abrasion. They also have to be compatible with the materials being moulded or cast. At the elevated temperatures involved these materials tend to become chemically reactive with the mould and die materials and this chemical erosion can cause early destruction of the mould and die cavity surface finish.

4.6 Forging dies

Figure 4.11 shows a typical set of forging dies, whilst Fig. 4.12 shows the temperature ranges for a variety of ferrous and non-ferrous metals and alloys. It can be seen that forging operations are carried out at very much higher temperatures than any of the processes described so far. However, because of the high temperatures involved oxidation of the surface of the workpiece occurs and generally, accuracy and finish is of a much lower order than for the processes described so far.

Strength

The cavity walls of forging dies are subjected to high compressive loading so the die material must have high compressive strength. At the same time the die material must have a high resistance to impact loading.

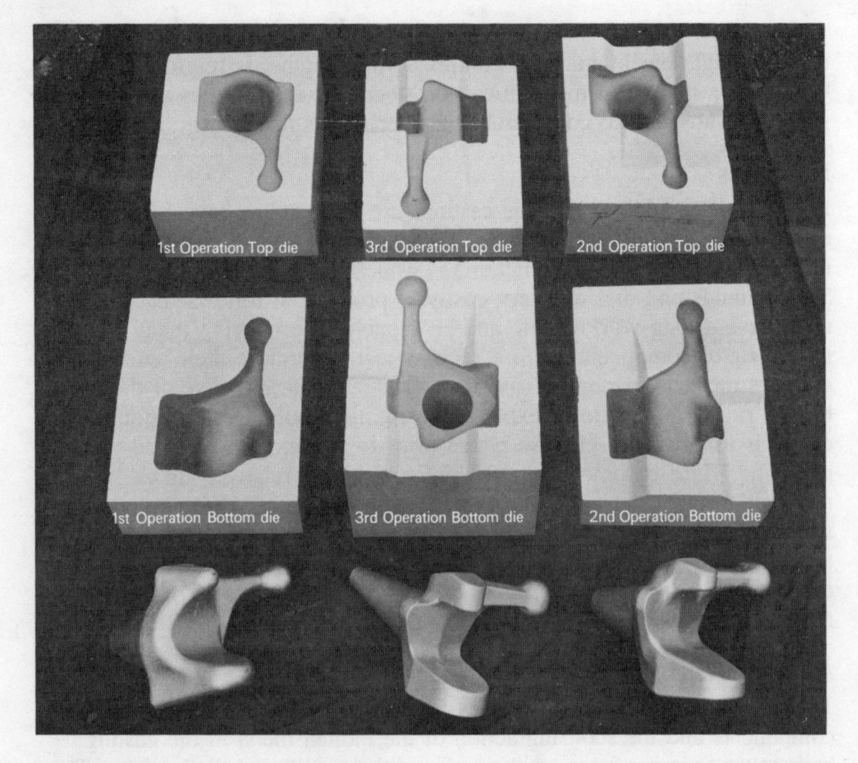

Fig. 4.11 Examples of drop forging

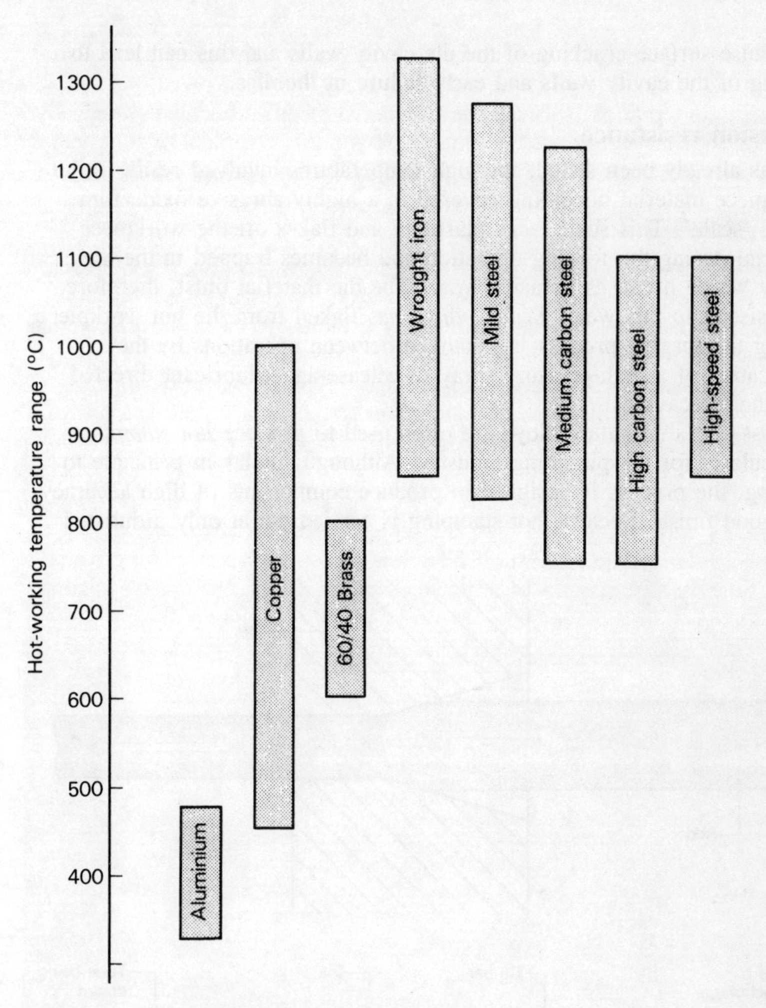

Fig. 4.12 Typical forging temperatures

Hardness

The cavity walls of forging dies are subjected to considerable abrasive wear, not only from the flow of the metal in the dies but also from the scale that exists on the surface of the heated workpiece material. Thus forging dies must be hard as well as strong and tough. This hardness must be maintained at the high temperatures at which the dies have to operate. However because of the mass of the dies and the fact that they are only in intermittent contact with the workpiece materials their working temperature is not as high as Fig. 4.12 would imply. The intermittent heating and cooling of the dies leads to thermal shock which can cause surface cracking of the die cavity walls and this can lead to flaking of the cavity walls and early failure of the dies.

Abrasion resistance

As has already been stated, the high temperatures involved result in the workpiece material becoming covered in a highly abrasive oxide film called 'scale'. This scale lacks plasticity and flakes off the workpiece material during the forging operation and becomes trapped in the die cavity where it causes abrasive wear. The die material must, therefore, be resistant to this wear. Scale, which has flaked from the hot workpiece during the forging process, is removed between operations by the application of a high-pressure spray of release-agent/lubricant directed into the die cavities.

Brass and aluminium alloys are often used to produce *hot-stampings*, particularly for the plumbing industry. Although similar in principle to forging, the process is designed to produce components of high accuracy and good finish. Because hot-stamping is carried out at only a dull-red

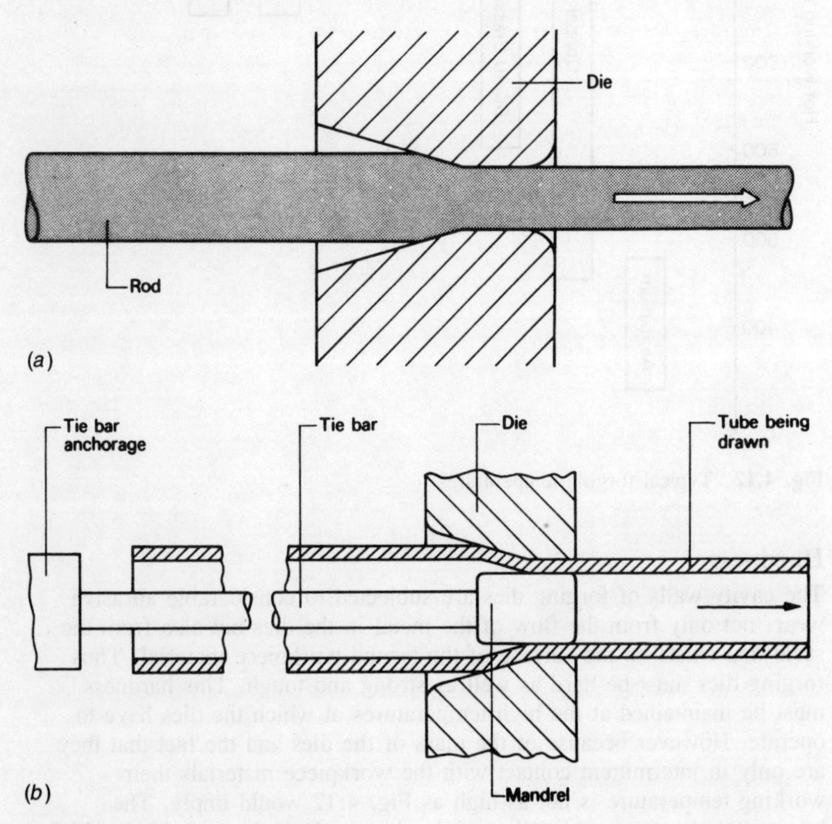

Fig. 4.13 Principles of metal drawing (*a*) wire drawing (*b*) tube drawing

heat, due to the lower melting temperatures of the alloys involved, scaling is less of a problem and can be easily removed by acid pickling. The properties of the die materials are similar to those required for forging.

4.7 Drawing and extrusion dies

Figure 4.13 shows the principles of wire and tube drawing whilst Fig. 4.14 shows the principle of hot extrusion. These processes produce wires, rods, tubes and sections with high length/diameter ratios.

Strength

Drawing and extrusion dies are subjected to high levels of hoop stress. At the same time they are subjected to very high levels of surface wear. Therefore such dies are usually made from hard and wear resistant

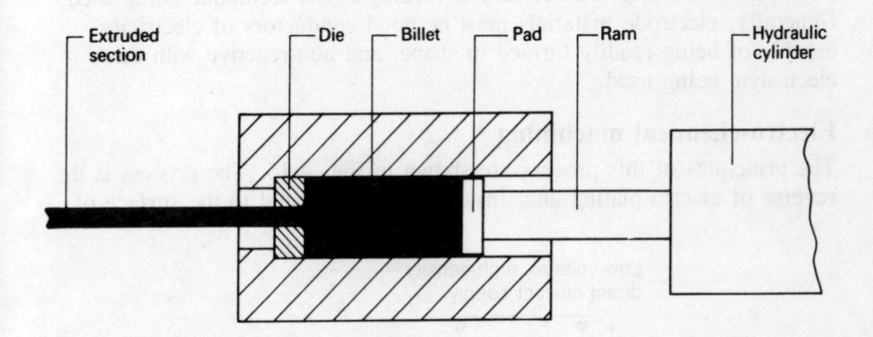

(a) **Commencement of extrusion stroke**

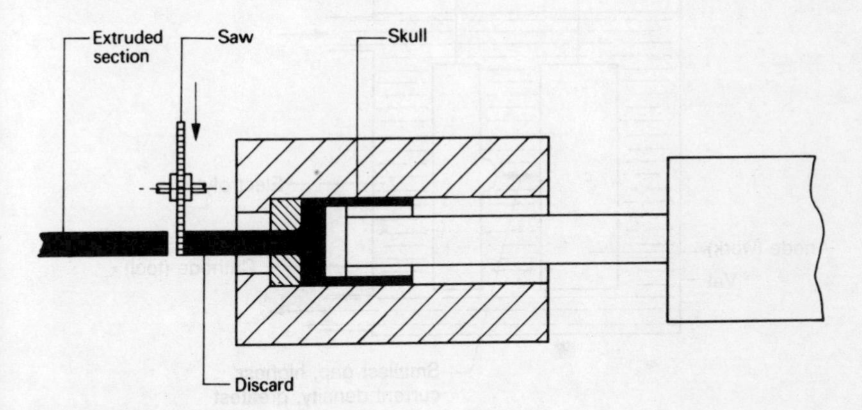

(b) **Completion of extrusion stroke**

Fig. 4.14 Extrusion

materials surrounded by a high tensile steel bolster or yoke which resists the hoop stresses.

Hardness

It has already been stated that drawing and extrusion dies need to be hard and wear resistant. This is because the metal being processed is in constant and moving contact with the working surfaces throughout the processing cycle and that very considerable contact pressures are involved. For fine wire drawing the die is often made from industrial diamonds inserted in a suitable steel bolster. Extrusion dies are usually made from alloy steel and have also to be heat resistant since the metal being extruded is at a dull red heat. Carbide inserts are also used.

4.8 Electrodes for ECM and EDM processes

Electrode materials for *electro-chemical machining* (ECM) and *electric discharge machining* (EDM) vary according to the technique being used. Generally, electrode materials must be good conductors of electricity, capable of being readily formed to shape, and non-reactive with the electrolyte being used.

Electro-chemical machining

The principles of this process are shown in Fig. 4.15. The process is the reverse of electro-plating and, instead of adding metal to the surface of

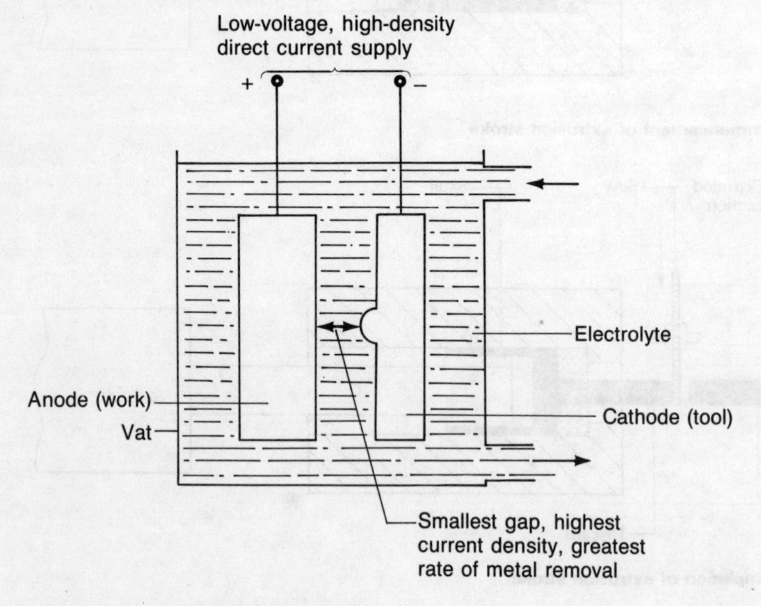

Fig. 4.15 Electro-chemical machining (ECM)

the workpiece, metal is stripped away. The tool never comes into contact with the workpiece and no wear takes place. The electrolyte is usually a solution of sodium chloride but a solution of sodium nitrate is sometimes used. The electrolyte is pumped through the system at a high rate to remove the reaction products (dross). These are filtered out and the electrolyte is re-used. The electrodes are usually made from brass, copper or bronze by any appropriate forming process. These materials have good electrical conductivity, have adequate rigidity, they are easily machined to shape, and they do not react with the electrolyte.

Electro-chemical grinding is a variation on the above process. the electrode is the grinding wheel and, unlike conventional grinding wheels, the abrasive (aluminium oxide or diamond particles) are metal bonded. The abrasive particles take no part in the forming process, acting merely as insulators to space the wheel at a constant distance from the workpiece and also to help remove the reaction products.

Electric discharge machining

The principles of this process are shown in Fig. 4.16. The surface of the workpiece material is eroded away by electric spark discharges. This process is also known as 'spark erosion'. The dielectric acts as an insulator to control the spark, as a coolant, and it also carries away the debris from the cutting zone. Tools for EDM are usually made from brass, copper, copper-tungsten alloy, or graphite. Tool wear is important since it affects the tolerances and the shape produced in the workpiece. Separate electrodes for roughing and finishing are often used. Tool wear

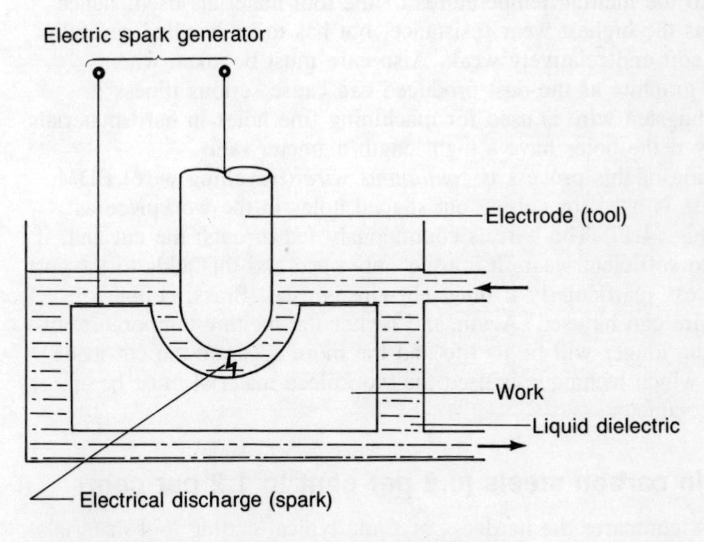

Fig. 4.16 Electrical discharge machining (EDM)

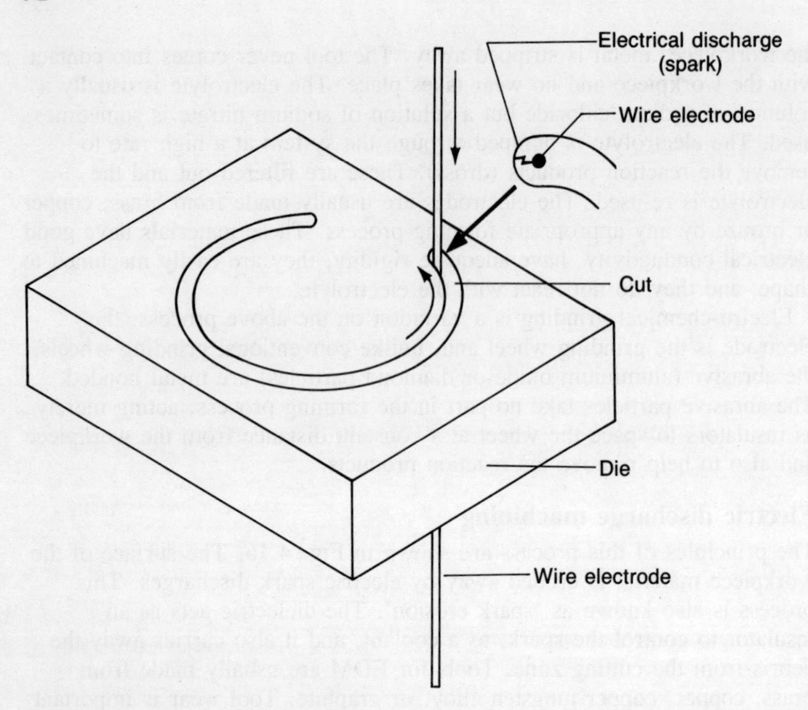

Fig. 4.17 Continuous wire EDM

is related to the melting temperatures of the tool materials used, hence graphite has the highest wear resistance, but has to be handled carefully since it is soft and relatively weak. Also care must be taken when machining graphite as the dust produced can cause serious illness if inhaled. Tungsten wire is used for machining fine holes in hard materials particularly if the holes have a high length/diameter ratio.

A variation of this process is *continuous wire* (travelling wire) EDM. This process is used for cutting out shaped holes in the workpiece as shown in Fig. 4.17. The wire is continuously fed through the cut and, if subjected to sufficient wear, it is used only once and this adds to the cost of the process, particularly if tungsten wire is used. Brass, copper or tungsten wire can be used. Again, the higher the melting temperature of the wire, the longer will be its life and the more accurate the cut made. No matter which technique is used the workpiece material must be an electrical conductor.

4.9 Plain carbon steels (0.8 per cent to 1.2 per cent)

Figure 4.18 compares the hardness of some typical cutting tool materials at various temperatures. It can be seen that plain carbon steels with a

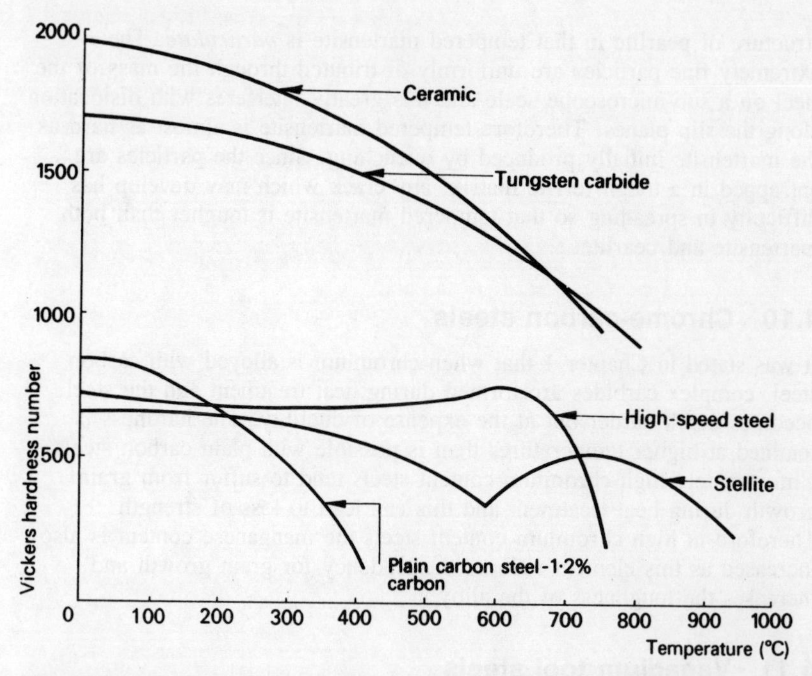

Fig. 4.18 Hardness/temperature curves for cutting-tool materials

high carbon content have a high initial hardness but that this hardness is rapidly reduced as the temperature increases (i.e. as the temper is 'drawn').

Steels in the lower carbon range are used where toughness rather than hardness is required and some typical applications are forging tools such as hot and cold sets, swages, fullers punches and drifts. Such steels are also used for making screwdriver blades, hammer heads, drifts, cold chisels, low grade spanners and similar hand tools.

Steels in the higher carbon range are used where a keen cutting edge is required. It can be seen from Fig. 4.18 that such steels are harder than alloy steels at room temperatures and this makes them very suitable for such applications as carpenters' chisels, knives, wood drills, and engineers' files where maximum sharpness is required but where the cutting temperature is low. However, Fig. 4.18 also shows that the hardness of plain carbon steels falls off rapidly as their temperature increases and this makes them unsuitable for machining operations under present-day production conditions.

The hardening and tempering of plain carbon steels had already been considered in detail in volume 1, and also in Chapter 2 of this text. The tempered martensitic structure of high-carbon steel tools is a two-phase structure of ferrite and iron carbide which differs from the lamellar structure of pearlite in that tempered martensite is *particulate*. The extremely fine particles are uniformly distributed through the mass of the steel on a sub-microscope scale and this greatly interferes with dislocation along the slip planes. Therefore tempered martensite is almost as hard as the martensite initially produced by quenching. Since the particles are entrapped in a tough ferrite matrix, any crack which may develop has difficulty in spreading so that tempered martensite is tougher than both martensite and pearlite.

4.10 Chrome-carbon steels

It was stated in Chapter 1 that when chromium is alloyed with carbon steel, complex carbides are formed during heat treatment and the steel becomes much harder but at the expense of ductility. The hardness is retained at higher temperatures than is possible with plain carbon steels. Unfortunately high chromium content steels tend to suffer from grain growth during heat treatment and this can lead to loss of strength. Therefore in high chromium content steels the manganese content is also increased as this element reduces the tendency for grain growth and increases the toughness of the alloy.

4.11 Vanadium tool steels

The addition of the alloying element, vanadium, to carbon-manganese steels and carbon-chrome steels restrains grain growth, increases the toughness of the steel and improves its 'hot hardness'.

4.12 Low tungsten steels

The addition of the relatively expensive metal, tungsten, results in the formation of hard and stable carbides when heat-treated. These carbides remain hard and stable at elevated temperatures, so that tools made from steels containing tungsten can operate continuously at high temperatures without loss of hardness.

4.13 High-speed steels

These are tool steel alloys designed to operate continuously at elevated temperatures. Their name derives from the fact that they can operate at very much higher cutting speeds than plain carbon and low tungsten steels. There are three groups of alloys:

(*a*) tungsten-chromium alloys, known as 'high-speed steels';
(*b*) tungsten-chromium-cobalt alloys, known as 'super-high-speed steels'; and
(*c*) molybdenum alloys, known as 'economy-high-speed steels' since the cheaper alloying element, molybdenum, is substituted for much of the tungsten and cobalt content.

The heat treatment of high-speed steels differs from other alloy tool steels inasmuch as they are hardened and then secondary hardened instead of being hardened and tempered. This secondary hardening actually increases the hardness as well as increasing the toughness of the steels, whereas tempering increases the toughness at the expense of some loss of hardness (see also Section 1.9).

High-speed tools can be given a very thin titanium nitride coating by chemical vapour deposition (CVD). Titanium nitride is a hard and heat resistant ceramic material. Twist drills are frequently treated in this manner to reduce wear of the flutes and the radial lands. Drills so treated have a characteristic golden colour.

4.14 Application of British Standard specifications to tool steels

Table 4.1 is derived from BS 4569 and shows some typical examples of the tool steels described in Section 4.13. Unlike BS 970 where the identification code is based upon the composition and properties of the steel, BS 4569 applies the number and letter code in a more arbitary manner. The coding is based upon that of the American Iron and Steel Institute (AISI) and the only difference is that the British code is prefixed by the letter B as shown in Table 4.2. Where a number follows the letter code, the number denotes a specific alloy. The number is applied arbitarily and in no way indicates the composition or properties of the alloy.

4.15 Stellite

This is a cobalt-based alloy, containing little or no iron, which can only be cast to shape or deposited as a hard facing material using an oxy-acetylene torch. It requires no heat treatment to make it hard, and can only be machined by grinding as it cannot be softened. Although slightly softer than high-speed steel, it retains its hardness even when the heat generated by the cutting process causes it to glow red-hot. It is considerably more expensive than high-speed and super-high-speed steels because of the large amounts of tungsten and cobalt present in the alloy. Stellite can be deposited, by welding, on to plain carbon and alloy steels to build up hard facings. It is also available in the form of ground tool 'bits' to fit standard tool holders. Stellite has sufficient strength and toughness to withstand positive rake cutting with the tool 'bit' supported as a cantilever. A typical composition is:

Cobalt	50%
Tungsten	33%
Carbon	3%
Various	14%

Table 4.1 Typical die and tool steels (derived from BS 4569)

Type of Steel	Composition C	Mn	Cr	Mo	V	Si	W	Co	Heat treatment	Applications
BD3	2.0	0.3	12.5	—	—	—	—	—	Heat slowly to 750–800°C and heat quickly to 960–990°C and quench in oil. Temper at 150–400°C for 30 to 60 minutes depending upon application	*High-carbon — high chromium (HCCR) Die Steel.* Widely used for high quality blanking and forming punches and dies for sheet metal pressings. Gauges, and thread rolling dies. Low shrinkage and distortion during heat treatment
BH12	0.35	—	5.00	1.50	0.40	1.00	1.35	—	Heat slowly to 800°C, soak and heat quickly to 1020–1050°C. Air cool, and temper at 540–620°C for up to 1½ hours depending upon section and application	*Hot-working die steel*, suitable for extrusion dies, hot stamping and forging dies. Hot pressing dies for copper and aluminium alloys

BT1	0.75	—	4.25	—	1.20	—	18.0	—	Heat slowly to 900°C, soak, and heat quickly to 1290–1310°C. Quench in oil or air blast. Double temper 565°C for 1 hour	*18% tungsten high-speed steel* for general purpose cutting tools for lathes, shapers, planers, milling machines etc. Threading taps, hacksaw blades, master gauges etc
BT6	0.80	—	4.75	0.50	1.50	—	22.0	12.0	Heat slowly to 900°C, soak, and heat quickly to 1300–1350°C. Quench in oil or air blast. Double temper at 565°C for 1 hour	*'Super' high-speed steel.* Cutting tools for machining hard and tough materials such as high duty alloy steels and cast irons
BM2	0.85	—	4.25	5.00	1.90	—	0.5	—	Heat slowly to 900°C, soak, and heat quickly to 1200–1250°C. quench in oil or air blast. Double temper at 565°C for 1 hour	'Economy' high-speed steel, developed during World War II when tungsten was in short supply. Similar in characteristics to BT1. Still used where shock resistance is required as it is less brittle than BT1. Useful for roughing and heavily scaled forgings and rustings

Table 4.2 Coding of tool and die steels

AISI Code	BS4659 Code	Description
A	BA	Medium-alloy steel; hardenable; but only suitable for processes which do not involve appreciable rise in temperature
D	BD	High carbon, high chromium die steel. This has better wear resistance than type BA, but can only be used for processes which do not involve appreciable rise in temperature
H	BH	Chromium or Tungsten based steel for hot working processes
L	BL	Low-alloy tool steels for special applications
M	BM	Molybdenum based steel for hot working processes
O	BO	Oil-hardening tool steels; only suitable for processes which do not involve appreciable rise in temperature
T	BT	Tungsten-based high speed steels
W	BW	Water-hardening tool steels

4.16 Cemented carbides

These are *cermet* composite materials which were introduced in Section 3.28. They consist of hard ceramic particles in a tough metallic matrix. Pre-formed tool tips made from metallic carbides are produced by a technique known as *sintering*. The production of sintered powder-metal compacts is discussed in *Manufacturing technology: volume 2*. Such tool tips are much harder than stellite and high-speed steels and they retain their hardness at very much higher cutting temperatures. Carbide cutting tools fall into four categories: tungsten carbide; mixed carbides (tungsten and titanium carbides); titanium carbides; and coated carbides.

Tungsten carbide

Tungsten carbide cutting tool tips are made from particles of tungsten carbide bonded together by a matrix of metallic cobalt. They are hard and brittle and are used mainly for cutting such materials as cast irons and cast bronzes which have a relatively low tensile strength, which form a discontinuous chip, and which have a highly abrasive skin. Because of its brittleness and low strength, 'straight' tungsten carbide is not suitable for roughing cuts on ductile, high-strength materials which form a continuous chip, although it can be used for light, finishing cuts using a high cutting speed and fine feed. Neither can it be used for interrupted cutting. Further, 'straight' tungsten carbide cutting tool materials tend to be porous and particles of the metal being cut can become embedded in the matrix. Although the metal being cut will not chip weld directly to the tungsten carbide, it will chip weld to the embedded particles of the metal being cut to form an undesirable built-up edge on the tool. Work piece materials such as cast iron, cast bronze and free-cutting brass which do not form a continuous chip and are, therefore, less susceptible to chip-welding can be successfully cut using tungsten carbide tools.

Mixed carbides

These are mixtures of titanium, tungsten, molybdenum and tantalum carbides in a matrix of cobalt. As well as being tougher and less susceptible to chipping, they have improved crater-wear characteristics and better hot-hardness than 'straight' tungsten carbide, and are suitable for taking heavy roughing cuts on such materials as high-strength alloy steels. They are also suitable for interrupted cutting. Mixed carbide tool tips are less porous than 'straight' tungsten carbide with the result that there is less tendency for particles of the workpiece material to become embedded in the matrix and, therefore, less tendency for chip-welding to occur.

Titanium carbide

Titanium carbide based tool tips have a matrix of nickel/molybdenum alloy. They have a higher wear resistance than tungsten carbide tool tips and mixed carbide tool tips but are not as tough. They are less susceptible to chip welding and are suitable for cutting hard materials at high speeds.

Coated carbides

These are more expensive than tungsten carbide and mixed carbides but have the advantage of being able to operate at significantly higher cutting speeds. They are generally 'straight' tungsten carbide tips which have been given a very thin coating of titanium carbide, titanium nitride, or aluminium oxide by chemical vapour deposition (CVD). The coating thickness is usually 5 to 9 μm. Alloy tool and die steels may also be coated by this process to increase their wear resistance. Multi-layer coatings are also available.

Titanium carbide coatings improve the wear resistance of the tool, whilst titanium nitride coatings reduce adhesion (chip-welding) and galling of the rake face. The use of titanium nitride greatly increases tool life. However, these coatings do not perform well at low cutting speeds as the coating tends to be chipped and removed by adhesion. Lubrication of the chip/tool interface is therefore important.

Aluminium oxide coatings have a low thermal conductivity and are therefore a good thermal barrier. Such coatings prevent the substrate of the tool tip when machining at very high cutting speeds becoming overheated.

4.17 Ceramics (oxides)

Ceramic tool tips are even harder than those made from metallic carbides and can also withstand higher operating temperatures without loss of hardness. Unfortunately, they are also more brittle. The ceramic material most widely used for cutting tools is finely ground aluminium oxide (Al_2O_3) particles together with chromium oxide, titanium oxide and titanium oxide bonded together by sintering. The aluminium oxide forms at least 70 per cent of the mix. Ceramic tips have a low transverse strength and are susceptible to edge chipping. Therefore they are only used for high-speed finishing operations where a low feed rate is used and cutting is uninterrupted. Unlike carbide tool tips, ceramic tips cannot be brazed to the tool shank but must be clamped in place. Figure 4.19 shows a typical toolholder suitable for both carbide and ceramic tips. The triangular tip is not usually resharpened but can be indexed and turned over to provide six alternative cutting edges before it is discarded. To reduce thermal shock, cutting fluids should not be used. The cutting tool tip should have negative rake geometry and the machine tool should be very rigid as vibration soon leads to tool failure.

4.18 Cubic boron nitride

This is the hardest material presently available except for diamond. Tool tips are made by bonding a layer of polycrystalline cubic boron nitride (CBN) to a substrate of cemented carbide. The boron nitride provides very high wear resistance and edge strength whilst the carbide provides the shock resistance and support. CBN is chemically inert to iron and nickel and is also resistant to oxidation at high temperatures. It is particularly suitable for machining hardened ferrous and high-temperature alloys.

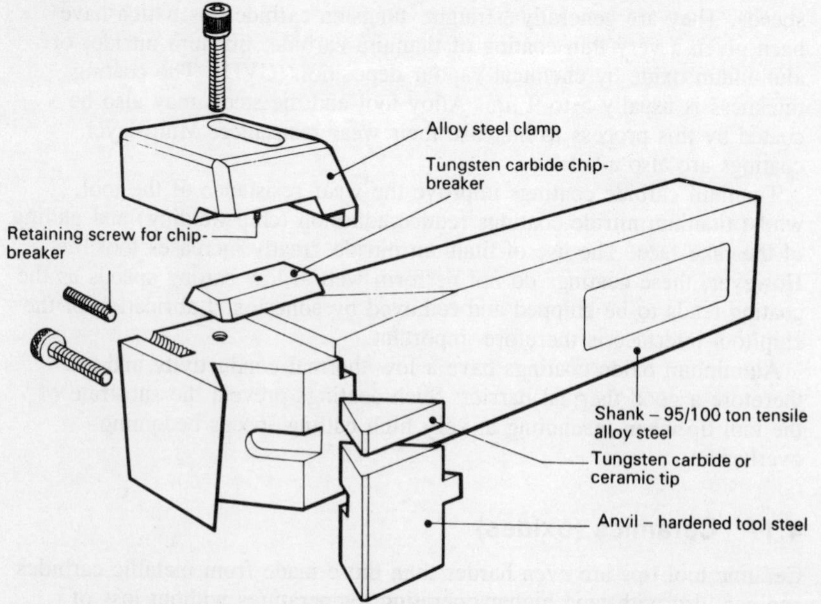

Fig. 4.19 Tool holder for carbide and ceramic tips

4.19 Diamond

Diamond — which is a crystalline allotrope of carbon — is the hardest substance known at present. In addition to its extreme hardness diamond has a high wear resistance, low coefficient of friction, low coefficient of thermal expansion, and high thermal conductivity. Because diamond is brittle, tool shape and mounting is important. Crystal orientation is also important to obtain optimum performance. The abrasive wear resistance of diamond can vary by at least tenfold depending upon crystal orientation. Wear in diamond tooling usually takes place by micro-chipping due to thermal stressing, and transformation to amorphous carbon due to frictional heating.

Diamond tools can be used satisfactorily at almost any speed, but are principally used for light uninterrupted cuts with high cutting speeds and low feed rates. Diamond tools are widely used for finish machining non-ferrous metals and alloys and abrasive non-metals to high dimensional accuracies and high surface finish. For example, the finish machining of motor car engine pistons where the aluminium alloy used cannot be satisfactorily ground. Unfortunately, diamond cannot be used for machining ferrous metals and alloys and nickel alloys as it has a high chemical affinity for these metals.

Gem stones are not used as they are too costly and chips of brown and black burt stones are generally used. However there is now an increasing tendency to use *polycrystalline diamond*. Small synthetic crystals are fused together by a high-pressure, high-temperature process, and bonded to a hard carbide substrate. The random orientation of the polycrystalline diamond prevents the propagation of cracks through the structure. The carbide substrate provides the toughness and support which the diamond layer required.

4.20 Abrasives

Abrasives are generally made from hard crystalline materials such as: aluminium oxide (emery); silicon carbide (corundum); cubic boron nitride; and diamond.

Abrasive wheels are made from large numbers of crystalline abrasive particles, called *grains*, held together by a *bond* to form a multi-tooth cutter similar to a milling cutter. Since an abrasive wheel has many more 'teeth' than a milling cutter, and because they are arranged in a random pattern, a ground surface usually has a very high standard of surface finish. However the reduced chip clearance between the grains of an abrasive wheel, compared with the space between the teeth of a milling cutter, results in a lower rate of material removal. Thus grinding is a finishing process. Figure 4.20 shows the dross from a grinding operation and it can be seen that this consists of a mixture of metal chips and blunt grains. The chips are remarkably similar to the chips produced by a milling cutter — proof that abrasive wheels cut metal from the work-

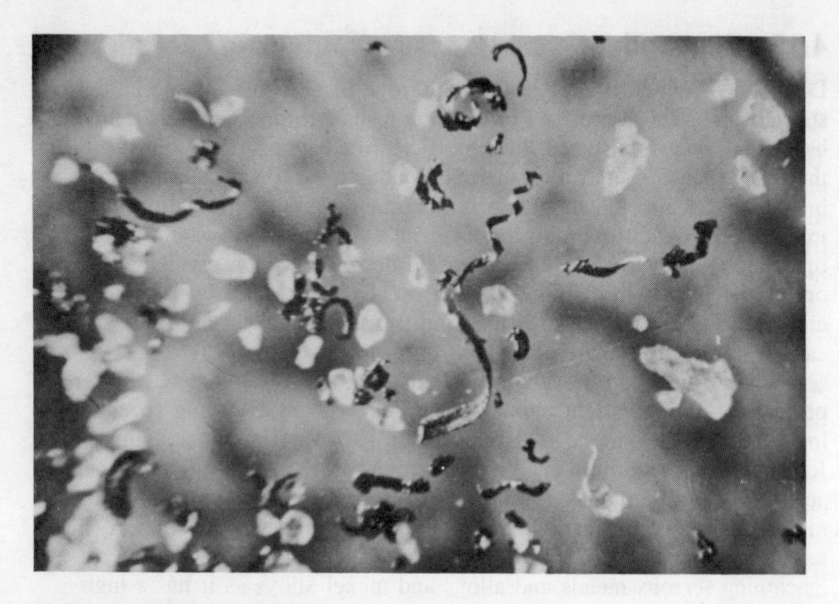

Fig. 4.20 Grinding wheel dross

piece and do not rub or burnish the surface. The blunt, spent grains are torn from the bond to expose new, sharp, *active* grains. When the abrasive grains become dull, the cutting forces acting upon them increase and the crystal particles shatter to expose new, sharp cutting edges, or the grains are ripped out of the bond wholesale to expose completely new, active grains. Figure 4.21 shows a comparison between the cutting actions of an abrasive grain and a milling cutter tooth.

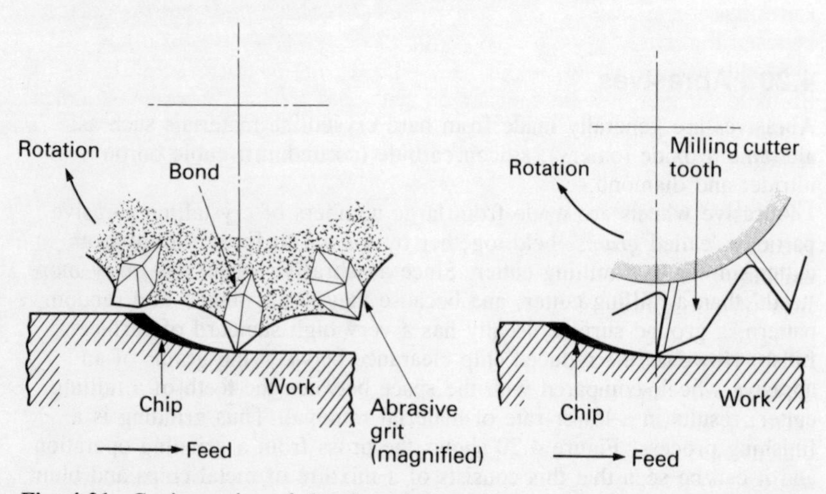

Fig. 4.21 Cutting action of abrasive wheel grains

4.21 Consideration of properties in the selection and application of cutting tool materials

Resistance to thermal softening

The mechanical energy required for cutting and forming engineering materials is invariably converted into heat energy which, in turn, results in an increase in temperature at the chip/tool interface. This may range from a gentle warming during manual cutting and forming operations to increasing the temperature of the chip to red heat when machining with carbide and ceramic tools. Therefore all tool materials require some degree of 'hot-hardness'.

Coefficient of friction

Since forming operations and cutting operations on ductile materials which form a continuous chip result in the workpiece material sliding over the tool material a low coefficient of friction is desirable. This not only reduces wear but reduces the forces required to perform the forming or cutting operation. In the case of metal cutting tools a low friction drag factor at the chip/tool interface can reduce the possibility of chip-welding occurring.

Wear resistance

Forming and cutting tools are costly, especially the large press tools used for car body panels and large plastic moulding dies, and the cost needs to be spread over a very large number of components. Therefore the tool materials must have a high resistance to wear, particularly if a high standard of product dimensional accuracy and surface finish is required.

Resistance to chemical reaction

Care must be taken to ensure that chemical reaction does not take place between the workpiece material and the tool material. For example diamond reacts with ferrous metals under cutting conditions and its tool life is lower than less hard materials such as coated carbides or ceramics. Again molten aluminium alloys react with some die steels and care is required in the choice of materials for pressure die-casting dies.

Electrical conductivity

Electrodes for electric discharge machining (EDM), and electro-chemical machining (ECM) require a high conductivity (low resistance) so that the flow of the electric current is not impeded by them.

Dimensional stability

The accuracy and surface finish of the workpiece is dependent upon the accuracy of the tooling, it is essential that the tooling material has a high dimensional stability. This applies particularly to large forming tools made from iron castings, forging dies, large die-casting dies and plastic moulds. After rough machining the tool materials should be stabilised by 'weathering' or by a normalising heat treatment to remove internal stresses. The slow release of these stresses under working conditions can cause distortion and dimensional inaccuracy of the tools. This particularly applies to hot forming processes such as forging, die-casting and plastic moulding where the repeated heating and cooling of the tool material can cause creep and distortion to occur. This is aggravated if the heating and cooling is not uniform. For hot forming processes the tool materials should have a low coefficient of expansion.

Machinability

This applies only to metallic tooling materials. Most of these materials can be softened and machined to shape using conventional techniques. They can then be finished to size by grinding after hardening by heat treatment. The exception is Stellite alloy which can only be machined by grinding. Carbides and ceramics cannot be machined except by grinding. They are preformed by compacting the powdered metallic carbide and the metallic matrix, and sintering prior to finish grinding.

Mechanical properties

The mechanical properties required by the material used for various typical tooling applications were discussed in Sections 4.2 to 4.8 inclusive. By analysing the stresses involved and choosing suitable materials to match these stresses considerable cost savings can be achieved together with longer tool life and lower maintenance and replacement costs. The use, for example, of wear pads of expensive die steel set into a yoke of low cost plain carbon steel is more cost effective than making a forming tool for pressed steel components from solid die steel. Not only will the tough yoke resist the working stresses of the tool better than the hardened die steel but, as wear takes place, the individual wear pads (die cheeks) can be replaced more easily and cheaply than having to replace the whole tool.

Size availability

Tool designers should always use standard components and standard size materials. This not only reduces the cost but by using components and materials which are readily available also reduces the lead time in getting the tools into production. Further, tool designers must also consider the 'ruling section' for the materials selected where heat treatment is involved if the required properties are to be achieved (see Section 2.5).

Cost

The tooling is only a means to an end, the main requirement is the production of components in the quantity, quality and time specified. Materials should be selected to keep the cost of the tooling to a minimum commensurate with adequate tool life between replacement. When choosing tool materials, the cost of manufacturing the tool should be considered as well as the cost of the individual materials. The cost of the tooling has to be shared out over the number of components to be produced (unit tooling cost). Simple, low cost tooling may increase the production cost unduly, whilst sophisticated tooling may increase the unit tooling cost beyond the saving in production cost. Usually a compromise has to be achieved between these two extremes and other economic parameters by the use of appropriate mathematical modelling.

5 Welded and brazed materials

5.1 Fusion welding

Fusion welding produces a bond between two components by the melting and solidification of the parent metal and filler metal at the joint interface. The principle of fusion welding is illustrated in Fig. 5.1, where it can be seen that the edges of the vee are melted and fused together with the molten filler metal. There are two fundamental methods of fusion welding and they are named after the heat source used, as shown in Fig. 5.2.

(*a*) *Oxy-fuel gas* (usually oxy-acetylene) in which a very high temperature flame is used to melt the metal being joined as well as melting the filler material.
(*b*) *Manual metallic arc* in which an electric current forms a very high temperature 'arc' between the metals being joined and an electrode made from suitable filler material.

5.2 Oxy-fuel gas welding

The principle of oxy-fuel gas welding is shown in Fig. 5.2(*a*). The flame from the welding torch provides the heat energy to melt (fuse) the parent metal and the filler rod which is made from a similar metal to that of the components being joined. Upon cooling the joint solidifies and the welded joint is complete.

For the production of satisfactory welded joints, it is essential to use the correct flame conditions. The appearance of the three basic flames are shown in Fig. 5.3.

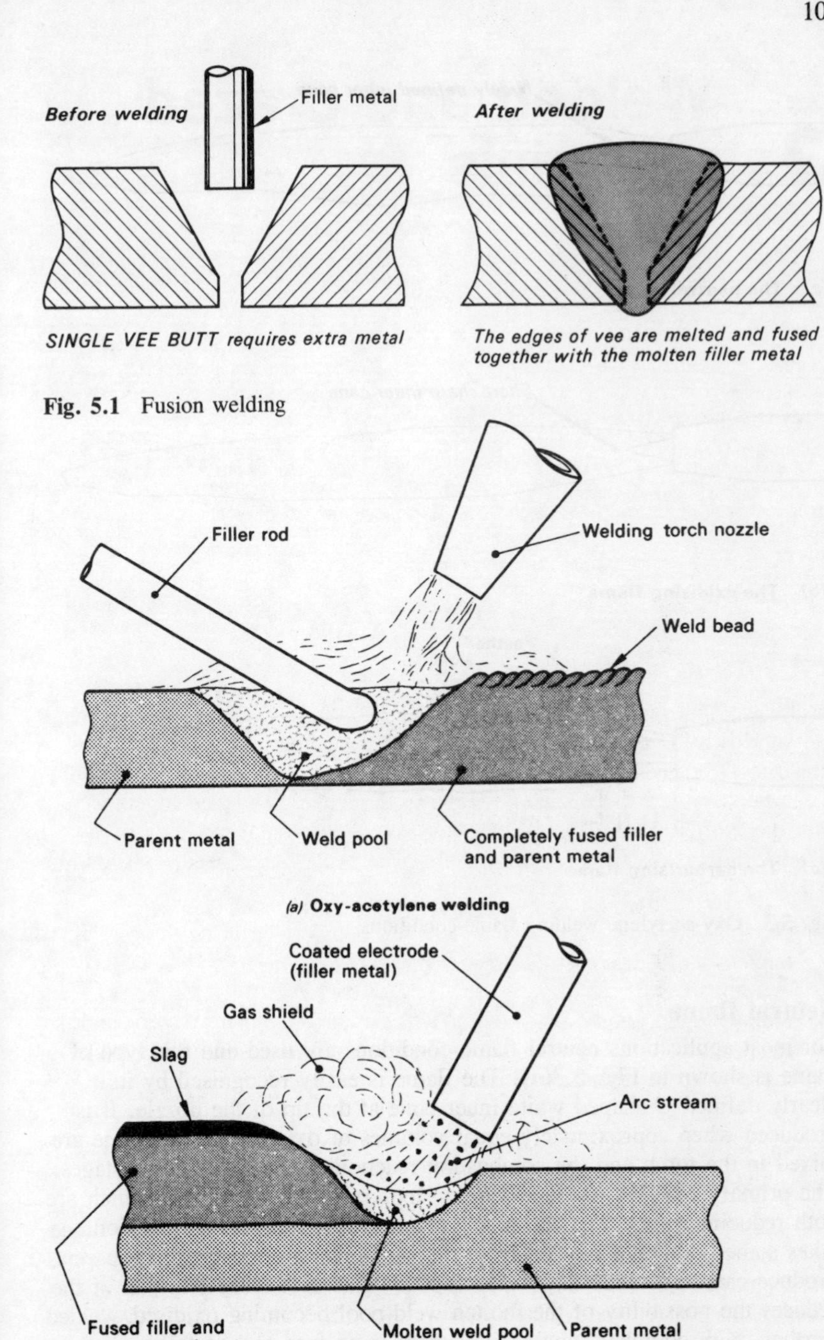

Fig. 5.1 Fusion welding

Fig. 5.2 Comparison of oxy-acetylene and metal-arc welding

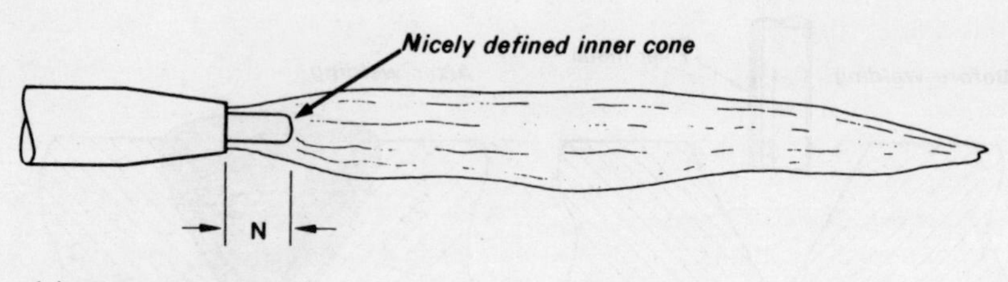

(a) **The neutral flame**

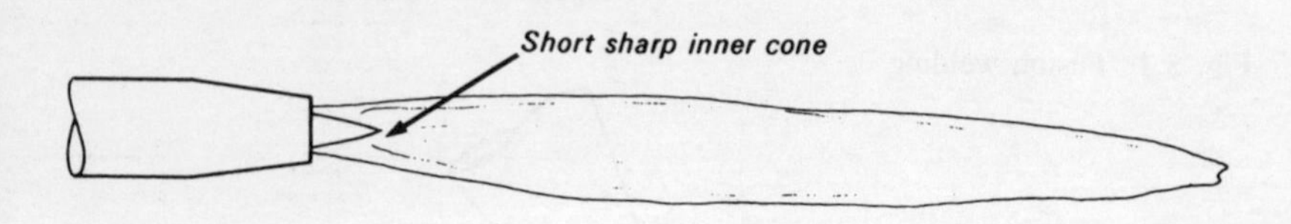

(b) **The oxidising flame**

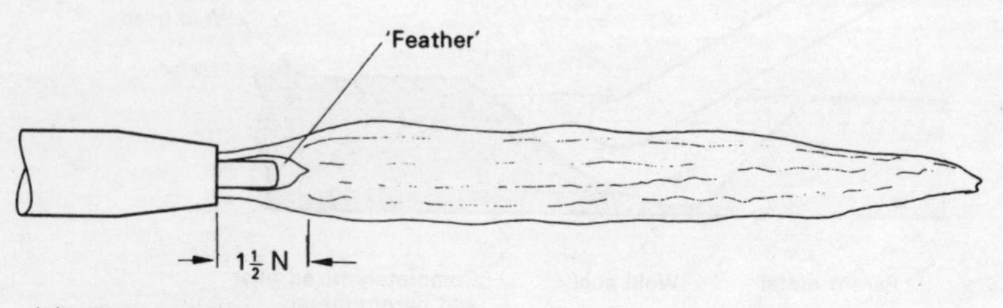

(c) **The carburising flame**

Fig. 5.3 Oxy-acetylene welding flame conditions

Neutral flame

For most applications neutral flame conditions are used and this type of flame is shown in Fig. 5.3(*a*). The flame is easily recognised by its clearly defined U-shaped white inner cone at the tip of the nozzle. It is produced when approximately equal volumes of oxygen and acetylene are mixed in the torch and the combustion reactions take place in two stages. The primary reaction produces carbon monoxide and hydrogen, which are both reducing gases. The secondary reaction for complete combustion takes some additional oxygen from the surrounding atmosphere to produce carbon dioxide and water vapour. This secondary reaction reduces the possibility of the molten weld pool becoming oxidised. Further, with complete combustion, there is no free carbon to be absorbed by the molten metal. Thus with a neutral flame virtually no change will take place in the composition of the weld metal.

Oxidising flame

This type of flame is shown in Fig. 5.3(*b*). It is obtained by first setting the torch for a neutral flame and then reducing the acetylene supply slightly at the torch. The oxidising flame is easily recognisable by the shorter and sharply pointed conical inner cone. Combustion is more noisy as the flame tends to 'roar'.

For complete combustion, one volume of acetylene requires two and a half volumes of oxygen. Thus, with the neutral flame, one and a half volumes of oxygen are taken from the atmosphere during the secondary reaction but, with the oxidising flame, all the oxygen necessary for combustion is taken from the cylinder and none is taken from the atmosphere. Therefore the hot weld metal will be rapidly oxidised by the atmospheric oxygen to which it is exposed.

When welding carbon steels with an oxidising flame, the excess oxygen tends to combine with the carbon in the metal to form bubbles of carbon monoxide in the molten weld pool. This leads to degradation of the joint resulting from decarburisation of the weld metal and gas porosity. Iron oxides will also be formed and will become trapped in the joint as undesirable and weakening inclusions.

Since the flame temperature is at a maximum with an oxidising flame, the metal may be overheated resulting in oxidation of the grain boundaries (burning) again leading to weakening of the joint. The only time that an oxidising flame is advantageous is when welding some brass, bronze and aluminium alloys.

Reducing or carburising flame

This flame is shown in Fig. 5.3(*c*) and is characterised by the 'feather' of incandescent carbon particles between the inner cone and the outer envelope. It is obtained by first setting the torch to neutral flame conditions and then slightly increasing the volume of acetylene at the torch. When the ratio of oxygen to acetylene becomes less than one to one, the flame condition is said to be 'reducing' or 'carburising'. This results in incomplete combustion which produces some free carbon. Thus ferrous weld metals become heavily carburised which results in major changes in the composition, hardness and ductility of the weld metal.

(*a*) The carbon content of the molten ferrous weld metal is increased by diffusion in a manner similar to that occurring during case hardening.
(*b*) If the critical cooling rate of the carburised metal is exceeded and the carbon content of the carburised metal exceeds 0.4 per cent, hardening will occur accompanied by a sharp reduction in ductility and toughness.

This effect can be used for the *flame hardening* of machine slide ways (see *Engineering materials: volume 1*). When carbon is dissolved in iron the melting temperature is lowered, and since carbon pick-up from the carburising flame is greatest at the surface of the metal, it follows that under these conditions the surface of the metal will melt before the body of the metal. This effect is exploited when cladding low- and medium-carbon steels with a hard surfacing material such as stellite.

5.3 Chemical reactions (gas welding)

The high temperatures encountered during welding not only cause structural changes in the metal, but they also promote chemical reactions between the metal being welded, the products of combustion, and the atmosphere. The higher the temperature is raised, the more rapid becomes the rate of absorption of the contaminating gases and the rate of reaction with these gases, particularly when the metal is molten.

Oxygen tends to react with the parent metal and filler metal to form metallic oxides. Metals which oxidise readily are likely to be difficult to weld and the use of protective fluxes may be necessary. Oxidised welds are undesirable and can be identified by the appearance of the surface which is irregular and pitted. Oxidation can result in:

(*a*) fusion becoming difficult;
(*b*) inclusions which may weaken the joint if the oxides formed are absorbed into the weld pool and remain entrapped in the solidified metal;
(*c*) lack of ductility in the joint (brittleness).

Nitrogen dissolves in many molten metals and alloys, and it may react with some of the constituents. If, when welding some steels, nitrogen is allowed to enter into the weld pool, the resulting joint will be porous and lacking in ductility.

Hydrogen can cause problems with many molten metals such as steel, aluminium and tough pitch copper as it is a powerful reducing agent. For example, tough pitch copper relies on the presence of particles of copper oxide to strengthen and harden the metal. Such particles would be reduced to metallic copper and water vapour by the hydrogen. The presence of hydrogen causes gas porosity which renders the joint weak and unsuitable for pipe joints and pressure vessels. As stated above, the removal of oxides by the hydrogen can directly reduce the strength and hardness of the metal. However, the water vapour formed during the reduction of the oxides can also cause gas porosity which further weakens the joint. The main sources of hydrogen are the products of combustion of the fuel gas which is a hydrocarbon or, in the case of electric arc welding, the electrode coating.

Water vapour tends to dissociate into hydrogen and oxygen when in contact with the molten metal. These *nascent* gases are highly reactive and exacerbate the effects described above.

5.4 The oxidation of welds

Some metals have such a high affinity for oxygen that the oxides on the surface of the metal reform as fast as they are removed. Although, as previously described, oxidised welds are generally unsatisfactory, this affinity for oxygen can be used to advantage in certain welding operations. For example, manganese and silicon, elements common to plain carbon steels, readily react with oxygen when the steel is in a molten condition. The reaction forms a thin layer of *liquid slag* which protects the weld pool from further oxidation. It also prevents the formation of gas pockets (cavities) in the joint. Steel welding wires (filler rods) have a high silicon and manganese content to ensure the protective slag is formed.

When welding other metals and alloys, removal and dispersion of the oxide film is not so easily achieved. Practical difficulties occur in welding when:

(*a*) the surface oxide forms a tenacious film;
(*b*) the oxide film has a very much higher melting point than that of the parent metal;
(*c*) the oxide film reforms very rapidly.

Fluxes are substances used to prevent oxidation and other undesirable chemical reactions from adversely affecting the quality of the joint. Table 5.1 lists some common fluxes and their applications when gas welding. The general requirements of fluxes are:

(*a*) to assist in removing the oxide film present on the surface of the parent metal in the joint zone;
(*b*) to prevent the oxide film reforming until the welded joint has been completed;
(*c*) to assist in removing any oxides which occur during welding by forming 'fusible slags' which float to the surface of the weld pool and do not interfere with the deposition and fusion of the filler material;
(*d*) to protect the weld pool from atmospheric oxygen and prevent the absorption and reaction of other gases (products of combustion), without obscuring the welder's vision, or hampering the manipulation of the molten pool;
(*e*) to lower the melting temperature of the oxide film on the parent metal below that of the parent metal itself. The flux should also have a lower melting temperature than that of the parent metal.

5.5 Manual metallic arc welding

The basic principles of manual metallic arc welding are shown in Fig. 5.4. The 'arc' is produced by a low voltage electric current jumping the air gap between the electrode and the metal being welded. The heat of the arc is concentrated on the edges of the two pieces of metal being joined. This causes the metal edges to melt and additional molten metal is transferred across the arc from the electrode which also melts and acts as a filler rod. The molten metal from the electrode (the *arc-stream*) is projected forcibly across the arc gap rather than flowing by gravity alone. This allows welding to take place vertically and overhead as well as

Table 5.1 Common applications of fluxes for gas welding

Metal or alloy	Flux	Remarks
Mild steel	—	No flux is required because the oxide produced has a lower melting point than the parent metal. Being less dense it floats to the surface of the molten weld metal as scale which is easily removed after welding. Use a neutral flame.
Copper	Borax base with other compounds	If Borax alone is used, a hard scale of copper borate is formed on the surface of the weld which is difficult to remove. Use a neutral flame.
Aluminium and aluminium alloys	Contains chlorides of lithium and potassium	Aluminium fluxes absorb moisture from the atmosphere, i.e. they are hygrosopic. Always replace the lid firmly on the container when the flux is not in use. The flux residue is very corrosive. On completion of the weld it is essential to remove all traces of this residue. This can be accomplished by scrubbing the joint area with a 5 per cent nitric acid solution or hot soapy water. Use carburising flame.
Brasses and bronzes	Borax type containing sodium borate with other chemicals	The flux residue is a hard glass-like compound which can be removed by chipping and wire brushing. Use an oxidising flame.
Cast iron	Contains borates, carbonates, and bicarbonates plus other slag-forming compounds	Oxidation is rapid at red heat, and melting point of the oxide is higher than that of the parent metal. For this reason it is important that the flux combines with the oxides to form a slag which floats to the surface of the weld pool and prevents further oxidation.

Flux removal: Many types of fluxes are corrosive to the metals or alloys with which they are used. Therefore it is important that residual flux be removed from the surface immediately after the welding operation. *Methods of removal generally employed include mechanical methods such as chipping and scratch brushing, rinsing or scrubbing with water, and use of acids or other chemicals.*

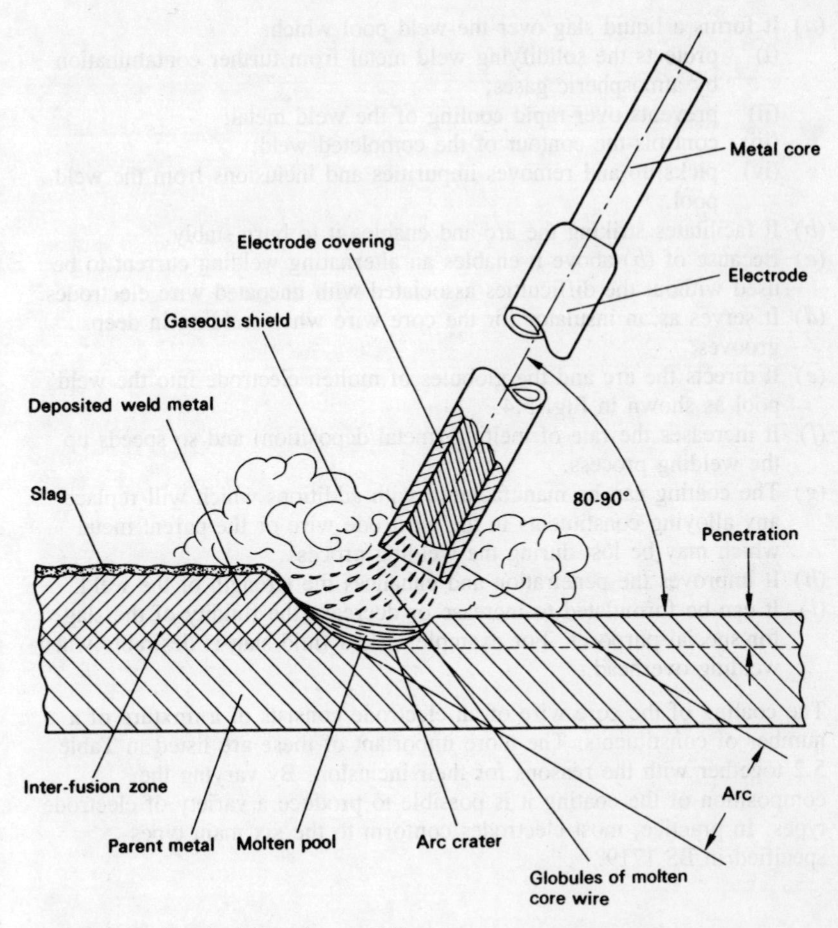

Fig. 5.4 Basic principles of arc welding

along horizontal joints. As soon as the arc is struck, the tip of the electrode begins to melt thus increasing the gap between the electrode and the work. It is necessary, therefore, to feed the electrode towards the work to maintain an arc gap approximately 3 mm in length. The electrode is moved at a uniform rate along the joint, melting the metal to be joined and adding additional metal into the joint as it passes.

The majority of electrodes used in manual metallic arc welding processes are *coated* electrodes. These consist of a core wire of closely controlled composition surrounded by a concentric coating of a solid flux which will melt uniformly with the core wire. The flux forms a partly vaporised and partly molten screen around the arc-stream and protects it from contamination by atmospheric gases. The flux coating which surrounds the electrode has several important functions apart from protecting the arc-stream. The more important of these are as follows.

(*a*) It forms a liquid slag over the weld pool which:
  (i) protects the solidifying weld metal from further contamination by atmospheric gases;
  (ii) prevents over-rapid cooling of the weld metal;
  (iii) controls the contour of the completed weld;
  (iv) picks up and removes impurities and inclusions from the weld pool.

(*b*) It facilitates striking the arc and enables it to burn stably.
(*c*) Because of (*b*) above it enables an alternating welding current to be used without the difficulties associated with uncoated wire electrodes.
(*d*) It serves as an insulator for the core wire when welding in deep grooves.
(*e*) It directs the arc and the globules of molten electrode into the weld pool as shown in Fig. 5.4.
(*f*) It increases the rate of melting (metal deposition) and so speeds up the welding process.
(*g*) The coating can be manufactured with additions which will replace any alloying constituents in the electrode wire or the parent metal which may be lost during the welding process.
(*h*) It improves the penetration and enhances the strength of the weld.
(*i*) It can be formulated to increase or decrease the fluidity of the slag for special purposes. For example, a less fluid slag is desirable when welding overhead.

The coating of the core wire of an electrode consists of a mixture of a number of constituents. The more important of these are listed in Table 5.2 together with the reasons for their inclusion. By varying the composition of the coating it is possible to produce a variety of electrode types. In practice, most electrodes conform to the six main types specified in BS 1719.

5.6 Shielding gases (arc welding)

An alternative to using coated electrodes is to use shielding gases to prevent oxidation of the weld pool. Although processes using shielding gases have higher material costs, labour costs are reduced as descaling is no longer required.

Tungsten inert gas welding (TIG)

This employs a tungsten electrode which is non-consumable. The atmospheric gases oxygen and nitrogen are excluded from the weld pool by a blanket of inert gas, such as *argon* or *helium*, which will not react with the molten metal. The principle of TIG welding is shown in Fig. 5.5(*a*). It can be seen that the arc is struck between the tungsten electrode and the parent metal and, unlike other arc welding processes, a separate filler rod is used in a similar way to gas welding.

Table 5.2 Electrode coating materials

Constituent	Remarks
Titanium dioxide	Available in the form of natural sands as **Rutile** containing 96 per cent titanium oxide. Forms a highly fluid and quick freezing slag. Is a good ionising agent.
Cellulose	Provides a reducing gas shield for the arc. Increases the arc voltage.
Iron oxide *and* manganese oxide	Used to adjust the fluidity and the properties of the slag.
Potassium aluminium silicate	Is a good ionising agent, also gives strength to the coating.
Mineral silicates *and* asbestos	Provides slag and adds strength to the coating.
Clays *and* gums	Used to produce the necessary plasticity for extrusion of the coating paste.
Iron powder	Increases the amount of metal deposited for a given size of core wire.
Calcium fluoride	Used to adjust the basicity of the slag.
Metal carbonates	Provides a reducing atmosphere at the arc. Adjusts the basicity of the slag.
Ferro-manganese *and* ferro-silicate	Used to deoxidise and supplement the manganese content of the weld metal.

Metal inert gas welding (MIG)

This is an automatic or semi-automatic process. The electrode is a bare wire which is continuously fed from a drum into the welding gun by means of an automatic electrode wire drive unit which senses the potential difference across the arc and maintains a constant length of arc gap. The shielding gases used may be *argon*, *helium* or *carbon dioxide*. The use of carbon dioxide considerably reduces the operating costs of the process and when this latter gas is used the process is often called 'CO_2 welding'. Unfortunately, carbon dioxide is not an inert gas and cannot be used with some non-ferrous metals and alloys. When it passes through the arc it tends to break down into carbon monoxide and oxygen. (The effects of these gases on the strength of welded joints was discussed in Section 5.4.) To ensure that the liberated oxygen does not contaminate the weld metal, *deoxidising* alloying elements are incorporated into the welding wire. These deoxidising elements combine with the oxygen to form a very thin and neat protective layer of slag on the surface of the completed weld and it does not have to be removed. Figure 5.5(*b*) shows the principle of MIG welding.

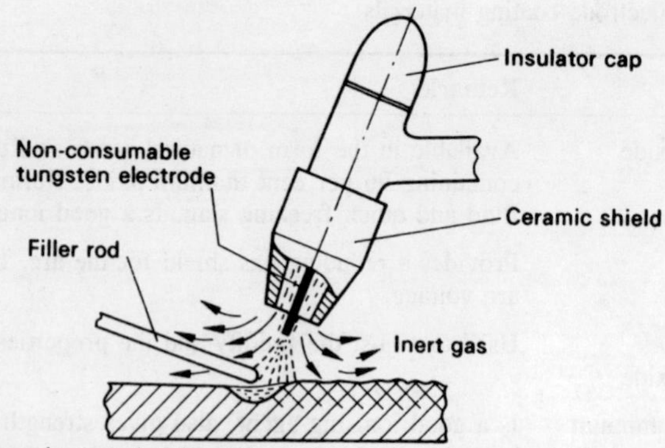

The atmosphere is excluded from the weld by shielding with an inert gas. No chemical reactions take place

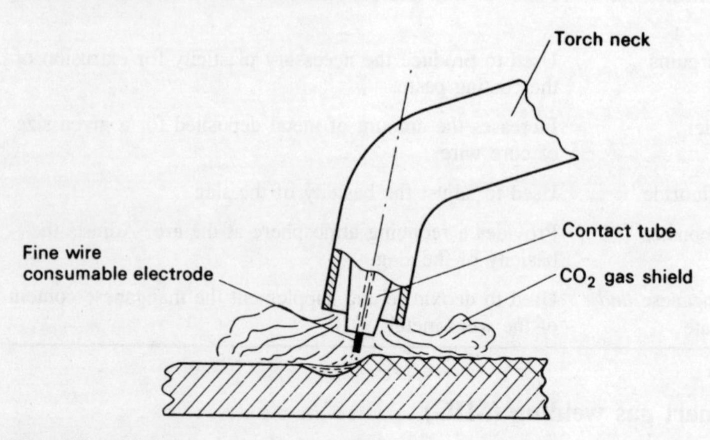

Fig. 5.5 Shielding gases in arc welding

5.7 Submerged arc welding

This is an automatic process used for arc welding thick plate where very heavy welding currents are required. The end of the bare wire electrode is submerged under powdered flux, thus preventing splatter and ultraviolet radiation which could be a hazard to persons working nearby. The principle of this process is shown in Fig. 5.6. The tube delivers the powdered flux to the joint from a hopper and surplus flux is removed by a vacuum cleaning unit and recycled to the hopper. The electrode wire is automatically fed into the joint to keep the arc constant. Much of the flux melts and rises to the top of the molten weld metal as a protective slag. The automatic flux supply and electrode feed mechanism is traversed along the joint on a small power driven 'tractor'. The process is most suitable for straight line welding and is widely used in ship construction and other heavy duty, large-scale applications.

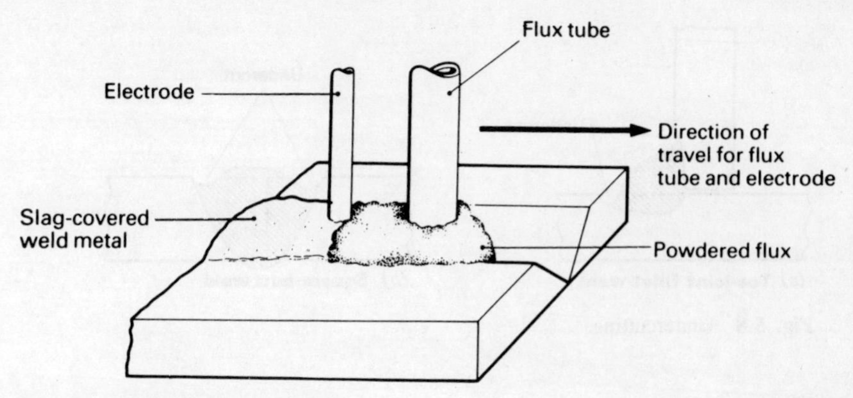

Fig. 5.6 Submerged arc welding

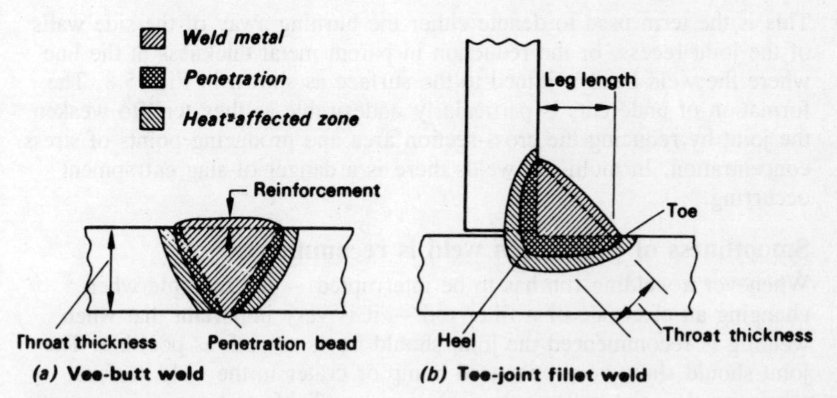

Fig. 5.7 Correctly formed welds

5.8 Welding defects

Some of the main factors to be considered when assessing the quality of a weld are:

(*a*) shape of profile;
(*b*) uniformity of surface and freedom from surface defects;
(*c*) degree of undercut;
(*d*) smoothness of join where weld is recommenced;
(*e*) penetration of bead and degree of root penetration;
(*f*) degree of fusion;
(*g*) non-metallic inclusions and gas cavities.

Figure 5.7 shows a correctly formed vee-butt weld and a tee-joint fillet weld. The effect of the heat affected zone on the joint strength will be considered in Sections 5.13 to 5.18 inclusive.

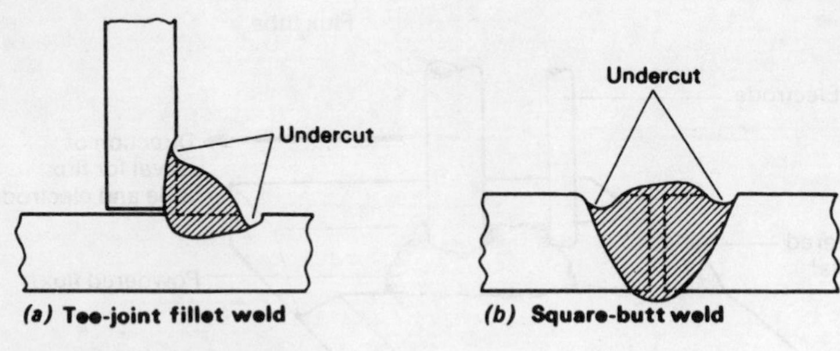

Fig. 5.8 Undercutting

Undercutting

This is the term used to denote either the burning away of the side walls of the joint recess, or the reduction in parent metal thickness at the line where the weld bead is joined to the surface as shown in Fig. 5.8. The formation of undercuts is particularly undesirable as they tend to weaken the joint by reducing the cross-section area and producing points of stress concentration. In multi-run welds there is a danger of slag entrapment occurring.

Smoothness of joint when weld is recommenced

Whenever a welding run has to be interrupted — for example when changing an electrode or a filler rod — it is very important that when welding is recommenced the joint should be as smooth as possible. The joint should show no pronounced hump or crater in the weld surface, otherwise the joint ends of the weld runs are liable to have poor strength. This is caused by crater cracks producing stress concentrations, and overlaps causing lack of fusion.

Surface defects

Surface defects in welded joints are generally due to the use of unsuitable materials and/or incorrect techniques. The weld surface should be free from porosity, cavities, and either burnt-on scale (when gas welding) or trapped slag (when arc welding).

Penetration

One of the more common causes of welded joints which are faulty is the lack of penetration. Examples of welds with unsatisfactory penetration are shown in Fig. 5.9. This results in a reduction in the cross-sectional area of the joint and corresponding lack of strength.

Lack of fusion

Lack of fusion is the failure to fuse together adjacent layers of weld or adjacent weld metal and parent metal as shown in Fig. 5.10. This

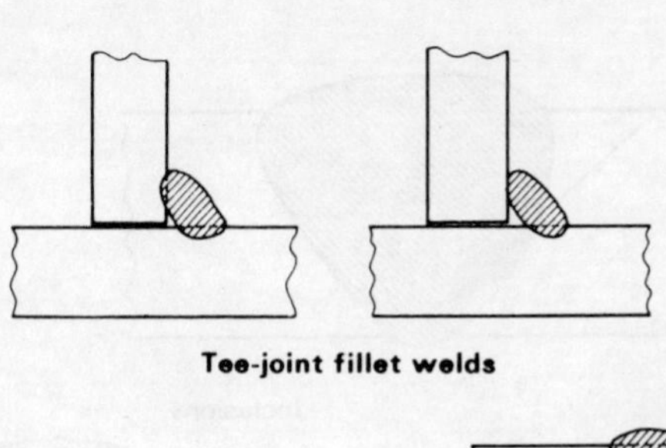

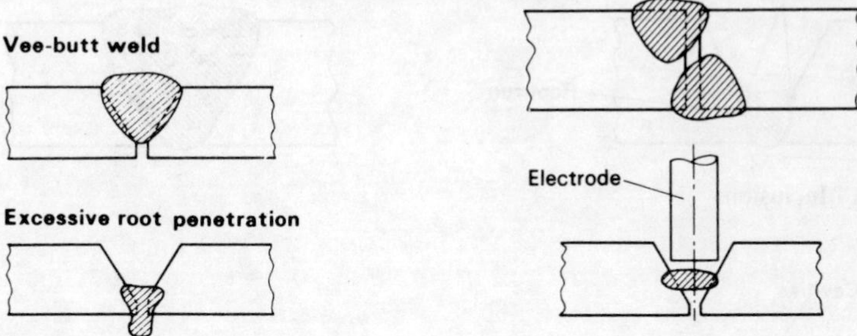

Fig. 5.9 Incorrect penetration

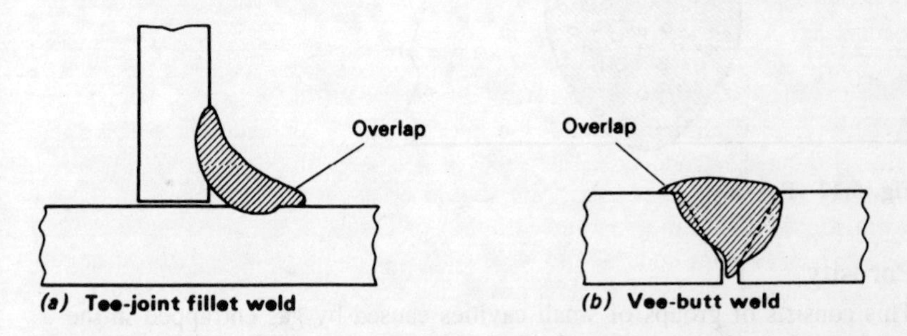

Fig. 5.10 Lack of fusion

condition is caused by failure to raise the metal to its melting point either because of lack of heat energy, or because of an insulating layer of oxide or other impurities on the joint surfaces of the parent metal.

Inclusions

Inclusions may be slag or other foreign matter entrapped in the weld metal. These inclusions originate from the slag formed by the electrode coating, from badly prepared and dirty joint surfaces, or from mill-scale on the surface of the parent metal. They may also originate from atmospheric contamination. Slag inclusion in multi-run welds may result from inadequate removal of slag after the initial root run so that it becomes entrapped by the subsequent runs. Figure 5.11 shows some typical problems involving slag inclusions. In all instances, inclusions are the result of faulty technique and lack of attention to detail.

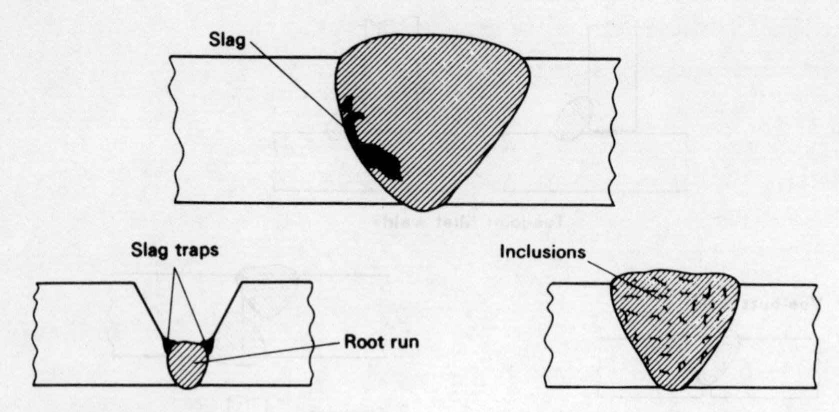

Fig. 5.11 Inclusions

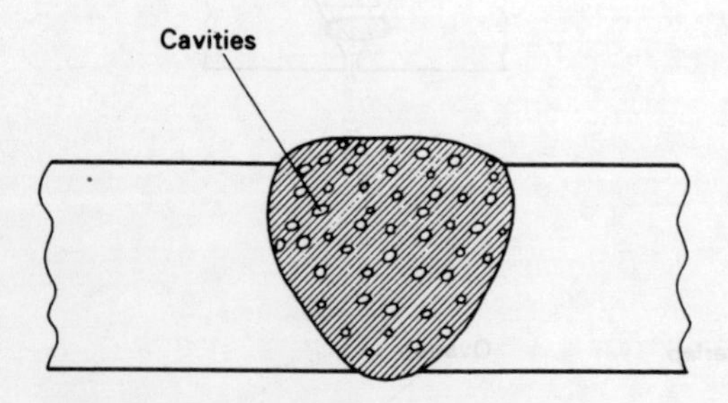

Fig. 5.12 Porosity

Porosity

This consists of groups of small cavities caused by gas entrapped in the weld metal as shown in Fig. 5.12. Large cavities just below the surface or in the surface of the joint are called 'blow-holes'. Table 5.3 lists some typical causes of porosity.

5.9 Resistance welding

The welding processes discussed so far in this chapter have been fusion processes; that is, the filler metal and the edges of the parent metal have been melted and allowed to run together. In resistance welding processes the metal is raised to just below its melting point and the weld is completed by the application of pressure as in forge welding.

Spot welding

This is the most common of the resistance welding processes. It is much quicker than riveting as a technique for joining sheet metal components and, since no holes are drilled in the components being joined, they are

Table 5.3 Miscellaneous causes of porosity

Cause	Remedy
1. High rate of weld freezing	Increase the heat input
2. Oil, paint, or rust on the surface of the parent metal	Clean the joint surfaces
3. Improper arc length, current, or manipulation	Use proper arc length (within recommended voltage range), control welding technique
4. Heavy galvanised coatings	Remove sufficient zinc on both sides of the joint
5. Use of oxidising flame	Use neutral flame for most welding operations

not weakened. The joint is produced by making a series of spot welds side by side at regular intervals. Such joints are not fluid tight and have to be sealed to prevent leakage or corrosion. Apart from ensuring that the joint faces are clean and free from corrosion, no special joint preparation is required.

The temperature of the components to be joined is raised locally by the passage of a heavy electric current, at low voltage, through the components as shown in Figs 5.13(*a*) and 5.13(*b*). When an electric current flows through a resistor the electrical energy is partially converted into heat energy and the temperature of the resistor is raised. Resistance welding uses this principle. Resistance to the flow of current occurs between the two surfaces of the components being joined, over the cross-section of the electrodes. Sufficient heat is generated to raise the components to the welding temperature at this spot. The current is then switched off and the weld is completed by squeezing the components tightly together between the electrodes. The electrodes, which are made from copper as this metal has high thermal and electrical conductivity, are water cooled to prevent them from over-heating. The complete cycle of events is controlled by an automatic timer.

No additional material filler metal is required to make a resistance welded joint. Since the process is akin to forge welding, the grain structure of the metal in the weld zone is in the wrought condition, rather than in the 'as cast' condition associated with fusion welding. This makes resistance welds stronger and more ductile than fusion welds of similar cross-sectional area.

Seam welding

The components to be joined are gripped between revolving, circular electrodes as shown in Fig. 5.13(*c*). The welding current is applied in

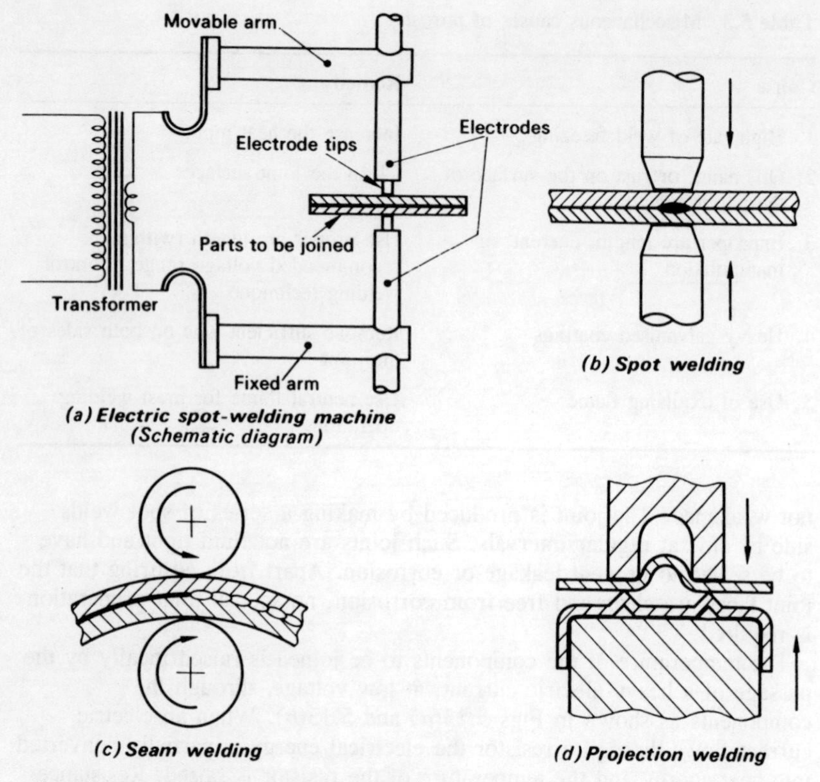

Fig. 5.13 Principles of resistance-welding

pulses and this results in a series of overlapping spot welds being made along the seam. This method of resistance welding is used for the manufacture of containers and fuel tanks.

Projection welding

With this process the electrodes act as locations for holding the parts to be joined. The joint is so designed that projections are performed on one of the parts as shown in Fig. 5.13(*d*). Projection welding enables the welding pressure and heated welding zone to be localised at predetermined points. This technique is largely used for small components which need to be accurately located.

Butt welding

The joints described so far in this section have been lap joints connecting sheet metal components. Butt welding is used for connecting more solid components such as carbon steel shanks onto the high-speed steel bodies of large twist drills. The principles of resistance butt-welding are shown

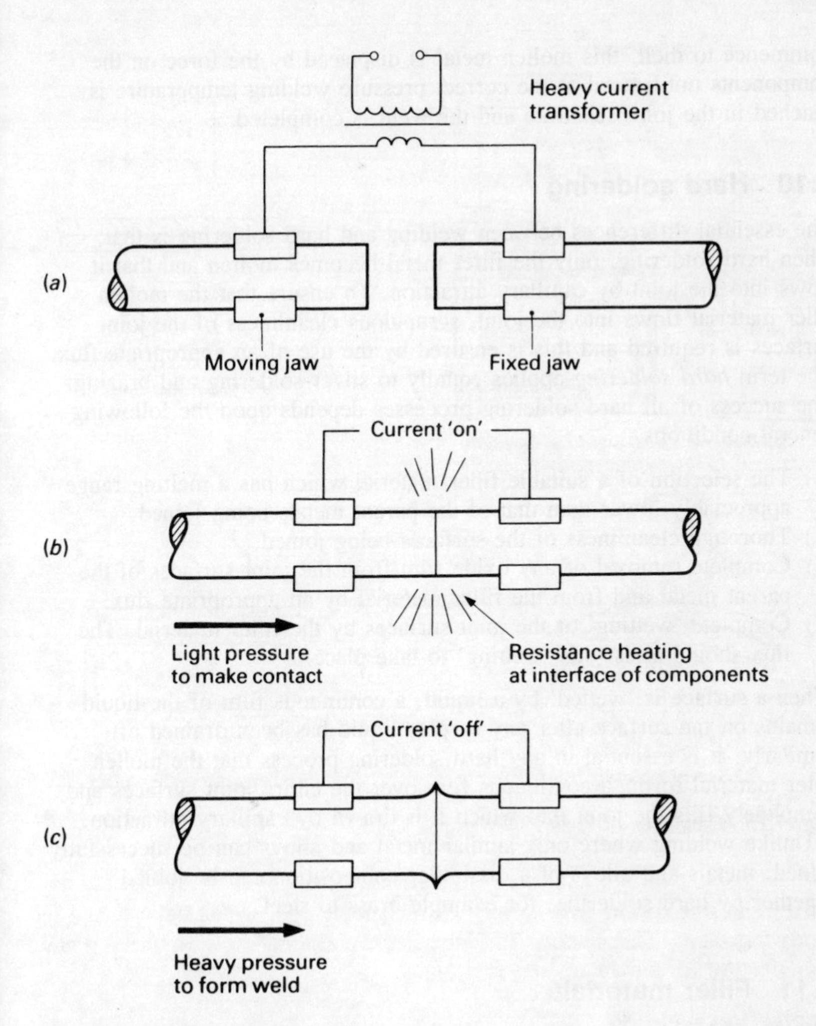

Fig. 5.14 Resistance butt-welding

in Fig. 5.14. The two ends of the rods are brought together with just sufficient force to ensure the current can flow without arcing. The resistance of the joint face ensures local heating will then take place on the passage of a heavy, low-voltage, electric current. When the metal in the joint zone has reached its welding temperature, the current is switched off and the axial force on the joint is increased to complete the weld.

As for all resistance welding processes, a sound weld is achieved by raising the temperature of the metal to just above the pressure welding temperature. Should the surfaces of the components at the joint interface commence to melt, this molten metal is displaced by the force on the components until metal at the correct pressure welding temperature is reached in the joint substrate and the weld is completed.

5.10 Hard soldering

The essential differences between welding and hard soldering is that, when hard soldering, only the filler metal becomes molten and that it flows into the joint by capillary attraction. To ensure that the molten filler material flows into the joint, scrupulous cleanliness of the joint surfaces is required and this is ensured by the use of an appropriate flux. The term *hard soldering* applies equally to silver-soldering and brazing. The success of all hard soldering processes depends upon the following general conditions.

(*a*) The selection of a suitable filler material which has a melting range appreciably lower than that of the parent metals being joined.
(*b*) Thorough cleanliness of the surfaces being joined.
(*c*) Complete removal of any oxide film from the joint surfaces of the parent metal and from the filler material by an appropriate flux.
(*d*) Complete 'wetting' of the joint surfaces by the filler material. The flux should assist this 'wetting' to take place.

When a surface is 'wetted' by a liquid, a continuous film of the liquid remains on the surface after any surplus liquid has been drained off. Similarly, it is essential in any hard soldering process that the molten filler material forms a continuous film over the entire joint surfaces and completely fills the joint into which it is drawn by capillary attraction.

Unlike welding where only similar metal and alloys can be successfully joined, metals and alloys of a dissimilar composition can be joined together by hard soldering, for example brass to steel.

5.11 Filler materials

Silver solders

These are more expensive than the common brazing alloys because they contain a high percentage of the precious metal silver. However, they offer the advantage of producing strong and ductile joints at much lower temperatures than those associated with brazing spelters and, consequently, have little heat-effect on the parent metals being joined. Silver solders are very free flowing which speeds up the process, results in a neat joint on fine and intricate work, and little finishing is required. Proprietary fluxes are required and these are supplied by the manufacturers of the silver solder.

Brazing brasses

The oldest filler material, from which the 'brazing' process gets its name, is a brass alloy containing equal amounts of copper and zinc. This was originally called 'brazing spelter' (spelter being the old name for zinc). This alloy melts at a much higher temperature than silver solder and phosphorus containing brazing alloys. Increasing the zinc content reduces the melting temperature range, thus enabling a joint to be made with components manufactured from ductile brass alloys which have a higher copper content. It might be thought that the use of a high zinc content brazing spelter would result in a relatively weak joint. However, during the brazing process some of the volatile zinc content is lost and the final composition of the filler material has a lower zinc content and a higher strength. When brazing with a copper-zinc alloy spelter, a paste of borax in water provides a suitable flux.

Brazing spelters containing phosphorus

Filler materials which contain phosphorus are usually referred to as self-fluxing. These alloys contain silver, copper and phosphorus or simply copper and phosphorus, the former possessing a lower melting temperature range. The phosphorus content lowers the melting range sufficiently for copper and copper-based alloys to be brazed without the parent metal melting. Copper and steel components can be brazed in air since the oxidation products form a liquid compound which acts as an effective flux. Copper-based alloys require a separate flux. Self-fluxing alloys are widely used for furnace brazing and other manufacturing techniques where the filler material is applied as 'preforms'.

5.12 Brazing techniques

Flame brazing

This technique may be used to fabricate almost any assembly on a small quantity basis. An oxy-propane gas torch is usually used and neat joints, requiring no finishing, can be produced by a skilled operative. It is also used where the assembly is too large to braze in a furnace. To ensure a sound joint, the correct flux must be applied, the joint must have a uniform capillary gap between the components, and the parent metal must be sufficiently hot to melt the spelter on contact after the flame of the torch has been momentarily withdrawn.

Furnace brazing

This technique is used when:

(*a*) the parts to be brazed can be pre-assembled or jigged to hold them in position;
(*b*) the brazing filler material can be preformed and pre-placed as shown in Fig. 5.15;
(*c*) when a controlled atmosphere is required.

The method of heating varies according to the application. The work to be brazed can be loaded into a muffle furnace so that the atmosphere can be controlled and the products of combustion will not affect the joint, or

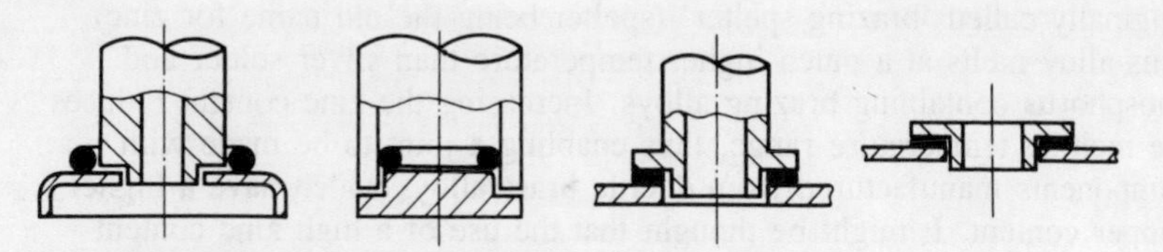

Fig. 5.15 Use of prepared filler metal

the work can be packed into sealed containers. Alternatively, the work can be passed continuously through the furnace on a conveyor.

Dip brazing (molten spelter)

The parts to be brazed are assembled together and submerged in a bath of molten filler material which is drawn into the fluxed joint by capillary attraction. It does not adhere to the surface of the work. The filler material is melted in a graphite crucible and a layer of flux is floated on the surface of the molten metal. Large assemblies have to be pre-heated before being lowered into the molten filler material.

Dip brazing (salt bath)

The molten salts, in a salt bath furnace, provide uniform heating of the work and prevent atmospheric contamination. Again, the work has to be pre-assembled, fluxed and the filler material applied as preforms.

Electric induction brazing

The component to be brazed is placed in the magnetic field of an induction coil through which is passed a high-frequency electric current as shown in Fig. 5.16. This induces eddy currents in the component which cause it to heat up. Thus the heat is generated within the component itself. Preformed and pre-positioned filler material is used together with a suitable flux if required.

Electric resistance brazing

In this process the heat required to melt the filler material is developed by:

(*a*) resistance at the joint interface (*direct heating*);
(*b*) resistance between the work and the electrodes (*indirect heating*).

The basic principle of resistance heating is that a heavy electric current at a low potential is passed through the assembly in such a way that a hot spot is generated at the joint interface as shown in Fig. 5.17. Heating can be localised and this ensures that the parent metal suffers no general loss of mechanical properties.

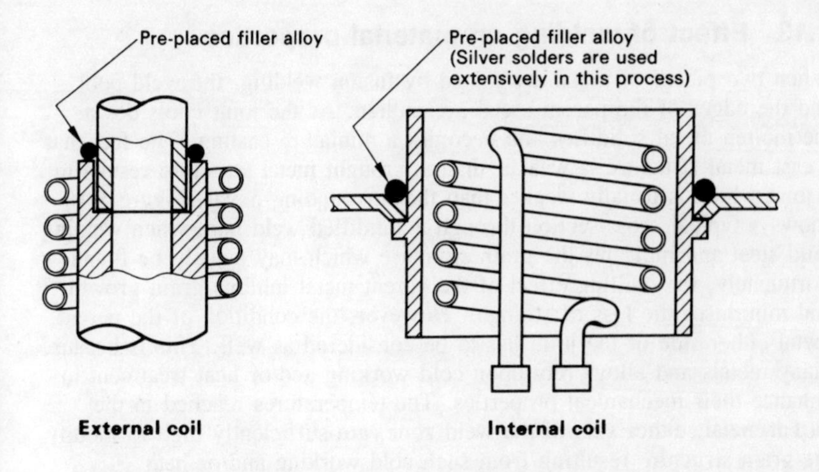

Fig. 5.16 Electric induction brazing

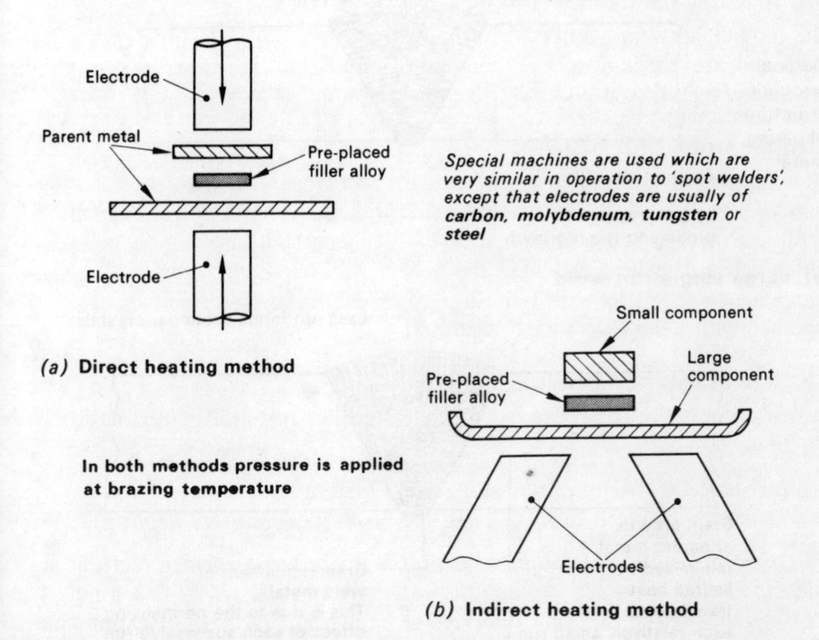

Fig. 5.17 Electric resistance brazing

5.13 Effect of welding on material properties

When two pieces of metal are joined by fusion welding, the weld pool and the edges of the parent metal are molten. As the joint cools down, the molten metal solidifies and becomes a miniature casting. The fact that a cast metal structure is weaker than a wrought metal structure results in a joint which is usually weaker than the surrounding metal. Figure 5.18 shows a typical cross-section through a solidified weld pool when welding mild steel and indicates the grain structure which may usually be found. Fortunately, the chilling effect of the parent metal inhibits grain growth and minimises the loss of strength. However, the condition of the parent metal either side of the joint has to be considered as well. This is because many metals and alloys rely upon cold working and/or heat treatment to enhance their mechanical properties. The temperatures reached in the parent metal, either side of the weld zone, are sufficiently high to modify the grain structure resulting from such cold working and/or heat treatment. The areas either side of the welded zone where structural modifications occur as a result of welding are referred to as *heat affected*

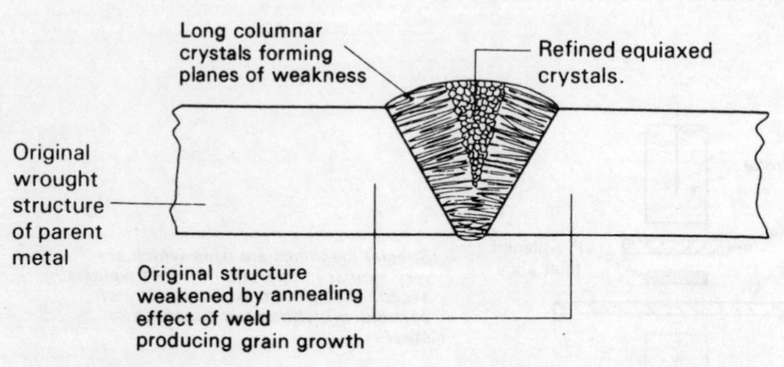

(a) **Large single-run weld**

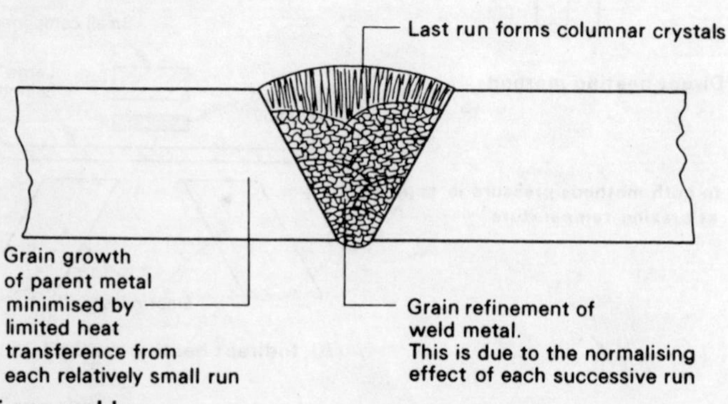

(b) **Multi-run weld**

Fig. 5.18 Structure of weld metal (mild steel)

zones. The upper limit of the temperature gradient across the heat affected zone is the temperature of the parent metal remote from the weld zone (room temperature). Figure 5.19 shows, schematically, the heat affected zone for a welded joint in mild steel, whilst Table 5.4

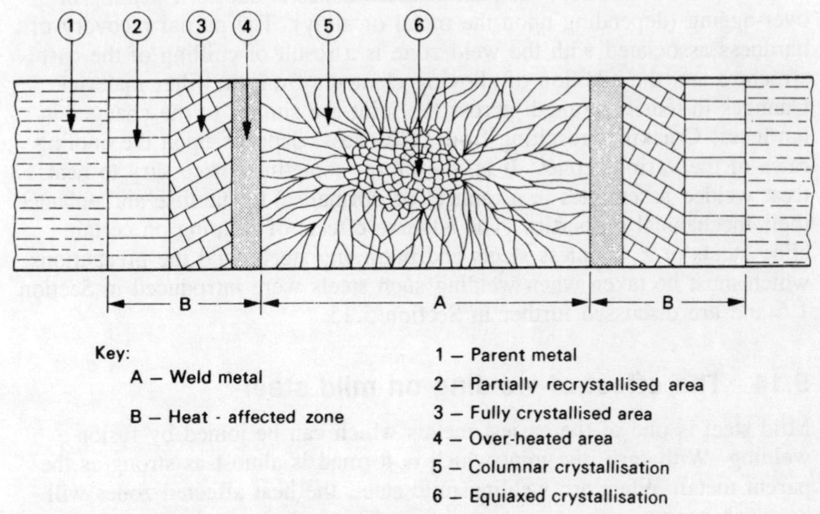

Fig. 5.19 Structure of a welded joint (schematic)

Table 5.4 Effect of temperature gradient on crystal structure of the metal

Temperature zones	**Remarks**
Fusion zone	Temperature reaches melting point. The cooling rate is in the order of 350°–400°C/min., which is the maximum quenching range. The weld is less hard than the adjacent area of the parent metal because of loss of useful elements (carbon, silicon and manganese).
Overheated zone	The temperature reaches 1100°–1500°C. Cooling is extremely rapid in the order of 200°–300°C/min. Some grain coarsening occurs.
Annealed zone	Here the temperature reaches slightly higher than 900°C. The parent metal has a refined normalised grain structure. The change is not complete because the cooling rate is still high, in the order of 170°–200°C/min.
Transformation zone	The temperature here is between 720°C and 910°C. These are the upper and lower critical temperatures between which the iron in steel transforms from a body-centred-cubic to a face-centred-cubic structure. The parent metal tends to recrystallise.

summarises the effect of the temperature gradient on the crystal structure of the metal.

The various zones discussed above can be identified by examining the variation in hardness across the welded joint. The rapid fall off in hardness associated with the heat affected zones is due to annealing or over-ageing (depending upon the metal or alloy). The partial recovery of hardness associated with the weld zone is a result of chilling of the cast structure and the addition of alloying elements from the filler material. Changes in tensile strength across the joint are similar to the changes in hardness. Cracking, resulting from brittleness, depends upon the cooling rates of the various zones. It is, therefore, sometimes necessary to heat treat welded assemblies (e.g. anneal or normalise) to stabilise and enhance their mechanical properties. The adverse effects of welding on certain alloy steels (e.g. stainless steels and maraging steels) and the precautions which must be taken when welding such steels were introduced in Section 1.6 and are discussed further in Section 5.15.

5.14 The effect of welding on mild steel

Mild steel is one of the easiest metals which can be joined by fusion welding. With care, the joint which is formed is almost as strong as the parent metal. When arc welding mild steel, the heat affected zones will be much narrower than in gas welding. This is because the heat of the arc is more localised and the temperature is raised more quickly. However, when the flame of a gas welding torch is applied to the edges of the components being joined, the time taken to raise them to the welding temperature causes appreciable heating of the parent metal and appreciable spread of the heat-affected zone. This not only softens the parent metal each side of the joint, but reduces the 'chill casting' effect on the weld deposit. Further, since the heat from the welding flame is applied longer and cooling of the joint is slower, grain growth is also more pronounced with a corresponding loss of strength. Thus, for thick plates, arc welding should be used. Figure 5.20 compares macrostructures of a single run weld in mild steel plate: (*a*) when gas welding and (*b*) when arc welding.

When a fusion weld is made in low carbon steel with the addition of filler material, the following structures will result.

(*a*) Metal which has been molten will make up a cast structure of deposited metal alloyed with parent metal.
(*b*) There will be a *fusion line* at the junction between the metal which has been melted and the parent metal which has not been melted.
(*c*) A *heat affected zone* which extends from the fusion line to that part of the parent metal which has not been heated sufficiently for structural changes to occur.
(*d*) In the heat affected zone there is an area, adjacent to the 'fusion line', which has cooled slowly and thus has a coarse grain structure. This results in loss of hardness and strength.

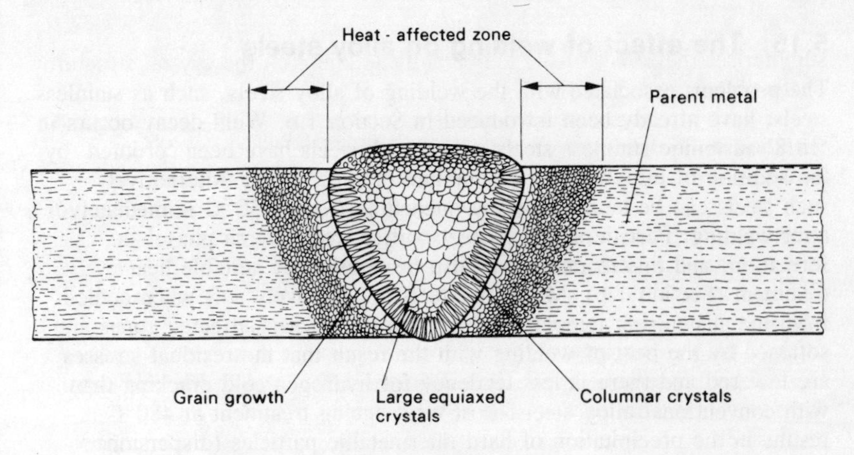

(a) **Oxy-acetylene weld in mild steel**

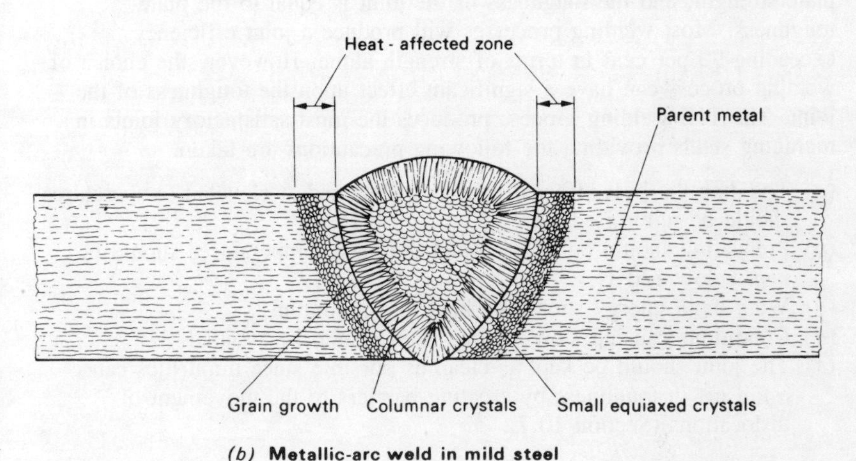

(b) **Metallic-arc weld in mild steel**

Fig. 5.20 Macrostructure of single run welds

(*e*) Progressing away from the weld zone through the coarse grain structure just described, the grains become smaller, and the zone where they become very small is called the *refined zone*. This metal has been heated to the transformation temperature just long enough for the metal to recrystallise but the metal has then cooled before grain growth could occur.

(*f*) Progressing away from the *refined zone* described in (*e*), there is a zone of *mixed structure*. In this zone, some of the grains have recrystallised and some have remained unaffected.

(*g*) Finally the last zone, remote from the weld zone, is the *unaffected zone*. This is where the parent metal has not been sufficiently heated to cause any structural change.

5.15 The effect of welding on alloy steels

The problems associated with the welding of alloy steels, such as stainless steels, have already been introduced in Section 1.6. Weld decay occurs in '18/8' austenitic stainless steels unless such steels have been 'proofed' by the addition of molybdenum or titanium (1.6 per cent). Alternatively, such steels can be heat treated by quenching from 1050°C immediately after welding. Ferritic stainless steels suffer from 'sigma phase' formation, and this tendency can also be reduced by suitable heat treatment (Section 1.6). Maraging steels are relatively soft when cooled from the austenising temperature. Therefore, the heat affected zones are softened by the heat of welding with the result that the residual stresses are lowered and there is less tendency for hydrogen cold cracking than with conventional alloy steels. Post-weld ageing treatment at 480°C results in the precipitation of hard intermetallic particles (dispersion hardening) and this raises the strength of the joint close to the original plate strength, and the toughness of the joint is equal to the plate toughness. Most welding processes will produce a joint efficiency exceeding 90 per cent in terms of strength alone. However, the choice of welding process can have a significant effect upon the toughness of the joint. The TIG welding process produces the most satisfactory joints in maraging steels providing the following precautions are taken.

(*a*) The time the heat affected zone is maintained at elevated temperatures should be minimised.
(*b*) Preheating should be avoided and interpass temperatures should be kept below 120°C.
(*c*) Weld input energy should be kept to a minimum.
(*d*) Conditions causing slow cooling rates should be avoided.
(*e*) The joint should be kept as clean as possible since impurities cause a fall off in toughness by creating barriers to the movement of dislocations (Section 10.7).

5.16 The effect of welding on copper

The various grades of copper available have already been discussed in volume 1. Ordinary tough pitch copper contains oxygen in the form of copper oxides. These oxides give the metal its increased strength. Unfortunately, the welding flame reacts with the oxide particles to produce steam. This results in *gas porosity* and weakness of the joint. This effect can be overcome by using a slightly oxidising flame and a filler rod containing phosphorus. Where it is known that copper components are going to be assembled by welding, one of the 'de-oxidised' grades of copper and a neutral welding flame should be used.

The high thermal conductivity of copper (seven times that of steel) is a disadvantage when welding. Heat energy is conducted away rapidly from the weld zone so that it is not only difficult to achieve complete fusion of the edges of the parent metal but grain growth in the parent metal is excessive. To achieve fusion a large jet has to be used in the torch. This is not only wasteful in the use of welding gases, but exacerbates the problem of softening and grain growth in the parent metal. Too small a jet causes the filler material to melt before the parent metal edges resulting in lack of penetration and planes of weakness at the joint edges.

Like most non-ferrous metal, copper depends upon cold-working to increase its strength and hardness. The heat conducted back from the weld pool anneals the parent metal resulting in general weakness in the vicinity of the joint. A typical cross-section through a welded joint in copper is shown in Fig. 5.21.

5.17 The effect of welding on aluminium and its alloys

Like copper, aluminium also has a high thermal conductivity and depends upon cold-working to improve its strength. Therefore the conditions discussed in Section 5.16 also apply to aluminium. Furthermore, aluminium oxidises very easily and has to be protected from atmospheric oxygen by the use of fluxes and a reducing flame setting. Unfortunately, aluminium and its alloys absorb hydrogen more readily than any other metal in the molten state. The hydrogen comes from various sources: incomplete combustion in the welding flame, the fluxes, and atmospheric moisture. As the weld cools, the dissolved hydrogen is expelled resulting

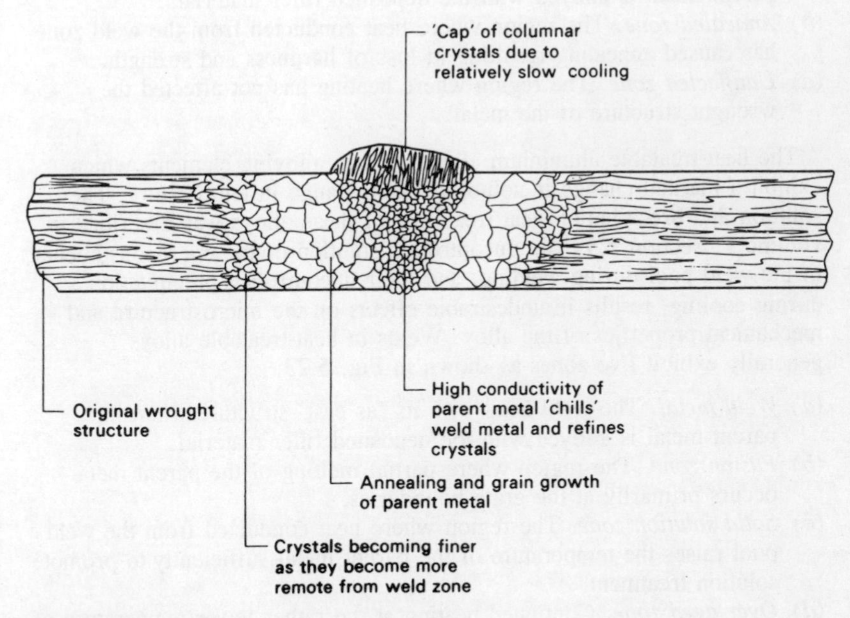

Fig. 5.21 Structure of weld zone (copper)

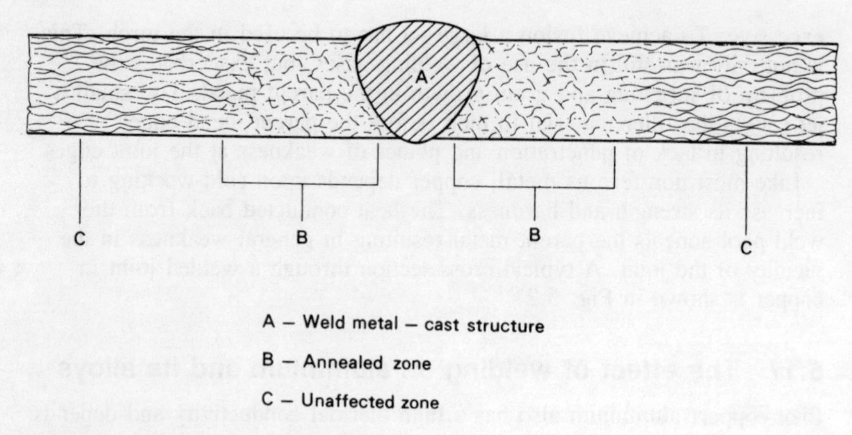

Fig. 5.22 Structure of weld zone (aluminium)

in 'gas porosity'. The fact that the conditions for preventing gas porosity and preventing oxidation conflict with each other is the main reason why aluminium and its alloys are such difficult metals to weld. There are usually three zones in welded joints in pure aluminium and non-heat-treatable aluminium allys. These are shown in Fig. 5.22.

(*a*) *Weld metal*. The weld bead with its 'as cast' structure, where the parent metal is alloyed with the deposited filler material.
(*b*) *Annealed zone*. The region where heat conducted from the weld zone has caused annealing resulting in loss of hardness and strength.
(*c*) *Unaffected zone*. The region where heating has not affected the wrought structure of the metal.

The heat-treatable aluminium alloys contain alloying elements which exhibit a marked change in solubility with change in temperature. The solution and precipitation heat treatment of these alloys was described in volume 1, Section 7.5. The uncontrolled solution of the microconstituents in the weld pool during welding, and their uncontrolled precipitation during cooling, results in undesirable effects on the microstructure and mechanical properties of the alloy. Welds in heat-treatable alloys generally exhibit five zones as shown in Fig. 5.23.

(*a*) *Weld metal*. The weld bead with its 'as cast' structure where the parent metal is alloyed with the deposited filler material.
(*b*) *Fusion zone*. The region where partial melting of the parent metal occurs primarily at the grain boundaries.
(*c*) *Solid solution zone*. The region where heat conducted from the weld pool raises the temperature of the parent metal sufficiently to promote solution treatment.
(*d*) *Over-aged zone*. Continued heating at the rather lower temperature of the next zone results in over-ageing and partial annealing occurring.

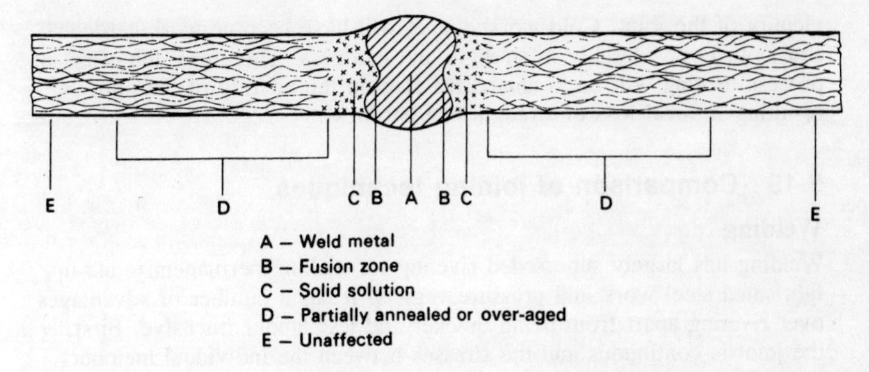

Fig. 5.23 Structure of weld zone (heat-treatable aluminium alloy)

The heat-treatable aluminium alloys rely upon precipitation age-hardening to increase their hardness and strength. However, over-ageing results in softening, grain growth, and weakness.

(*e*) *Unaffected zone*. The region where heating has not affected the structure.

The five zones are generally quite evident in welds made in heat-treatable alloys in which copper and zinc are the major alloying elements. Alloys of the magnesium-silicon type exhibit structural changes in the heat-affected zone which are somewhat different, with the principal heating effect being over-ageing. In these alloys the over-aged and partially annealed zones are of much greater widths.

The speed of welding has a marked effect upon the properties of welds in heat-treatable alloys. High welding speeds not only decrease the width of the heat-affected zones, but they minimise effects such as grain boundary precipitation, over-ageing and grain growth. Heat-treatable aluminium alloys may, in some cases, be heat-treated after welding to bring the heat-affected zones back to their original strength and structure. Where this is not possible the 'as cast' strength must be accepted. For example, an aluminium alloy containing 4 per cent copper has a permissible axial stress after heat treatment of 130 MPa (before welding), whereas the permissible axial stress in the 'as cast' condition is only 50 MPa (after welding).

5.18 The effect of brazing on cold-worked, low-carbon steel

As explained in Section 5.10, brazing processes do not involve such high temperatures as welding and the joint edges of the parent metal are not melted. Nevertheless, the process temperature is above that required to stress relief anneal cold-worked low-carbon steels. Further, the parent metal is often heated long enough for grain growth to occur in the vicinity of the joint. Cold-worked steel will have become work-hardened and gained considerably in strength and hardness. Therefore, the effect of process heating will be to anneal the steel in the vicinity of the joint resulting in local loss of strength and hardness.

5.19 Comparison of joining techniques

Welding

Welding has largely superseded riveting for making permanent joints in fabricated steel work and pressure vessels. It has a number of advantages over riveting apart from being quicker and less labour intensive. First, the joint is continuous and the stresses between the individual members

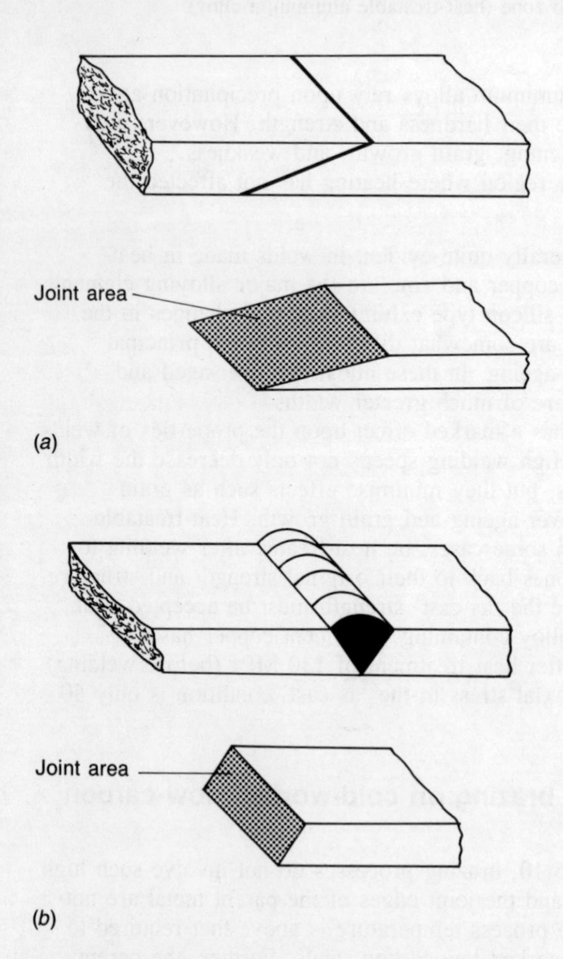

Fig. 5.24 Comparison of brazed and welded joints

are transmitted more uniformly. Second, the joints are fluid tight and therefore ideal for pressure vessels. Third, the joint's surfaces are smooth, more pleasing in appearance and easier to maintain on site (painting). Welding does have some disadvantages compared with riveting: the equipment is more expensive, more energy intensive, and relatively dangerous (particularly compressed gases). The labour required for welding is more skilled and joint defects are more difficult to detect. Riveting does not affect the properties of the metal being joined whereas the heat-affected zone of a welded joint can lead to weakness, to weld decay and eventually to fatigue failure. Further, cracks in welded plates can run throughout the structure, crossing joint lines, and causing catastrophic failure, whereas cracks in riveted structures will only run to the nearest rivet hole where the strain energy will be reduced to a value too low to sustain crack propagation (Section 10.10).

Brazing

Considering the relatively low strength of brazing spelter compared with welding filler material, brazed joints are remarkably strong. In fact they can even be stronger than welded joints under some conditions. This is because of the greater joint area of properly designed brazed joints, as shown in Fig. 5.24, compared with the joint area of a corresponding welded joint, resulting in reduced stress per unit area in the filler material. Further, as previously explained (Section 5.18), although the process temperature for brazing is sufficient to cause some modifications of the material structure, these are far less than the structural modifications associated with the heat-affected zone of a welded joint, again enhancing both the strength of brazed joints and their fatigue resistance.

6 Corrosion prevention

6.1 Dry corrosion

The corrosion of metals and preventative treatments were introduced in volume 1. The basic principles of a number of corrosion mechanisms will now be considered in greater detail.

Dry corrosion occurs as the result of a metal-gas reaction either at ambient temperatures indoors, or during processing at elevated temperatures as when hot-forging or hot-rolling metals. The corrosion film produced is the result of an oxidation reaction. Chemical oxidation reactions occur when a metal is converted from its elemental atomic form to the ionic form of its compound (M to M^{n+}). This loss of electrons in the metal occurs whenever it forms a compound by a chemical oxidation, for example:

$$4Al + 3O_2 \rightarrow 2Al_2O_3 \quad \text{and} \quad Fe + S \rightarrow FeS$$

Thus oxidation reactions can occur with gases other than oxygen. However, in the practical world of engineering, the term oxidation invariably refers to reactions between metals and atmospheric oxygen, and reactions with other gases are named accordingly, e.g. sulphidation, nitridation.

A typical example of dry corrosion at room temperature is the formation of the transparent, protective film of alumina which forms on aluminium products in dry air indoors. After polishing, such films usually achieve a maximum thickness of 1 to 5 nm in about two to four weeks after which any further growth becomes negligible. The corrosion products are found not only on the surface of the metal but also just below the surface. This is due to the reaction gases diffusing into the atomic structure of the metal. The higher the surface temperature of the metal the greater will be the agitation of the metal atoms and the easier it is for diffusion to occur. This results in a more rapid reaction and a thicker oxidation layer.

At higher temperatures the tempering colours on quench hardened and tempered carbon steel tools are an example of dry oxidation. As the tempering temperature is increased, the oxide film becomes thicker and the colour of the film becomes darker.

At processing temperatures above 570°C, a *scale* commences to form on ferrous metals, that is, the oxide film exceeds 1 μm in thickness and the film becomes more complex as shown in Fig. 6.1. The ability of an oxide film to protect the metal beneath it from further corrosion depends upon many factors. If the film is porous or subject to flaking then it will offer little protection as new metal is constantly being exposed to attack. The oxide film can often be improved by the addition of alloying elements. For example, when chromium is added to carbon steel it forms a continuous barrier layer of chrome oxide (Cr_2O_3) which opposes the migration and diffusion of the metal and oxygen atoms at the surface of the metal. Other alloying elements which help to improve the protection offered by the oxide layer when added to steels and cast irons are aluminium, silicon and nickel.

Table 6.1 shows the effect of adding some of these alloying elements to iron. It also shows that small additions of a third or fourth element to the alloy can be more effective than simply increasing the quantity of the second element (e.g. compare Fe + 1% Si with Fe + 3% Si and Fe + 2% Si + 2% Cr). When alloying reduces the oxidation rate it is often possible to use the material at higher temperatures. For example, the temperature at which a low carbon steel commences to form a scale is only 480°C compared with an 18/8 stainless steel which does not commence to scale until it has been heated to 900°C.

Further effects of oxidation in metals and alloys will now be considered. 'Growth' in grey cast irons occurs if the castings are raised to the temperature at which they become austenitic. The volume increase associated with this transformation allows the infusion of atmospheric

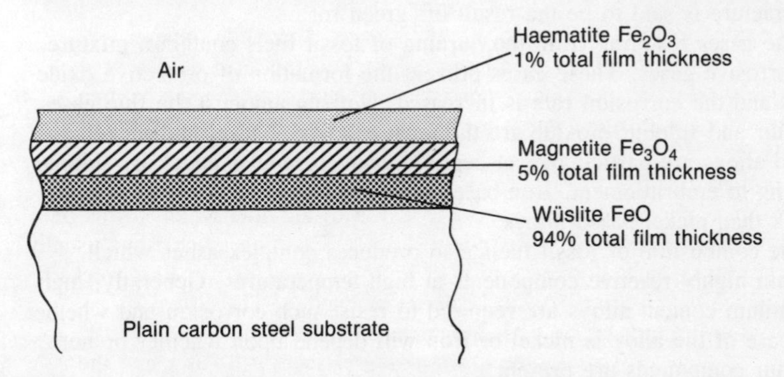

Fig. 6.1 Oxide film on plain carbon steels above 570°C

Table 6.1 Effect of alloying elements on the corrosion (oxidation) rate of iron

Composition %				Average weight gain
Fe	Si	Al	Cr	mg/cm^2
100	—	—	—	15.50
99	1	—	—	1.00
99	—	1	—	1.50
99	—	—	1	4.00
97	3	—	—	0.50
95	—	5	—	0.20
86	—	—	14	0.01
98	1	1	—	0.03
98	1	—	1	0.50
96	2	—	2	0.02

Oxidation of iron in dry air at 700°C.

oxygen which reacts with the flake graphite to form carbon monoxide. Some of this gas becomes trapped in the casting as it cools and causes a permanent increase in volume or 'growth' to occur resulting in warping and loss of strength. This effect can be overcome by using alloy cast irons such as *silal* or *nicrosilal* for castings which have to be used at elevated temperatures.

Intergranular corrosion of nickel-chromium alloys can occur at temperatures of 1000°C or above in the presence of hydrocarbon gases or even residual mineral oils or greases on the surface of the metal. Some of the chromium in the alloy reacts to form non-protective compounds such as chromium carbide (Cr_2C_6). This reduces the amount of chromium present in the alloy and oxidation of the alloy becomes easier. The strength of the alloy is reduced and it becomes more brittle. If the metal fractures, green chromic oxide (Cr_2O_3) colours the broken surfaces and the fracture is said to be the result of 'green rot'.

Flue gases resulting from the burning of fossil fuels contain a mixture of corrosive gases. These gases prevent the formation of protective oxide films and the corrosion rate is increased. Notably amongst the flue gases, sulphur and sulphur dioxide are the most reactive. Nickel and nickel based alloys are particularly susceptible to intergranular attack by sulphur leading to embrittlement. Iron-base alloys are less susceptible to sulphur attack than nickel based alloys.

The combustion of fossil fuels also produces complex ashes which contain highly reactive components at high temperatures. Generally, high chromium content alloys are required to resist such corrosion and whether the base of the alloy is nickel or iron will depend upon whether or not sulphur compounds are present.

Catastrophic oxidation occurs when no protective oxide film is formed and the reaction continues until no metal is left. For example, a steel containing 1 per cent chromium and 1 per cent molybdenum is completely destroyed when heated to 650°C in air. The molybdenum forms a volatile oxide which continually breaks down any protective oxides formed by the chromium. The chromium content has to be increased to at least 9 per cent to prevent this happening and the presence of nickel in the alloy also helps.

6.2 Wet corrosion

Wet corrosion occurs where a liquid, usually water, is present. For corrosion to occur, solid, liquid or gaseous contaminants must be dissolved in the water to form an *electrolyte* because wet corrosion is an electrochemical reaction. In addition, molten metals, molten salts and organic solutions can also cause wet corrosion.

A common example of wet corrosion is the rusting of ferrous metals. Iron will not rust in dry air nor will it rust in pure water from which any dissolved oxygen has been removed. However, in a moist atmosphere rusting will occur as the distinctive, reddish-brown rust film of ferric hydroxide builds up on the surface of the metal.

$$4Fe + 6H_2O + 3O_2 \rightarrow 4Fe(OH)_3$$

It is the presence of the dissolved oxygen in the water which makes this reaction possible. This reaction is achieved as follows and is typical of the electrolytic processes of wet corrosion generally.

Elemental atomic iron becomes ionised by the loss of two electrons (this is the oxidation mechanism described in Section 6.1) and enters the solution as ferrous ions.

$$Fe \rightarrow Fe^{2+} + 2e^- \text{ (2 electrons).}$$

These ferrous ions are further oxidised to ferric ions.

$$Fe^{2+} \rightarrow Fe^{3+} + e^- \text{ (1 electron).}$$

At the cathodic area of the reaction the electrons which flow from the anodic area, as the result of the ionisation of the iron, are intercepted by the oxygen atoms present.

$$4e + O_2 + 2H_2O \rightarrow 4(OH)$$

The ferric ions then combine with the hydroxyl groups so that ferric hydroxide (rust) is formed and the overall electronic equilibrium is maintained.

Wet corrosion also commonly occurs when two dissimilar metals come into contact in the presence of an electrolyte and an electrical cell is formed. This type of corrosion is also known as *galvanic corrosion* or *bimetallic corrosion* and it results in one or other of the metals being eaten away. Metals can be arranged in a special order called the *electrochemical series*. Some of the metals used in engineering are listed

Table 6.2 Electrochemical series

Metal	Electrode potential (volts)	
Sodium	−2.71	*Corroded (anodic)*
Magnesium	−2.40	
Aluminium	−1.70	
Zinc	−0.76	
Chromium	−0.56	
Iron	−0.44	
Cadmium	−0.40	
Nickel	−0.23	
Tin	−0.14	
Lead	−0.12	
Hydrogen (reference potential)	0.00	
Copper	+0.35	
Silver	+0.80	
Platinum	+1.20	
Gold	+1.50	*Protected (cathodic)*

in the order of the electrochemical series in Table 6.2 and it should be noted that, in this context, hydrogen gas behaves like a metal. If any two metals in the table come into contact in the presence of an electrolyte the more electro-negative metal will be attacked and eaten away whilst the more electro-positive metal will be protected. This is shown in Fig. 6.2.

(*a*) In the case of galvanised iron (zinc-coated low carbon steel), any porosity in the coating or damage to the coating results in the zinc corroding away whilst protecting the steel from rusting. Thus the zinc is said to be *sacrificial*. The iron is only protected as long as some zinc remains in the vicinity of the surface discontinuity. Once the zinc is destroyed, rusting will commence. Figure 6.2(*a*) shows the conventional current between the two metals as flowing from the iron to the zinc. Thus the electron current flows from the anodic zinc to the cathodic iron and, as a result, the Zn^{2+} ions enter solution in the electrolyte. The current flow will cause hydroxyl ions to form at the surface of the iron as previously described. The Zn^{2+} ions and the OH^- ions will react to form a white deposit of zinc hydroxide on the iron.

$$Zn^{2+} + 2(OH) \rightarrow Zn(OH)_2 \downarrow$$

(*b*) In the case of tin-plate, the mild steel substrate is eaten away if the tin film is broken. Hence cut edges should always be tinned and bend lines should be marked with a soft pencil and not with a scriber.

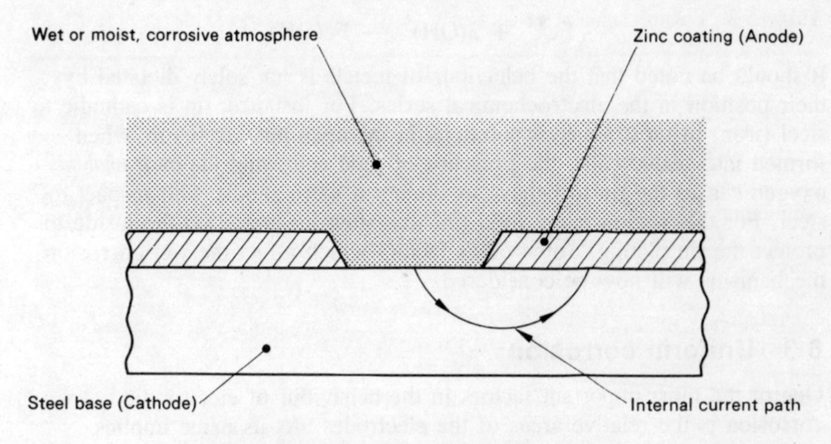

(a) **Protection by a sacrificial coating**

Coating is eaten away whilst protecting the base

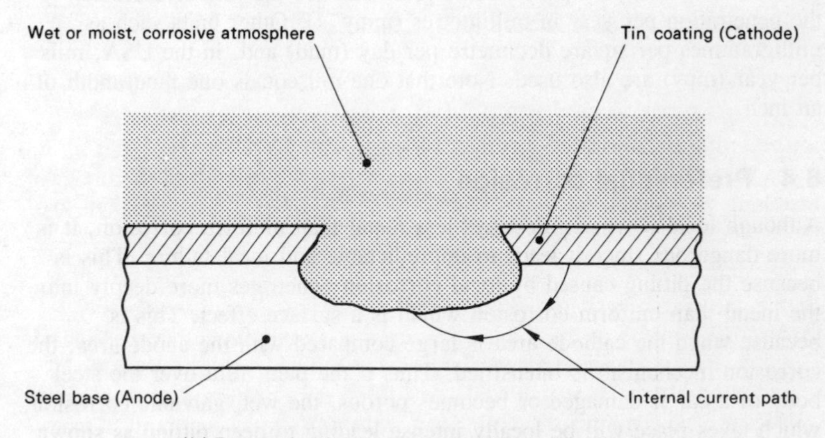

(b) **Protection by a purely mechanical coating**

Coating only protects the base if intact.
If coating is damaged, base is eaten away
quicker than if coating were not present.

Fig. 6.2 Galvanic corrosion

Figure 6.2(*b*) shows the conventional current flow from the tin to the iron. Thus the electron current flow is from the anodic iron to the cathodic tin and, as a result, the Fe^{3+} ions will enter solution in the electrolyte. Hydroxyl groups will form at the surface of the tin cathode as previously described. The Fe^{3+} ions and the OH^- ions will react to form ferric hydroxide (rust)

$$Fe^{3+} + 3(OH)^- \rightarrow Fe(OH)_3 \downarrow$$

It should be noted that the behaviour of metals is not solely dictated by their position in the electrochemical series. For instance, tin is cathodic to steel (iron) in most aqueous solutions in the open air. However, when formed into sealed cans, the presence of food acids and the lack of oxygen causes the tin to behave as though it were anodic with respect to steel. For this reason many food tins are often lacquered on the inside to protect the tin plating. Some other factors which affect the wet corrosion mechanisms will now be considered.

6.3 Uniform corrosion

One of the most important factors in the behaviour of electrolytic corrosion is the relative areas of the electrodes. As its name implies uniform corrosion occurs where the anodic and cathodic regions have approximately equal areas. Metal loss is greatest with uniform corrosion and it may be specified as the corrosion rate from which the service life of the material may be predicted. The SI unit for the rate of corrosion is the penetration per year in millimetres (mmy^{-1}). Other units such as milligrammes per square decimetre per day (mdd) and, in the USA, mils per year (mpy) are also used. Note that one mil equals one thousandth of an inch.

6.4 Preferential corrosion

Although local corrosion destroys less metal than uniform corrosion, it is more dangerous since it leads to unpredictable and local failure. This is because the pitting caused by local corrosion penetrates more deeply into the metal than uniform corrosion which is a surface effect. This is because when the cathode area is large compared with the anode area, the corrosion mechanism is intensified. Thus if the paint film over the steel body of a car is damaged or becomes porous, the wet, galvanic corrosion which takes place will be locally intense leading to deep pitting as shown in Fig. 6.3.

Where bolts or rivets are used to join metal components together the following rules should apply.

(*a*) The rivets or bolts should be of the same material as the components being joined.

(*b*) If the bolts or rivets have to be of a different metal they should be more 'noble' in the electrochemical series so that they are cathodic relative to the metal of the components being joined. That is, steel bolts or rivets of small surface area will soon dissolve away if used to join brass, bronze or copper components whereas brass, bronze or copper bolts or rivets of small surface area will be unaffected when joining steel components.

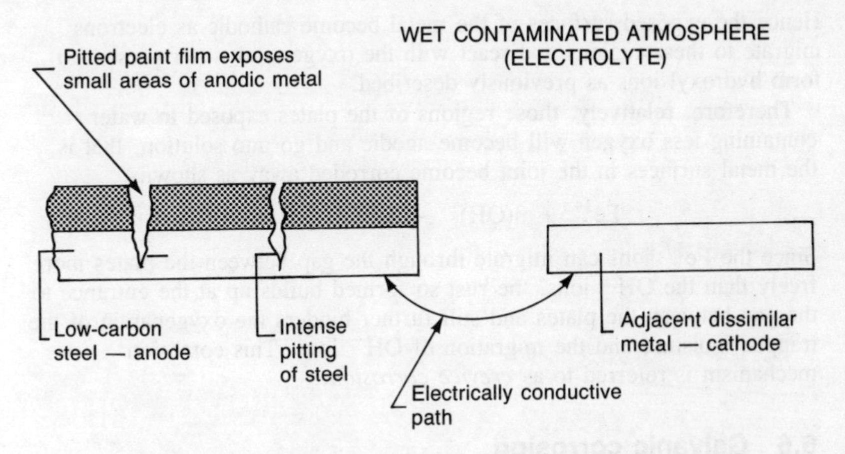

Fig. 6.3 Preferential corrosion

6.5 Crevice corrosion

Figure 6.4 shows two plates of the same material riveted together (the effect would be the same if they were bolted together) and immersed in oxygenated water. It might be thought that where the plates are in contact with each other little corrosion could take place. However, variations in the amount of oxygen present from one region of the joint to another results in corrosion occurring where it is least expected. This is because the water drawn into the joint between the plates by capillary attraction becomes starved of oxygen as the reactions proceed whilst those regions of the plates exposed to free water have an adequate supply of dissolved oxygen. Further, it is much more easy for electrons to migrate out of the joint than it is for replacement oxygen atoms to move into the joint.

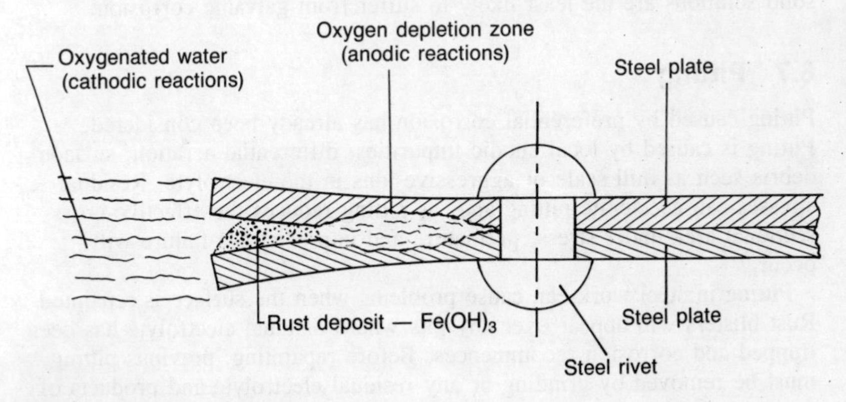

Fig. 6.4 Crevice corrosion

Hence the exposed surfaces of the metal become cathodic as electrons migrate to these regions and react with the oxygenated water present to form hydroxyl ions as previously described.

Therefore, relatively, those regions of the plates exposed to water containing less oxygen will become anodic and go into solution, that is, the metal surfaces in the joint become corroded away as shown.

$$Fe^{3+} + 3(OH)^{-} \rightarrow Fe(OH)_3 \downarrow \text{(rust)}$$

Since the Fe^{3+} ions can migrate through the gap between the plates more freely than the OH^{-} ions, the rust so formed builds up at the entrance to the gap between the plates and still further hinders the oxygenation of the trapped moisture and the migration of OH^{-} ions. This corrosion mechanism is referred to as *crevice corrosion*.

6.6 Galvanic corrosion

Galvanic corrosion between two dissimilar metals was shown in Fig. 6.2. In Fig. 6.2(*b*) it was the iron which was corroded away rather than the tin coating because the iron was anodic in that instance. Similarly, hot rolled or forged steel corrodes (rusts) more quickly than bright steel because the scale is not only porous and offers no protection, it is cathodic to the metal beneath it.

Galvanic corrosion on the microscopic scale can also occur due to impurities. These impurity particles may be anodic or cathodic relative to the metal in which they are found. Where the impurity is cathodic the metal surrounding it will be eaten away, whilst where the impurity is anodic the impurity is eaten away.

Again, galvanic action can take place within the metal structure. For example, lamellar pearlite consists of alternate laminations of ferrite and cementite. In the presence of oxygenated water the ferrite is anodic and eaten away whilst the cementite (iron carbide) is cathodic and unaffected.

Thus metals which are very pure or metals which form homogeneous solid solutions are the least likely to suffer from galvanic corrosion.

6.7 Pitting

Pitting caused by preferential corrosion has already been considered. Pitting is caused by local anodic impurities, differential aeration, surface debris such as mill scale or aggressive ions in the electrolyte. Residual stresses can accelerate pitting and, as pitting reduces the effective cross-sectional area under stress, premature and unpredictable failure will occur.

Pitting in steel work can cause problems when the surface is repainted. Rust blisters will appear over any pits where residual electrolyte has been trapped and corrosion recommences. Before repainting, previous pitting must be removed by grinding or any residual electrolyte and products of corrosion in the pits must be neutralised chemically. For example, resprayed car body shells frequently blister very quickly if they are not properly prepared before painting.

6.8 Intergranular corrosion

The effects of impurities on local corrosion have already been discussed in Section 6.7. Since impurities in metals frequently migrate to the grain boundaries as the metal solidifies, they cause intergranular corrosion resulting in embrittlement and weakness.

However, intergranular corrosion can occur without impurities being present, possibly due to the irregular arrangement of ions at the grain boundaries, and the fact that the grain boundaries are regions of higher energy levels. This is particularly so when the material has been severely cold worked. Under suitable wet corrosion conditions the grain boundaries become anodic relative to the body of the grains which exhibit cathodic characteristics. Under such conditions the anodic boundaries are attacked and this attack is intensified by the fact that there is a small anode area compared with the large cathode area (Section 6.4).

An example of intergranular corrosion is the 'season cracking' of α-brass after severe cold-working. Corrosion follows the grain boundaries, reducing the effective cross-sectional area of the metal, until it can no longer sustain the applied load, and failure occurs. Stress relief by a low temperature annealing process following the cold-working allows the atoms and ions to re-align themselves sufficiently to reduce the energy levels at the grain boundaries and corrosion is eliminated for all practical purposes.

The intergranular corrosion of high-alloy steels occurs not only when such alloys are welded but also when castings of such alloys are cooled slowly or when they have been held at the 500°C to 800°C temperature range too long during heat treatment (temper-brittleness). Slow cooling through this temperature range results in the formation of chromium carbide at the grain boundaries, leaving the surrounding areas deficient in chromium and unable to form the chromium oxide necessary to prevent corrosive attack, as shown in Fig. 6.5.

Intergranular corrosion can be largely overcome in various ways depending upon the limitations of service conditions. The metal can be reheated to 1100°C after processing to dissolve the carbides. It is then quenched to prevent the carbides precipitating out again. Alternatively, a low carbon stainless alloy ('stainless iron') such as BS 304S12 could be used to lessen the risk of carbide precipitation. Alloys such as BS 321S20 containing titanium or BS 347S17 containing niobium are said to be stabilised or 'proofed'. Titanium and niobium both form carbides more readily than chromium and leave the chromium content undepleted. Further, the carbides formed do not congregate at the grain boundaries.

Unfortunately, the temperatures achieved during arc welding may exceed 1100°C resulting in the titanium or niobium carbides being dissolved in the heat affected zone. As the weld zone cools, chromium

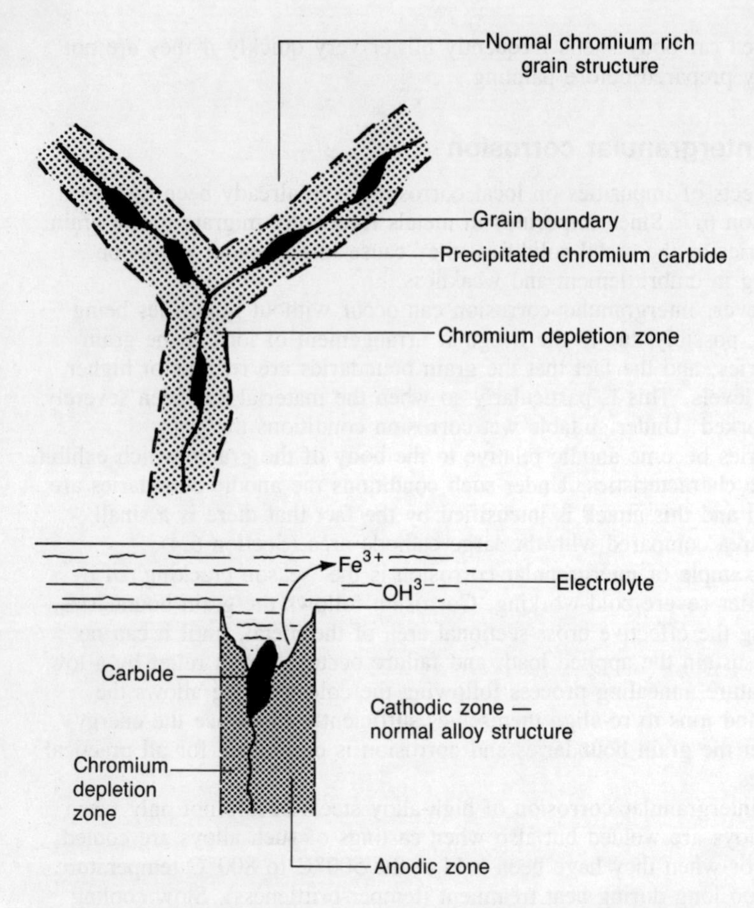

Fig. 6.5 Intergranular corrosion of high chromium alloy steels

carbide is precipitated out rendering the joint liable to 'weld-decay'. Under these conditions precipitation occurs in a narrow band either side of the weld itself and is referred to as 'knife-line attack', as shown in Fig. 6.6. This can be overcome by reheating the weld zone to just above 1000°C and allowing the assembly to cool naturally. This avoids the distortion and stresses associated with quenching.

Although stainless steels are normally unreactive and corrosion resistant in the presence of oxygenated water and many other aqueous solutions because of their natural oxide film, they corrode rapidly in the presence of sea water. This is because the chlorine ions in the sea water are particularly aggressive towards stainless steels. They cause the break down of the protective oxide film at localised regions causing pitting and activation of the alloy. This is aggravated where welding has occurred, resulting in catastrophic failure even after the precautions described previously have been taken.

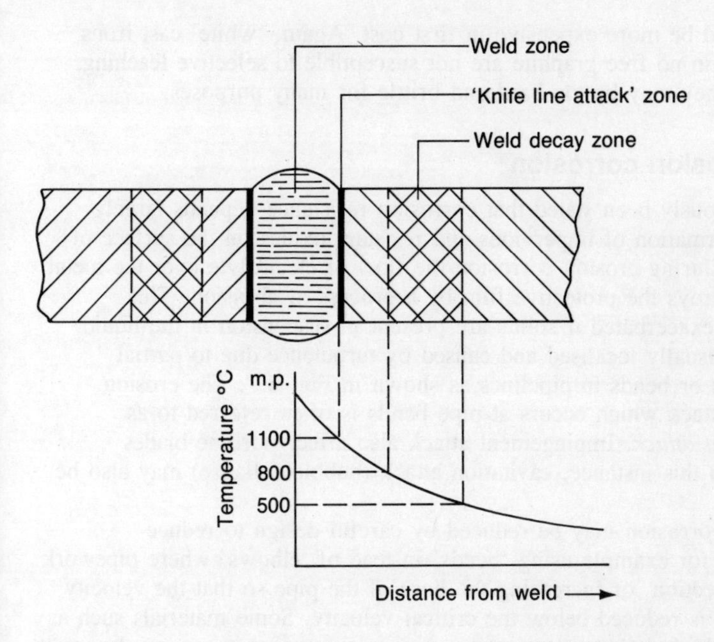

Fig. 6.6 Intergranular corrosion zones for welded high chromium alloy steels

6.9 Selective leaching

This is the electrolytic attack of one element in an alloy by another in the presence of an aqueous electrolyte. For example, *dealuminiumification* where the aluminium content of aluminium-copper alloys is corroded out to leave a spongy mass of copper. Again, there is the *dezincification* of brass where the zinc content is corroded out leaving a spongy mass of copper. In both these examples the anodic aluminium and the anodic zinc are destroyed in the presence of the cathodic copper in what is a simple electrolytic reaction. This reaction is particularly rapid when chloride ions are present as in brine and sea-water. The residual copper has no mechanical strength and, because of its spongy nature, is itself increasingly susceptible to further corrosion.

Selective leaching can also occur between metals and non-metals in an alloy, for example the *graphitisation* of grey cast iron. Here, the uncombined flake graphite is cathodic and the anodic iron surrounding the flakes is corroded away as rust to leave a spongy mass with no mechanical strength.

The only way selective leaching can be overcome is to replace the materials under attack with materials which are resistant to the aqueous solutions present, for example replacing the simple, binary brass alloy with Admiralty brass which contains tin and arsenic in addition to zinc. Alternatively, the brass alloy may be replaced by a cupronickel alloy which would be even more resistant to salt solutions and sea-water, but which would be more expensive in first cost. Again, 'white' cast irons which contain no free graphite are not susceptible to selective leaching. However, they may be too hard and brittle for many purposes.

6.10 Erosion corrosion

It has previously been stated that corrosion resistance depends largely upon the formation of impervious and resistant oxides on the surface of the metal. During erosion corrosion the flow of electrolyte over the metal surface destroys the protective film by a process of abrasion. This abrasion is exacerbated if solids are present in suspension in the liquid. Erosion is usually localised and caused by turbulence due to partial obstructions or bends in pipelines as shown in Fig. 6.7. The erosion corrosion attack which occurs at pipe bends is often referred to as *impingement attack*. Impingement attack also affects turbine blades although, in this instance, cavitation attack (bubble collapse) may also be present.

Erosion corrosion may be reduced by careful design to reduce turbulence, for example using 'bends' instead of 'elbows' where pipework changes direction, or increasing the bore of the pipe so that the velocity of the fluid is reduced below the critical velocity. Some materials such as nickel-based alloys have a greater resistance to erosion corrosion than

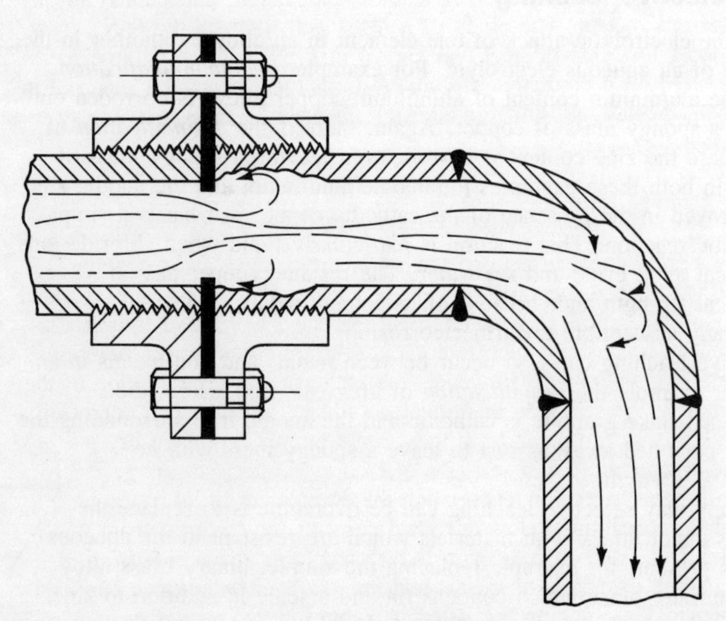

Fig. 6.7 Erosion–corrosion caused by turbulence

iron-based alloys. If iron-based alloys have to be used, then stainless steels and cast irons such as 'ni-resist' and high silicon alloys are reasonably resistant to erosion corrosion providing chloride ions are not present. In closed systems, such as central heating systems, erosion corrosion can be reduced by the use of an 'inhibitor' in the circulating fluid. This is a chemical which encourages the formation of protective films on the metal. However, the use of an inhibitor does not excuse bad design and the system should be as free from turbulence as possible.

Another form of erosion corrosion is *cavitation attack*. This is often present in the vicinity of ships' propellers, turbine blades and pump impellers. The pressure drop at the trailing surface of modern high speed blades can fall below the vapour pressure of the fluid in which it is operating so that the liquid actually 'boils' on the surface of the metal. A vapour bubble so formed bursts so rapidly that the pressure wave hammering on adjacent metal surfaces can reach intensities of 1.5 GN/m. Since more than one million bubbles can form and collapse in one second over a relatively small area, the cumulative effect damages the surface of the blades and any other adjacent metal. Careful choice of materials which can resist such attack accompanied by careful design to reduce cavitation to a minimum is the only way to overcome this problem.

6.11 Corrosion fatigue

Mechanical fatigue failure is exacerbated by corrosion, particularly in the case of ferrous metals. The effect of aerated sea-water on the S-N diagram for a low carbon steel was shown in Fig. 13.10 in volume 1.

The *damage ratio* compares the fatigue limit for a particular material in dry air with the fatigue limit for the same material in a corrosive environment.

$$\text{Damage ratio} = \frac{\text{fatigue limit in dry air}}{\text{fatigue limit in corrosive environment}}$$

For a low carbon steel in aerated sea-water the damage ratio is about 0.2 compared with a damage ratio of 1.0 for a material which is not corroded in such an environment, for example pure copper. Corrosion fatigue can be reduced by controlling such factors as solution composition, aeration, pH value and temperature. However, the only successful resolution of the problem lies in the selection of a material which is resistant to corrosion in its working environment. Note that unlike fatigue failure in dry air where the component under test is subjected to cyclical stressing, corrosion fatigue can occur even when the component is subjected to constant load conditions below the yield point of the material from which it is made. This latter phenomenon is referred to as *stress-corrosion cracking*. Some examples of metals and the environments which can lead to stress-corrosion cracking are listed in Table 6.3.

Table 6.3 Material/environment combinations which can lead to stress corrosion cracking

Material	Environment
Aluminium alloys	Aqueous chloride solutions
Copper alloys	Ammonia
Low-carbon steels	Hot concentrated alkaline solutions
Magnesium alloys	Aqueous chloride solutions
Nickel alloys	Molten caustic soda
Stainless steels	Aqueous chloride solutions
Titanium alloys	Aqueous chloride solutions

Common source of aqueous chloride solutions is sea-water.

6.12 Hydrogen damage

Atomic (nascent) hydrogen produced by such processes as electro-plating, electrolytic cleaning and polishing, pickling and by some micro-organisms can cause 'blistering' and embrittlement of a metal. Particularly susceptible are high-strength steels. Although the hydrogen is generated at the surface of the metal, some of the relatively small hydrogen atoms diffuse readily between the larger metal atoms and recombine as molecular hydrogen to form pockets of hydrogen gas. These gas pockets cause internal stress and surface blistering and embrittlement of the material.

6.13 Biological corrosion

Micro-organisms do not attack metals directly, but often exude and thrive upon substances which do attack metals. For example, the anaerobic organisms which may be found in sea-water and soils thrive on nitrogen, hydrocarbons (e.g. methane caused by vegetable decay) and sulphur compounds. In turn, they exude hydrogen and corrosive compounds.

Organic substances can also attack metals and careful selection of materials is particularly important in food-processing plant. (Note: some metals which are corrosion resistant are also toxic and cannot be used in such situations.) An example of organic attack is the dark-green *verdigris* which builds up on copper and which must not be confused with the pale-green *patina* formed by inorganic chemical action and which is a protective coating. Copper utensils used for food preparation should, therefore, be heavily tin plated.

6.14 Factors affecting aqueous corrosion

Having discussed the mechanisms of aqueous corrosion, the factors affecting the rate of such corrosion can be summarised as follows.

Metal composition and structure

The position of a metal in the electrochemical series is a good indication as to its relative rate of corrosion. Metals at the anodic end of the table will corrode more rapidly than those at the cathodic end of the table. Metals with a high degree of purity tend to corrode less rapidly than the same metals when impurities are present since electrolytic reactions take place between the anodic metal and the cathodic impurities in which the metal is eaten away. Metals with coarse grain structures generally corrode more rapidly than those with fine grain structures. This is because a coarse grain structure is more susceptible to ionic diffusion. Alloys can be developed to be resistant to a wide range of corrosive environments. However, it should be remembered that ferrous based alloys, such as stainless steels, which are resistant to most rural, urban and industrial environments, are attacked in coastal and marine environments due to the presence of chloride ions. The metallurgical structure of the material can also affect its corrosion. For example, the electrolytic attack on the iron content in lamellar pearlite and the dezincification of brass were discussed in Sections 6.6 and 6.9 respectively.

Environment

Air consists of oxygen (20 per cent), nitrogen (nearly 80 per cent) and other gases, such as argon, helium and neon in very small amounts. The oxygen is responsible for both the corrosion of ferrous metals when it is dissolved in water (rain) and the formation of protective oxides on metals which are corrosion resistant. When the oxides of nitrogen and the carbon dioxide produced by the burning of fossil fuels become dissolved in rain water to form acid solutions, the range of metals attacked and the rate of attack is increased. Atmospheric corrosion depends upon a number of local factors. In rural communities corrosion is generally due to oxygenated rain water alone, although prevailing winds may carry urban and industrial pollutants from neighbouring areas. In urban areas, additional pollutants such as carbon dioxides and oxides of nitrogen are present from the burning of fossil fuels. In industrial areas, sulphur compounds and ammonia compounds may also be present. In coastal and marine environments, highly reactive chlorides will also be present. The greater the concentration and range of pollutants present, the greater will be the rate and range of corrosive damage.

Surface defects

Damage to protective coatings allows the aqueous solutions responsible for corrosion to come into contact with the structural metal and corrosion can occur. As has been previously explained, the rate of corrosion is increased because of the small exposed anode area to cathode area ratio which is present under such conditions.

Structural design

The following factors should be observed during the design stage of a component or fabrication to reduce corrosion to a minimum.

(*a*) Materials which are inherently corrosion resistant should be chosen or, if this is too costly, then an anti-corrosion treatment should be specified.
(*b*) The design should avoid crevices and corners where moisture and silts may become trapped. Adequate drainage and ventilation should be provided.
(*c*) The design should allow for easy washing down and cleaning.
(*d*) Joints which are not continuously welded should be sealed, for example, by the use of mastic compounds.
(*e*) Where dissimilar metals have to be joined their electrolytic compatibility should be established at the design stage. Further, they should be insulated from each other either by the use of a suitable adhesive in the case of a permanent joint, or by suitable bushings and washers made from insulating materials, or by the use of non-metal connections.

Applied or internal stress

Chemical and electrochemical corrosion is intensified when a metal is in a stressed condition. Internal stresses are usually caused by cold-working and, if not removed by stress relief annealing, result in corrosive attack along the crystal boundaries. This is because distortion of the grain structure facilitates the diffusion and migration of ions between heavily cold-worked zones of the component which have anodic characteristics and less heavily cold-worked zones which have cathodic characteristics. Hence, the more severely cold-worked and stressed zones of the component will be subjected to corrosive attack and destruction. The 'season cracking' of an α-brass after severe cold-working is an example of such corrosion, which can be overcome by stress relief annealing after cold-working.

Temperature

All chemical and electrochemical reactions have a critical temperature below which the reaction cannot take place. Hence, metal is preserved from corrosion in very cold climates as met with at the polar regions. Also corrosion is negligible in desert areas for, although the ambient temperature is above the critical reaction temperature, the atmosphere is too dry for 'wet corrosion' to occur. The worst conditions for corrosion are found in the tropics where high temperatures combine with high humidity and engineering products need to be 'tropicalised' if they are to have a reasonable service life. Even in temperate zones corrosion is a constant problem.

Aeration

It has already been shown that dissolved oxygen in water is essential for many forms of corrosion. Hence the rate of corrosion will be greater in aerated water than in stagnant water. Consider the legs of the piers found at many coastal resorts. It can be seen that corrosion is heaviest and

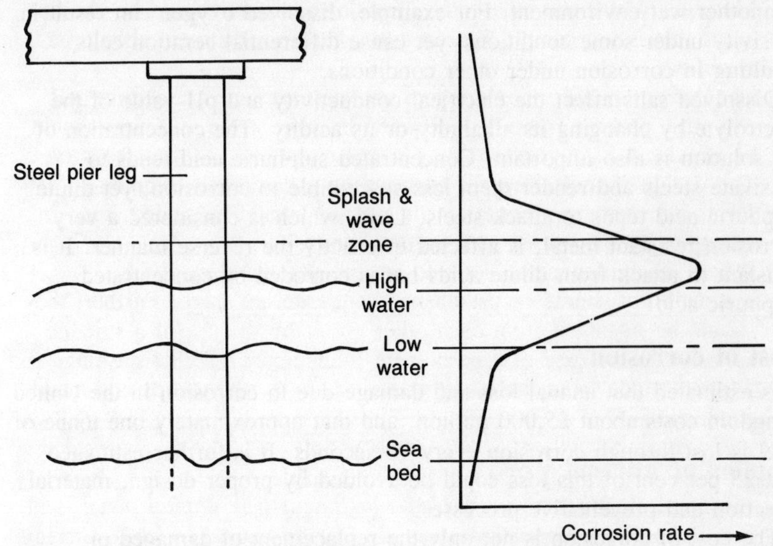

Fig. 6.8 Effect of partial immersion on rate of corrosion

fouling by marine organisms is most prevalent from the low tide water-mark through the high tide water-mark into the splash and wave zone. This is shown in Fig. 6.8. Below the low tide water-mark little aeration takes place, but the agitation of the surface of the sea increases aeration and this is locally increased still further by the turbulence and splashing which occurs around any obstruction such as the legs of the pier. This accounts for the increase in the rate of corrosion in this zone.

Chemistry of the electrolyte

The moisture necesssary to form solutions which can act as electrolytes can be classified as: water vapour; rain water; fresh (natural) water; brackish water; and sea-water. Water vapour and rain water will contain dissolved atmospheric pollutants (gases and dusts) and will generally be acidic. Fresh water from underground springs and rivers will have a variety of contaminants depending upon the rain which has fallen into it, the rock layers through which it has passed, the soils over which it has passed, and any debris or effluents which have been allowed into it. Deep sea-water is fairly constant in composition, but coastal waters can vary widely in composition depending upon changing concentration due to evaporation where it is shallow, effluent discharge from industry and sewage treatment plants, and fresh water dilution at the estuaries of large rivers. Thus the chemical composition and corrosive effects of aqueous electrolytes can vary extensively and anti-corrosion treatment suitable in one wet environment can be quite unsuitable or even counter-productive in another wet environment. For example, dissolved oxygen can result in passivity under some conditions, yet cause differential aeration cells resulting in corrosion under other conditions.

Dissolved salts affect the electrical conductivity and pH value of the electrolyte by changing its alkalinity or its acidity. The concentration of the solution is also important. Concentrated sulphuric acid tends to passivate steels and render them less susceptible to corrosion, yet dilute sulphuric acid tends to attack steels. Lead, which is considered a very corrosion resistant metal, is affected in exactly the reverse manner. It is resistant to attack from dilute acids but is corroded by concentrated sulphuric acid.

Cost of corrosion

It is estimated that annual loss and damage due to corrosion in the United Kingdom costs about £5,000 million, and that approximately one tonne of steel is lost through corrosion every 90 seconds. It is further estimated that 25 per cent of this loss could be avoided by proper design, material selection and preventative processes.

The cost of corrosion is not only the replacement of damaged or destroyed equipment, but also such factors as preventative maintenance, loss of production due to unexpected failure and compensation when plant failure leads to destruction, environmental contamination, injury and death. Therefore, the prevention of corrosion and its effect on safety, performance and cost is of prime importance to engineers. Some aspects of corrosion prevention will now be considered.

6.15 Corrosion prevention

The importance of choice of material and the effect of alloying elements and heat treatment (e.g. composition, change of phase and impurities) on corrosion problems have already been considered in this chapter. Also, metals which resist corrosion and corrosion resistant coatings were introduced in volume 1. Some further techniques for the prevention of corrosion will now be considered.

Anodic protection

There are two fundamental techniques of anodic protection.

(*a*) The use of *sacrificial anodes*. For example, the proximity of a manganese bronze propeller (cathodic) to the low-carbon steel hull of a ship (anodic) in highly agitated and aerated sea-water should result in the rapid corrosion and destruction of the hull. This tendency is largely eliminated by bolting large slabs of zinc onto the hull near the propellers. Since the zinc is anodic to both the manganese bronze propellers and the steel hull of the ship, the zinc will corrode sacrificially whilst protecting the hull. The zinc anodes are replaced from time to time.

(*b*) *Anodic passivation* may be achieved by two techniques:

(i) *Galvanically* (but not sacrificially — see above). An electrolytic cell is created between the metal to be protected and a more noble metal by plating (e.g. platinum on stainless steel) or by alloying. The alloying additions need only be small and the addition of only 0.5 per cent of such metals as platinum or rhodium to titanium, or chromium to carbon steels, is sufficient. This technique is only satisfactory in oxygen-free conditions. The corrosion rate for titanium when boiled in dilute sulphuric acid is 100 mm/y compared with 1 mm/y for a titanium-platinum alloy.

(ii) *Impressed e.m.f.* Previously it has been stated that when a material is anodic in an electrolytic cell it is corroded away, whilst the cathodic material is protected. Therefore, it appears strange that to make a material increasingly anodic can protect it. It should be noted that this technique can only be applied to very few metals capable of forming *passive* surface oxides, that is, surface oxides which are unreactive and protective. For example, low-carbon steel and stainless steel are increasingly corroded as they are made more anodic until, at a critical e.m.f., a surface oxide forms which is passive and the corrosion rate drops significantly. For example, BS 304 stainless steel requires the impressed e.m.f. to be increased until a current density of 50 A/m^2 is achieved in order to form a passive oxide film. Once *passivation* has been achieved, the impressed e.m.f. is lowered so that the current density is reduced to 0.04 A/m^2, which is sufficient to maintain passivation.

Cathodic protection

Unlike anodic protection where corrosion is not eliminated but reduced to an acceptable level, cathodic protection can prevent corrosion completely. Further, whilst anodic protection can only be applied to a limited range of materials, cathodic protection can be applied to most metals. A typical example of cathodic protection is shown in Fig. 6.9. Buried iron pipes would normally be anodic compared with the surrounding moist soil and would quickly corrode. However, when subjected to an impressed e.m.f., which is electro-negative relative to the surrounding soil, the iron pipe behaves as though it was cathodic relative to the soil and does not corrode. A low-voltage d.c. generator is used to provide the impressed current.

Inhibitors

Corrosion inhibitors are chemicals which reduce or prevent anodic or cathodic reactions. For example, chromates (CrO_4^{2-}) and nitrites (NO_2^-) are oxidising anodic inhibitors which are used in anti-corrosive primers in paint systems. They promote the formation of passive oxide films on the painted metal. Nitrite inhibitors are also incorporated in protective oils and greases and in cutting fluids. Care must be taken when using anodic

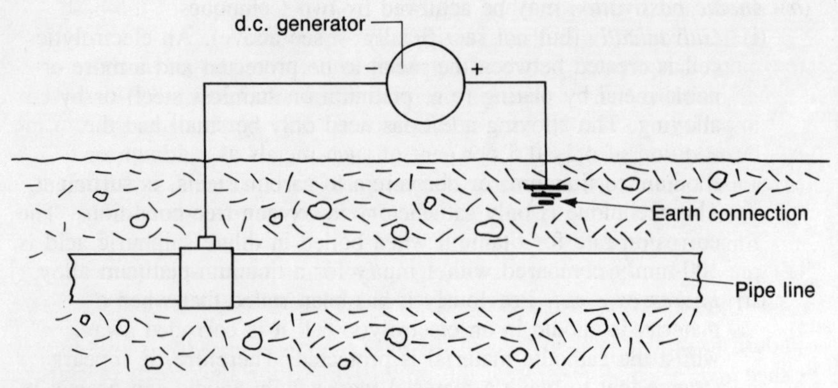

Fig. 6.9 Cathodic protection-impressed current

inhibitors. If the concentration falls below the minimum for a given system, severe pitting can occur because of the high cathode to anode area ratio of the unprotected zones.

Vapour phase inhibitors (VPI) and volatile corrosion inhibitors (VCI) are used in confined places and in packing materials (e.g. VPI impregnated papers). Silica gel may also be used to absorb any moisture present and so prevent wet corrosion.

Cathodic inhibitors reduce the rate of electron production at the anode of an electrolytic cell indirectly by forming a protective barrier at any cathodic sites. Unlike anodic inhibitors, there are no minimum concentration levels for cathodic inhibitors and they do not require such skilled monitoring for safety. For example, if the concentration level of a cathodic inhibitor falls below the minimum critical level, only mild uniform corrosion will occur and there will be no aggressive pitting. However, cathodic inhibitors are not so efficient as the anodic types because the deposit formed is more soluble and less adherent.

Mixtures of inhibitors tend to reinforce each others advantages whilst reducing their individual disadvantages. Since the primary inhibitor in the mixture becomes more effective at low concentrations, there is less likelihood of pitting when the more efficient anodic inhibitors are used.

6.16 Protective coatings (preparatory treatments)

Corrosion resistant metals and corrosion resistant metallic coatings were discussed in volume 1, and some inorganic and organic coatings were briefly introduced. These inorganic and organic coatings will now be considered in greater depth. First, however, it is necessary to examine the various preparatory techniques which are essential for the satisfactory protection of the substrate. These techniques are equally applicable both for first time treatment and for subsequent, remedial treatment.

Most component surfaces are contaminated with one or more of the following:

(*a*) oxide and hydroxide films resulting from the reaction of the base material with atmospheric oxygen and moisture;
(*b*) metal salt deposits, such as sulphates and carbonates of the base or its cladding, as the reaction products when attacked by the dissolved impurities in rain or surface water (e.g. 'acid rain' resulting from the burning of fossil fuels);
(*c*) soils in the form of grease, dust and dirt together with swarf and grinding wheel and polishing wheel dross from machining and finishing processes;
(*d*) previous protective coatings (e.g. paint films which need to be stripped to provide a sound base for replacement).

Failure to prepare the surface of the base material correctly results in either *lack of adhesion* so that the protective coating flakes off, or *self-perpetuating corrosion* resulting in destruction of the base material under the protective film. For example, if all traces of rust are not removed from steel-work before painting, rusting will continue under the paint film and will show itself by the paint film blistering and lifting off the base metal (e.g. evidence of the onset of body-rot in motor vehicles). Some common preparatory processes will now be considered.

6.17 The removal of existing coatings

Organic coatings, such as paint films, can be removed by use of a chemical solvent or a propane torch to soften the coating followed by mechanical scuffing of the surface using a manual or power scraper.

Anodic films can be removed by mechanical means or, more usually, by etching and chemical polishing after solvent or chemical cleaning.

Chromate films can be removed by chemical stripping and cleaning.

Phosphate films can be removed by chemical stripping and cleaning.

6.18 The removal of corrosion products

In addition to the removal of deteriorated or obsolete protective coatings it is also important to remove corrosion products and scale before applying any decorative or protective coating.

Acid pickling in hydrochloric or sulphuric acid is used to remove rust and scale. The acid cannot distinguish between the oxide and the metal and will often attack the metal if the oxide film is thick. This results in uneven pickling and pitting of the metal surface. This attack of the metal surface can be prevented by the addition of an *inhibitor* chemical to the pickling bath. It is essential to wash and neutralise the pickled metal and treat it with a temporary protective film such as lanolin or oil.

Conversion coatings. Where it is not possible to remove corrosion products completely, prior to painting, proprietary chemical compounds — in liquid form — may be applied to convert and passivate the residual corrosion products, for example the conversion of rust to a phosphate coating which acts as a protective film as well as a key for painting.

6.19 The removal of miscellaneous debris

Wire brushing

A rapidly rotating coarse wire brush can be used to dislodge loose debris and soils from structural steelwork before painting or repainting. The surface left by brushing provides a key for the first paint coat and helps it to adhere to the metal. Fine wire brushing can also be applied to aluminium and its alloys both as a decorative finish and as a key prior to painting.

Shot and vapour blasting

Fine particles are blasted against the metal surface at high velocity using compressed air. This is used to remove soils, previous coatings and scale from structural steelwork on site as well as smaller components under factory conditions. It is also used for cleaning and descaling forgings and sand castings.

Flame descaling

This depends upon the difference in expansion between the scale, soils and other surface debris and the base metal when locally heated. This process is used for cleaning heavily rusted and soiled steelwork before initial and maintenance painting. The surface is heated with an oxy-fuel gas torch fitted with a specially designed nozzle which gives a broad fan-shaped flame. The rapid expansion of the scale compared with the cooler metal substrate causes the scale, debris and rust to flake off. Any entrapped moisture is driven off as steam and helps in the stripping process.

Abrasive finishing

This can range from the use of portable grinding machines where heavy corrosion layers need to be removed on site to polishing and decorative finishing.

6.20 The removal of oil and grease (degreasing)

The presence of greases and oils prevents wetting of the surface to be treated and must be removed before any pre-treatment or finishing process can be applied.

Solvent degreasing

Trichlorethylene and perchlorethylene are still widely used in vapour degreasing plants despite the highly toxic nature of these chemicals. Oil and grease removal is effective, but inorganic soils are only removed by the washing action of the condensed liquid.

Kerosene (paraffin) will dissolve most oils and greases. It is usually blended with oil-soluble surface-active agents (surfactants) and becomes imulsifiable. Such systems have the advantage over vapour degreasing in as much that soils and residues can be rinsed away from the metal surfaces by the detergent and flushing action of the liquid.

Alkali cleaning

This is used where degreasing is followed by electro-plating as any residual solvent film leads to poor adhesion of the plating. Further, alkalis do not have the toxicity of chlorinated hydrocarbons or the flammability of kerosene. Alkali detergents range from washing soda and caustic soda to sophisticated blends of silicates, phosphates, carbonates and surface active agents. Alkali solutions are used at temperatures of 80°C to 90°C. It is important that work so treated is thoroughly rinsed so as to avoid 'carry over' into the plating baths, where the presence of alkalis would be highly undesirable. Alkali cleaning must NOT be used with aluminium-based or zinc-based alloys, unless suitable buffered solutions are used, as these metals suffer from alkali attack.

6.21 Protective coatings (inorganic)

These consist of ceramic materials, as discussed in Section 3.17, applied over metallic components. The ceramic coating acts as a barrier to corrosive and erosive agents. Such coatings are susceptible to thermal and mechanical shock and, therefore, can only be applied to rigid components and structures for a limited range of applications. Such applications could be the lining of chemical and water storage tanks, the protection of pipework and the insulation of rigid electrical conductors.

Vitreous finishes are often used for the protection of low-carbon steel cooking utensils and domestic appliances as an alternative to using more expensive stainless steel. Unfortunately, such coatings are very easily chipped. The vitreous finish consists of applying an opaque powdered glass slurry, called a *slip*, to the metal surface to be protected. The coated component is then dried and *fired* in a kiln so that the glass matrix melts and flows evenly over the metal surface. This finish is often referred to as 'vitreous enamelling' and must not be confused with organic enamelling which is a paint process.

6.22 Protective coatings (organic)

Organic coatings can be divided into three general categories:

(*a*) bitumastic coatings;
(*b*) plastic and elastomer coatings;
(*c*) paint films (Section 6.23).

Bitumastic coatings

Carbon-based coatings such as bitumen, pitch and tar are used to form barriers against the absorption of moisture. Such materials are used to protect underground pipes either as a direct coating or by wrapping the pipes in impregnated woven material such as hessian. Underground electric mains armoured cables are also protected in this way. Bitumastic paints are also available for protecting underground steel structures and the steelwork on ships.

Plastic and elastomer coatings

Plastics and rubbers, when used as protective coatings, can be functional as well as being corrosion resistant and decorative. The wide range of these materials available for coating purposes provides the designer with means of achieving:

(*a*) abrasion resistance;
(*b*) cushion coating (up to 6 mm thick);
(*c*) electrical and thermal insulation;
(*d*) flexibility over a wide range of temperatures;
(*e*) non-stick properties;
(*f*) permanent protection against weathering and atmospheric pollution, subject to the inclusion of anti-oxidants and ultraviolet filter dyes;
(*g*) reduction in maintenance costs;
(*h*) providing an impervious, protective barrier to a wide range of chemicals which would otherwise be potentially corrosive to the metal substrate;
(*i*) the covering and sealing of mechanical joints, welds and porous castings.

It must be remembered that neither the rubber nor the plastic coating in any way passivates the metal component which it is covering. Therefore, it is important that the component is treated before coating so that corrosion does not occur under the coating due to residual impurities. It is also important that the plastic or rubber coating is not broken at any point, otherwise moisture will seep in between the coating and the component by capillary attraction unless the coating is securely bonded to the component.

There are many processes by which plastic coatings may be applied, for example *fluidised bed dipping*, as shown in Fig. 6.10. The plastic powder is kept in a state of agitation by compressed air passing through the porous bed from the plenum chamber. In its state of agitation, the powder offers little resistance to the immersion of the pre-heated workpiece and adheres to it to form a homogeneous skin. *Liquid plastisol dipping* is used for coating the workpiece with PVC. The plastisol consists of the resin powder held in suspension in a plasticiser and no dangerous solvent is used. The heated workpiece is dipped into the thixotropic (non-drip) liquid plastic and a film forms on the heated surface. For larger work, *spraying* can be used, the plastic powder being sprayed onto the heated surface of the workpiece in a similar manner to paint spraying.

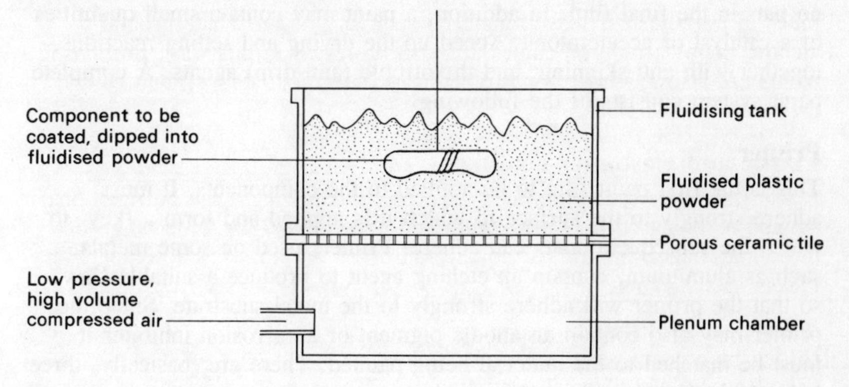

Fig. 6.10 Fluidised bed dipping

6.23 Paint films

Painting is widely used for the protection and decoration of metallic components and structures. It is the easiest and cheapest coating which can be applied with any degree of permanence and, by careful choice, painting can provide a wide range of protective properties. Paint films can be used as sealants over such finishes as galvanising, sherardising and phosphating. This is particularly useful in urban and industrial environments where the sulphur products in the atmosphere destroy the sacrificial zinc coating of galvanised or sherardised steelwork.

Paints may generally be described as consisting of finely divided solids (pigments) in suspension in a liquid (binder or 'vehicle') which dries or sets to provide a coherent film over the metal surface. Usually a paint is made up of three main constituents.

Binder

This contains the film forming component in a volatile solvent. The binder is a natural or synthetic resinous material and reflects the essential properties of the paint: its durability; protective ability; flexibility; and adhesion.

Pigment

This provides the paint with its opacity and colour. Further, some pigments have special properties and act as corrosion inhibitors, fungicides, insecticides, etc.

Solvent or thinner

This controls the consistency of the paint and its application. Since the solvent is volatile and evaporates once the paint has been spread it forms no part in the final film. In addition, a paint may contain small quantities of a catalyst or accelerator to speed up the drying and setting reactions, together with anti-skinning, and thixotropic (anti-drip) agents. A complete paint system consists of the following.

Primer

This is the first paint film to be applied to the components. It must adhere strongly to the surface to which it is applied and form a 'key' to which the subsequent coats can adhere. Primers used on some metals, such as aluminium, contain an etching agent to produce a suitable 'key' so that the primer will adhere strongly to the metal substrate. Since the primer may also contain an anodic pigment or a corrosion inhibiter it must be matched to the material being painted. There are, basically, three types of primer.

(*a*) Primers containing metallic pigments which are anodic and sacrificial to the metal being painted, for example, zinc-rich primers for steel. This prevents under-rusting of the primer until the anodic pigment is exhausted.

(*b*) Primers containing pigments which are inhibitors, such as chromates, phosphates, and red lead, are used when painting steel-work. These pigments dissolve in any moisture permeating the paint film to form solutions which stifle the corrosion process in the presence of aerated water. The red lead oxide reacts with the oil binder of the paint to form lead azelate which is a strong corrosion inhibitor. Therefore red lead oxide can only be used in oil-based paints. Note that chromates and red lead oxide are toxic.

(*c*) Primers which have high adhesion and chemical resistant properties, for example two-part epoxy primers. These offer no anodic protection nor any inhibition to corrosion reactions but merely act as a barrier. Therefore the metal being protected must be free from impurities and carefully prepared before application of the primer so that corrosion does not occur beneath the primer.

Putties or fillers

These are applied by a putty knife or a spatula to fill surface defects in castings or dents and defects in sheet metal. After setting the putty is sanded down smooth ready for undercoating.

Undercoat

One or more undercoats are used to built up the thickness of the paint film, to give opacity to the colour and to provide a smooth surface for the finishing coat. To this end, undercoats should be thoroughly 'flatted down' between each coat. Highly pigmented undercoats decrease the permeability of the paint to oxygen, and the use of laminar pigments reduces and delays the penetration of moisture.

Finish or top-coat

This is not only decorative because of its high gloss, but provides most of the corrosion resistance of the system. This is because the finish or top-coat contains a varnish which seals the undercoats and prevents the absorption of moisture. The varnish content is usually tough and abrasion resistant, being based upon acrylic or polyurethane rubbers.

6.24 Types of paint

Paints can be broadly classified, by the manner in which they dry, into four groups.

Group 1

In this group, atmospheric oxygen reacts with the binder causing it to polymerise into a solid film. This reaction is speeded up by forced drying at 70°C. Paints which dry by oxidation include the traditional linseed-oil based paints, the oleo-resinous paints, and the modern general purpose air-drying paints based on oil-modified alkyd resins. Note: paints based upon drying-oil type binders must not be used in alkaline environments or the binder will soften and dissolve by saponification. Therefore such paints cannot be used in the presence of cathodic protection systems, for example over galvanised steel-work.

Group 2

These paints are based upon amino-alkyd resins which do not cure (set) at room temperature but have to be 'stoved' at 110–150°C to promote the polymerisation reaction. When set, such paints are tougher and more resistant to abrasion than air drying paints. Further, the drying cycle is much quicker than for air-drying paints. Paints in this group are used for motor car bodies.

Group 3

In this group, polymerisation is achieved by the addition of an activator or hardener. Since this is stored separately and only added to the paint immediately before use, such paints are referred to as 'two-pack' paints. Polymerisation (hardening) commences as soon as the hardener is added to the paint. At first this will be slow and give ample time for application of the paint but, as soon as the paint is spread, a solvent (thinner) commences to evaporate increasing the relative concentration of the hardener. When this increase in concentration reaches a critical level, rapid polymerisation occurs and the paint is soon 'touch-dry'. However, it does not attain its full mechanical properties and resistance to damage for a few days. Paints in this category are based upon polyester, polyurethane, and epoxy resins. The tendency nowadays is to use 'one-can' paints. The hardener is added at the time of manufacture but below the critical concentration level. Polymerisation cannot, therefore, occur until the paint has been spread and the volatile solvent has evaporated, increasing the concentration of hardener to above the critical level.

Group 4

These are the laquers, that is, paints which dry by simple evaporation of the volatile solvent with no hardening or polymerisation reactions taking place. For example, cellulose nitrate dissolved in acetone with a pigment in suspension. As soon as the acetone volatilises (evaporates), a dry film of coloured cellulose nitrate covers the surface to which the lacquer has been applied. Since both the base and the solvent are highly flammable and the fumes given off during volatilisation are toxic great care must be taken in their use.

6.25 Application of paints

The success of a paint system depends upon the satisfactory preparation of the surface being treated and the correct application of the paint. Brush painting is labour intensive and the skill required depends upon the quality of finish required. Except for maintenance and remedial treatment it is little used in the engineering industry.

Spraying

There are a number of paint spraying processes but, the most commonly used, employs compressed air to atomise the paint in a spray gun and project it onto the component being coated. It is a quick and relatively simple process requiring relatively low-cost equipment. Further, it is versatile and can accommodate frequent colour changes. It gives consistently high standards of finish but paint and solvent wastage is high due to overspray and bounce.

Dipping

In this process the work to be coated is suspended in a bath of paint and then lifted out and allowed to drain off. The surplus paint drains back into the bath and little is wasted. The drying process is usually accelerated by 'stoving' at elevated temperatures. In order to maintain consistency of the paint bath it is neither usual nor desirable to use air drying paints and paints specially formulated for dipping should be used exclusively. Dipping plants usually operate on a continuous conveyor system where the components pass through the dipping bath, then over a drainage area, and finally through a stoving tower before cooling and being off-loaded from the conveyor with the paint film set and ready for handling. Dip painting is highly productive and the labour costs are low. However the capital cost is high, especially for large components, and close control is required if consistent results are to be achieved.

6.26 Hazards of paint application

The hazards of industrial paint application fall into two main categories.

(*a*) Explosion and fire hazards resulting from the use of flammable solvents and the formation of flammable dust particles as the spray mist dries in the atmosphere.
(*b*) Toxic and irritant effects due to the inhalation of paint mist (wholly or partially solidified) and solvent fumes.

These hazards are particularly related to spraying and stoving processes, but the storage of paints also presents special problems. The local Health and Safety Inspector and the fire authorities should be consulted before painting on an industrial scale is undertaken.

It is essential when spray painting to provide an efficient means of extraction to remove excess spray mist and solvent fumes. Spray booths serve the double purpose of removing spray mist and fumes from the working area and then treating the exhausted air so that it is cleansed before being released back into the atmosphere. The lighting, fans and other electrical equipment associated with spray painting booth and equipment have to be to *Buxton Approved Standards* for flame and explosion proof fittings.

Stoving

The main hazards associated with stoving ovens results from:

(*a*) the use of unsuitable paints having volatile and flammable solvents which are liable to ignite at the stoving temperature;
(*b*) the accumulations of explosive dusts and gases in the fume extraction ducts. Stoving ovens and their extractors should have pressure relief vents so that any explosion is carried upwards and away from the working area.

6.27 Testing protective coatings

As for any other manufacturing process the quality of any decorative and/or protective coatings and finishes can only be maintained by a rigorous programme of inspection and testing. To detail such tests is beyond the scope of this chapter since the tests vary for each type of coating and its application. Full information on the causes of corrosion, its prevention, and the testing and inspection of protective and/or decorative finishes can be obtained from such sources as:

(*a*) the manufacturers of the chemicals and materials used;
(*b*) the Institution of Corrosion Science and Technology;
(*c*) the Department of Industry Committee on Corrosion;
(*d*) the National Physical Laboratory; and
(*e*) the British Standards Institute.

The efficiency of a protective coating is dependent upon a number of factors and these will now be outlined briefly.

Type

Permanent protective coatings must be selected to satisfy the service requirements of the component, that is, they must be economic to apply yet satisfy such design criteria as appearance and level of corrosion prevention (passivation). If corrosion resistance is the primary objective, the process selected will depend upon the material from which the component has been made and the environment in which it is to operate.

Adhesion

The satisfactory adhesion of a protective finish to the metal substrate is extremely important. Any lack of adhesion will result in the protective film flaking or peeling away resulting in exposure of the substrate and its corrosion. Satisfactory adhesion is largely dependent upon the correct preparation of the surface to be protected.

Thickness

An unnecessarily thick protective coating is a waste of relatively expensive corrosion resistant material and a waste of processing time. A coating which is too thin will not present an adequate barrier to the corrosive environment and corrosion will occur. Uneven coating can result in both waste and inadequate protection.

Uniformity

The composition of any coating must be uniform otherwise, even if the thickness is constant and correct, the efficiency of the protective coating may vary and allow local corrosion such as pitting.

Chemical stability

Except where coatings are deliberately sacrificial, they must be inert to or become passivated by the reactive agents in the environment against which they are to provide protection. Further, the process by which the protection is applied must not itself affect the substrate to which it is applied. For example, the processing temperature must not cause temper brittleness, neither should any hydrogen released during electro-deposition be allowed to cause embrittlement.

6.28 Remedial measures

These consist of the replacement of a deteriorated or obsolete protective coating with a replacement coating to the original or an improved specification. Where corrosion and/or mechanical damage has occurred, a combination of the preparatory treatments previously discussed will be required, together with the replacement or repair of the structure itself, after which the replacement coating can be applied by a method appropriate to the size and type of the plant or structure, its situation and its environment.

6.29 Effects of finishing processes on material properties

All remedial and preparatory processes affect the mechanical properties of the metal being processed. For example, shot blasting enhances the mechanical properties of metals by putting the surface into a state of compression which improves the fatigue performance of the metal. On the other hand, machining, grinding, or scratch-brushing the surface reduces the fatigue performance of the metal since they leave the surface of the metal in tension. However, negative rake machining and polishing can enhance the metal properties as these processes tend to leave the surface of the metal in compression.

Chemical processing such as pickling in acid can lead to hydrogen embrittlement particularly in the case of high strength alloy steels. Chemical etching and chemical polishing can also lead to hydrogen embrittlement. Hence particle blasting and mechanical polishing is preferable for highly stressed components. Even a light vapour blast after chemical treatment is all that is required to restore the fatigue properties of the metal.

Many finishing processes are carried out above ambient temperatures and such processes can have the effect of impairing the mechanical properties of the metal from which the workpiece is made. For example, low-carbon steels become brittle when heated to 200°C for any prolonged period of time, yet many finishing processes are carried out around this temperature. Similarly, aluminium alloys are particularly susceptible to processing at temperatures between 100°C and 150°C, yet this is the temperature for force drying paint. Grinding and polishing and machining processes can also raise the temperature of the metal surface to a level which can adversely affect the mechanical properties of the metal.

Finally, it should be noted that there is inevitably some chemical interaction at the interface between the cladding and the substrate. This usually results in some loss of fatigue strength. However, the protection from corrosive fatigue often far outweighs the slight lowering of the mechanical fatigue performance.

7 Electrical properties of materials

7.1 Atomic structure

The structure of the atom was considered in Section 2.2 of volume 1. As a reminder, Fig. 7.1 shows a copper atom. It can be seen that it consists of a positively charged nucleus surrounded by orbiting electrons. Not all the electrons surrounding the nucleus are at the same energy level and it is useful to arrange them into 'shells'. In an *atom*, the number of positively charged protons in the nucleus is electrically balanced by an equal number of negatively charged electrons. The neutrons in the nucleus have the same mass as the protons but carry no electrical charge. If the electrical balance between the protons and the electrons is disturbed, the atom becomes an *ion*. Diagrams such as Fig. 7.1 are referred to as *Bohr models*.

Reference to Fig. 7.1 shows that the first *quantum shell* contains only two electrons; the second shell a maximum of eight electrons the third shell a maximum of eighteen electrons and, although the fourth shell can have a maximum of thirty two electrons, in the copper atom there is only one electron. Note that no matter what the maximum number of electrons in a particular quantum shell, when that shell is the outermost or *valency shell* it can never hold more than eight electrons.

The Bohr model concept oversimplifies the situation as it implies that all electrons in the shell are equal and that they follow strictly defined orbits in a single plane, whereas they do not. The electron is no longer considered to be a clearly defined particle following a precisely defined orbit, but more as a diffuse 'mist' of electrical charge. Further, it is impossible to ascertain the exact position of the electron because any known measuring technique disturbs the wave characteristics of the electron. Thus only probabilities of electron position can be predicted by

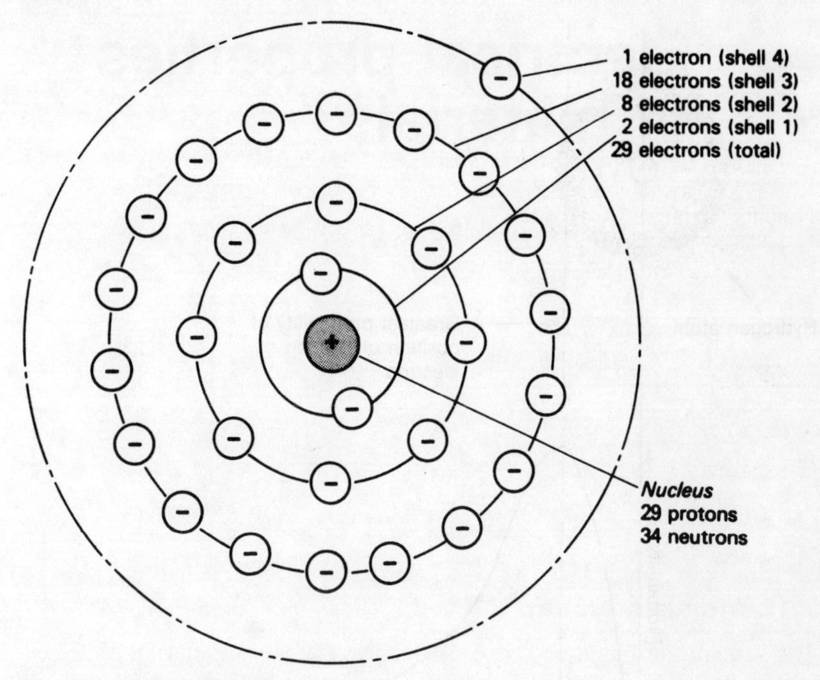

Fig. 7.1 Bohr model of a copper atom

calculation. Consider a hydrogen atom where one electron orbits around a nucleus of one proton. Figure 7.2 shows the probability of the actual electron positional range. The greatest probability is that the electron will be located at a radius 0.53 Å from the nucleus. (Note that when measuring distance, Å = Angstrom unit (0.1 nm). Hence in Fig. 7.2 the position of the electron is shown not as a clearly defined path but as a shaded zone of probable positions with the shading darkest where the probability is greatest.

The Bohr model also breaks down in showing all the electrons in a given shell as lying in the same plane and having the same energy levels. *Pauli's exclusion principle* states that there are definite rules governing the energy levels and probable positions of the electrons as they orbit the nucleus. Further, there cannot be more than two interacting electrons with the same orbital quantum number, and even these are not identical since they have inverted magnetic behaviour (opposite 'spins').

Reference back to Fig. 7.1 shows that the first shell has only two electrons and, therefore, complies with *Pauli's exclusion principle*. However, the subsequent quantum shells have many more electrons and, to comply with Pauli's exclusion principle, each quantum shell has to be divided into *subshells* with the electrons orbiting three-dimensionally so that there are no more than two electrons in any one orbit. The electron

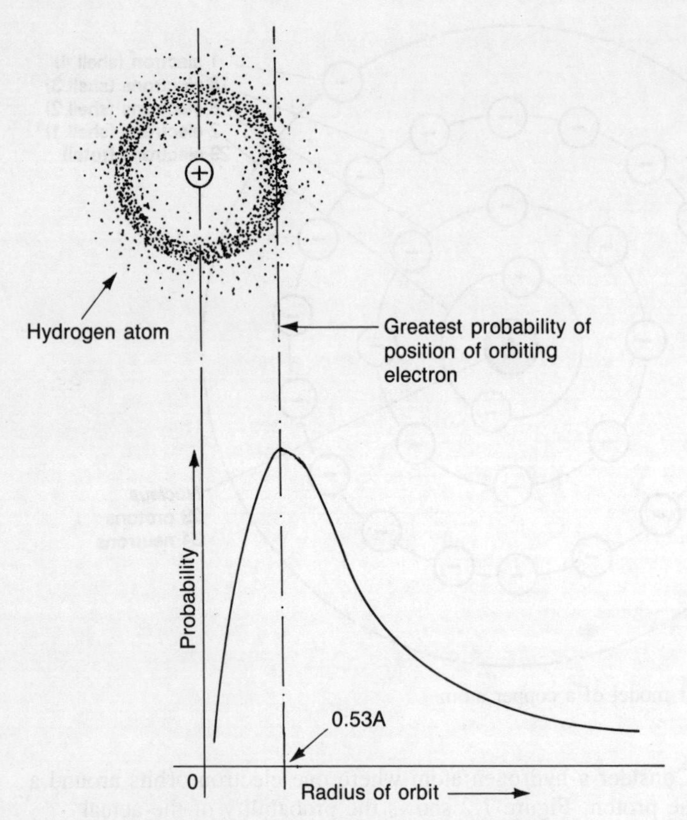

Fig. 7.2 Possible position of single hydrogen electron

notation for part of the first three periods of the periodic table is shown in Table 7.1.

Each electron in an atom has its own set of quantum numbers denoted by the symbols n, l, m_l and m_s which are defined in Table 7.2. The maximum number of electrons in any one shell is $2n^2$. The method of calculating the possible combinations of quantum numbers for any given shell is shown in Example 7.1.

Example 7.1
Using the data given in Tables 7.1 and 7.2, determine the maximum number of electrons for any given shell, together with the possible combinations of quantum numbers for the K shell and for the L shell.

(a) K shell
From Table 7.1, n = 1. Therefore, using the expression $2n^2$, the

maximum number of electrons that can be present in the shell is $2 \times 1^2 = 2$.

Since $l = n-1$, $l = 1-1 = 0$, and the possible combinations of quantum numbers will be:

l	m_l	m_s	
0	0	$+\frac{1}{2}$	subshell 1s (2 electrons)
0	0	$-\frac{1}{2}$	

Thus the K shell can only contain a maximum of 2 electrons with opposing spins. This satisfies the requirements of hydrogen and helium and also the requirements of the Pauli exclusion principle.

(b) L shell
From Table 7.1, $n = 2$. Therefore, using the expression $2n^2$, the maximum number of electrons that can be present in the shell is $2 \times 2^2 = 8$.

Since $l = n-1$, $l = 2-1 = 1$ and l can be 0 or 1. The possible combinations of quantum numbers will be:

l	m_l	m_s		
0	0	$+\frac{1}{2}$	subshell 2s (2 electrons)	
0	0	$-\frac{1}{2}$		shell L 8 electrons
1	0	$+\frac{1}{2}$		
1	0	$-\frac{1}{2}$		
1	−1	$+\frac{1}{2}$	subshell 2p (6 electrons)	
1	−1	$-\frac{1}{2}$		
1	+1	$+\frac{1}{2}$		
1	+1	$-\frac{1}{2}$		

Thus the L shell can contain a maximum number of 8 electrons (4 pairs of opposing spins — one pair per orbit). This satisfies the requirements of Table 7.1 and the requirements of the Pauli exclusion principle.

7.2 Energy levels

In any single atom the electrons can occupy a number of discrete energy levels which become closer together the more remote they are from the nucleus. Electrical conductivity is concerned only with the electrons in the outermost (valency) shell. A copper atom has twenty-nine electrons arranged in four main shells (seven subshells). The subshells 1s, 2s, 2p, 3s, 3p, and 3d are filled but the valency subshell 4s has only one electron in it and could hold another electron according to Pauli's exclusion

Table 7.1 Electron notation for elements in first three periods of the periodic table

Element	Atomic number Z	Electron configuration						
		Shell K n = 1	Shell L n = 2			Shell M n = 3		
		1s	2s	2p	3s	3p	3d	4s
H	1	1						
He	2	2						
Li	3	2	1					
Be	4	2	1					
B	5	2	2	1				
C	6	2	2	2				
N	7	2	2	3				
O	8	2	2	4				
F	9	2	2	5				
Ne	10	2	2	6				
Na	11	2	2	6	1			
Mg	12	2	2	6	2			

Al	13	2	2	6	2	1		
Si	14	2	2	6	2	2		
P	15	2	2	6	2	3		
S	16	2	2	6	2	4		
Cl	17	2	2	6	2	5		
Ar	18	2	2	6	2	6		
K	19	2	2	6	2	6		1
Ca	20	2	2	6	2	6		2
Sc	21	2	2	6	2	6	1	2
Ti	22	2	2	6	2	6	2	2
V	23	2	2	6	2	6	3	2
Cr	24	2	2	6	2	6	5	1
Mn	25	2	2	6	2	6	5	2
Fe	26	2	2	6	2	6	6	2
Co	27	2	2	6	2	6	7	2
Ni	28	2	2	6	2	6	8	2
etc.								

Sub-shell 4s commences to fill up before sub-shell 3d is full.

Table 7.2 Quantum number symbols

Symbol	Description
n	This is the *principal quantum number* for any given shell and it can have an integer value of 1, 2, 3, etc.
l	This is the *angular momentum quantum number* and its integer value ranges from 0 to $n-1$.
m_l	This is the *magnetic quantum number* and its integer value can only be -1, 0, $+1$.
m_s	This is the *spin quantum number* and it can only have values of $+\frac{1}{2}$ or $-\frac{1}{2}$.

principle. The electrons in each subshell occupy specific energy levels in which the s-level is lower than the p-level which, in turn, is lower than the d-level and so on. Further, shell K has a lower energy level than shell L and so on. Thus when an electron moves up from one energy level to a higher energy level it has to be given a quantum (or 'packet') of energy. When it moves down to a lower energy level it gives up a quantum of energy in the form of a photon.

When single atoms bond together to form a solid mass, they influence each other and the electrons are forced to seek alternative energy levels so that no more than two electrons occupy the same quantum state, in accordance with Pauli's exclusion principle. Within any crystal there will be a vast number of atoms and it is not possible to consider individual energy *levels*. Instead, it is necessary to consider densely filled energy *bands*. The concept of energy levels and energy bands is shown diagramatically in Fig. 7.3.

Since copper atoms have only one valency electron (only half the 4s band is filled), the energy required to raise a valence electron to an empty state is negligible. Therefore the valency electrons for copper can move freely within the crystal and conduct electricity (See section 7.3). Similarly all metals with an odd number of electrons in their valency bands have good conductivity. Although bivalent metals such as beryllium, calcium and magnesium have their valency bands completely filled, these are only 's' bands and some electrons transfer easily to the next energy band which are 'p' bands and this endows such metals with good electrical conductivity. On the other hand, the transition metals with the 3d band incompletely filled show relatively high resistivity because the electrons in the 3d band overlap with the electrons in the 4s band causing scattering.

In materials, such as the diamond allotrope of carbon, the valency band is filled by the four outer shell electrons and a wide energy gap exists between the valency band and the next possible energy band. In fact, at

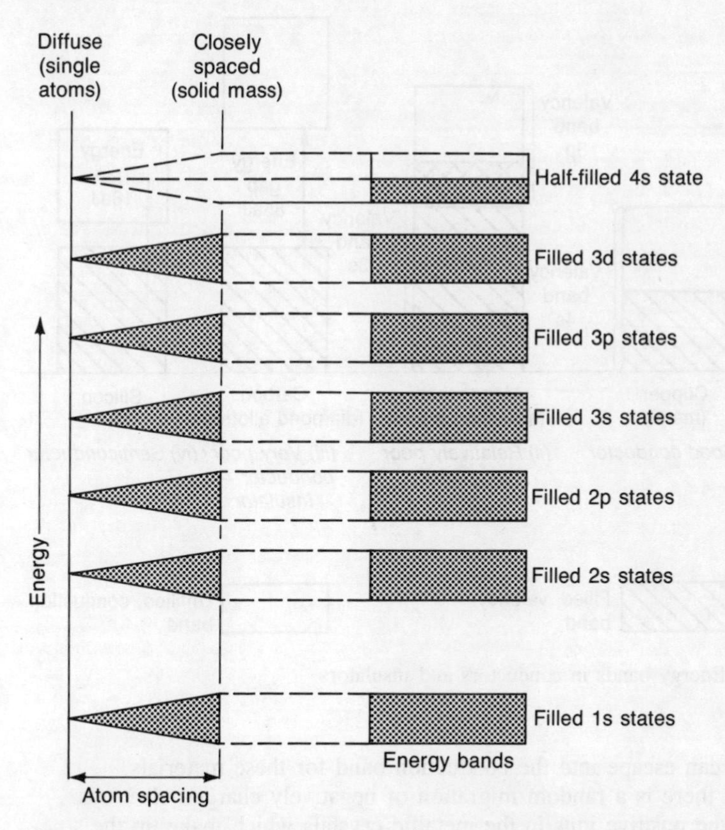

Fig. 7.3 Energy states of the electrons in a copper atom

20°C, about 8.5×10^{-19} J of energy is required by each electron to move from the 'valency band' to the 'conduction band' (note: 10^{-19} J = aJ). This accounts for diamond having such a high resistance to the flow of an electric current that it can be classified as an insulator. Graphite, which is also an allotrope of carbon, has a much lower resistivity than diamond and is classified as a high resistance conductor.

Some materials, known as *semiconductors* also have energy gaps between their completely filled valency bands and the next available energy or conduction bands. However, as can be seen from Fig. 7.4, the energy required to promote an electron from the valency band to the conduction band for semiconductor materials is very much less than for diamond.

7.3 Electrical conduction

It has already been shown that the high electrical conductivity of metals such as copper is due to the ease with which the outer, valency shell

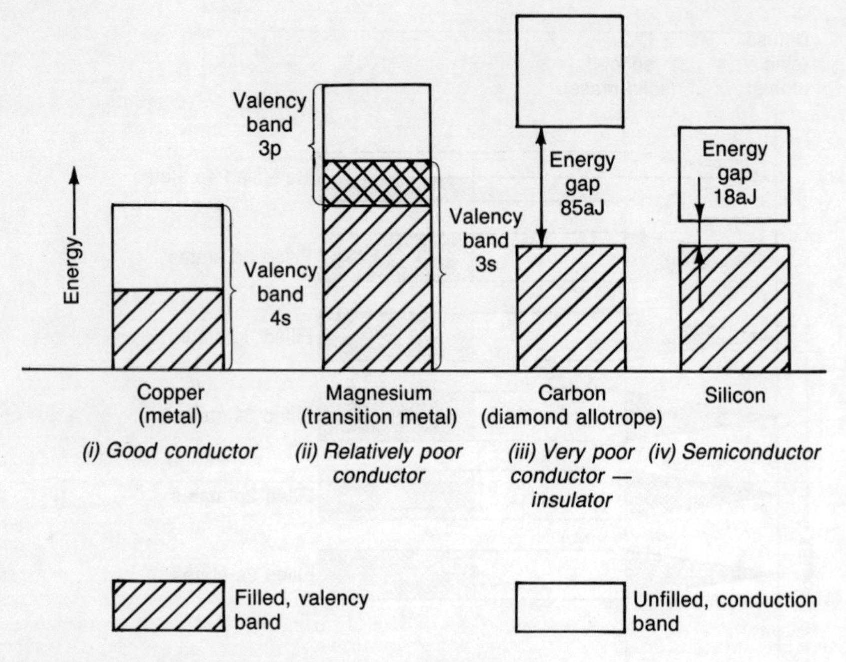

Fig. 7.4 Energy bands in conductors and insulators

electrons can escape into the conduction band for these materials. Normally there is a random migration of negatively charged electrons amongst the positive ions in the metallic crystals which make up the material. (A positive ion is an atom which has lost one or more electrons and has a residual positive charge.) However when such a material forms part of an electrical circuit the random migration of electrons becomes a directed flow as shown in Fig. 7.5. Since like charges attract and unlike charges repel, the negative electrons will flow away from the negative pole of the cell and towards the positive pole of the cell. Hence the *electron current* flows from negative to positive whilst the traditional concept of *conventional current* flow is from positive to negative.

The negative charge of one electron is too small for practical purposes and a much larger unit is used called the *coulomb*.

$$1 \text{ coulomb (C)} = 6.3 \times 10^{18} \text{ electrons.}$$

An electric current of 1 ampere (A) is said to flow in a circuit when a charge of one coulomb passes any point in that circuit in one second.

An electric current can also flow in non-metallic materials by the drift of positive and negative ions. This occurs in electrolytes — see corrosion cells.

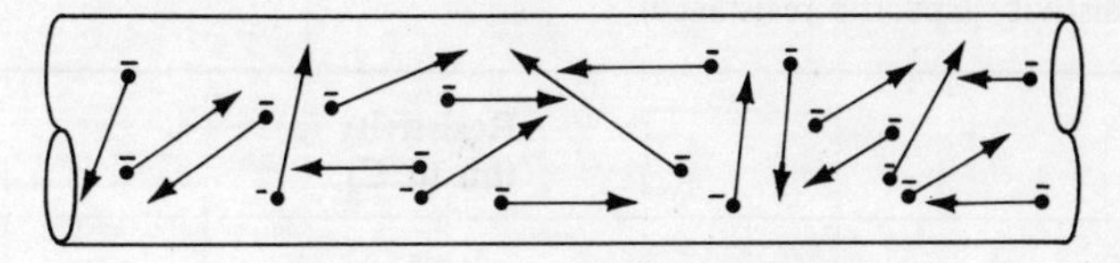

(i) Random movement of electrons in a conductor

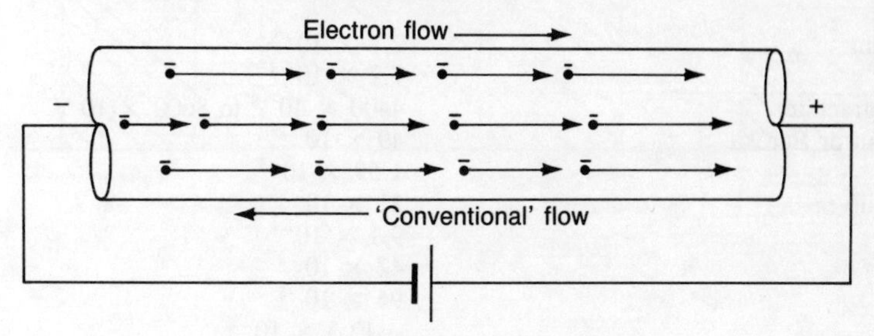

(ii) Directed movement of electrons in a conductor which forms part of an electric circuit

Fig. 7.5 An electric current as a flow of free electrons

7.4 Conductor materials

It has already been shown that metals are good conductors of electricity because of the relative ease (low energy requirement) with which their valency electrons can be moved from the valency energy band to the conduction band. It follows, therefore, that anything which interferes with the free movement of the valency electrons results in increased energy being required to make the transition and lowers the conductivity of the material. Since an electric current is a directed flow of electrons amongst the positive ions in a metallic crystal, anything which impedes this flow will also lower the conductivity of the material.

The *conductance (G)* of a conductor, measured in *siemens (S)*, is the reciprocal of the *resistance (R)* of that conductor measured in *ohms* (Ω).

$$G = 1/R$$

Similarly, the *conductivity* of a material, (sumbol σ) is the reciprocal of the resistivity of that material (symbol ρ).

$$\sigma = 1/\rho$$

The resistivity of a material may be defined as the resistance between any two opposite faces of a unit cube of that material. Table 7.3 lists the resistivity for a number of typical engineering materials at 0°C. Care must be exercised in the use of such a table as the resistivity of a

Table 7.3 Resistivity (specific resistance)

Material	Resistivity (ρ) Ωm (0°C)
Polythene	$\sim 10^{16}$
Germanium (intrinsic) Silicon (intrinsic)	$\sim 10^{-3}$
Aluminium	2.7×10^{-8}
Brass	7.2×10^{-8}
Carbon (graphite)	4400×10^{-8} to 8600×10^{-8}
Constantan or Eureka	49×10^{-8}
Copper	1.59×10^{-8}
German silver	21×10^{-8}
Iron	9.1×10^{-8}
Manganin	42×10^{-8}
Mercury	94×10^{-8}
Nickel	$\sim 12.3 \times 10^{-8}$
Tin	13.3×10^{-8}
Tungsten	$5.35\ 2\ 10^{-8}$
Zinc	5.75×10^{-8}

material is affected by such factors as temperature, impurities and distortion of the crystal lattice by hot- or cold-working. Example 7.2 shows how the resistance of a conductor may be calculated.

Example 7.2
Calculate the resistance of a copper conductor 4 m long and having a cross-sectional area of 1 mm. Temperature 0°C

$$R = \frac{\rho l}{A}$$

where: R = resistance (Ω)
ρ = resistivity (Ωm)
l = length of conductor (m)
A = cross-sectional area (m^2)

From Table 7.3, $\rho = 1.59 \times 10^{-8}$ Ωm at 0°C. Therefore:

$$R = \frac{1.59 \times 10^{-8} \times 4}{1 \times 10^{-6}}$$

$$= 6.36 \times 10^{-2}\ \Omega$$

$$= \underline{\underline{0.0636\ \Omega}}$$

Temperature

Any increase in temperature of a conductor produces greater thermal agitation of the metallic ions as they vibrate about their mean positions. Therefore the chance of collision between the electrons and ions becomes greater and the mean free path is effectively reduced. This restricts the flow of electrons and reduces the conductivity of the material. The effect of temperature on the resistivity of some typical conductor materials is shown in Fig. 7.6. It should be noted that the temperature effect on graphitic carbon is the opposite to the temperature effect on the metals. The decrease in resistivity with temperature rise which occurs within graphitic carbon is a phenomonon it shares with all other non-metals.

The temperature coefficient of resistance of a material is the change in resistance of 1 ohm at 0°C per degree of temperature rise (symbol α_0*).* Table 7.4 lists the temperature coefficients for a number of typical engineering materials. Note the negative coefficient for graphitic carbon which has been referred to previously. The reference temperature is rarely 0°C and the method of calculating the effect of temperature change is shown in Example 7.3.

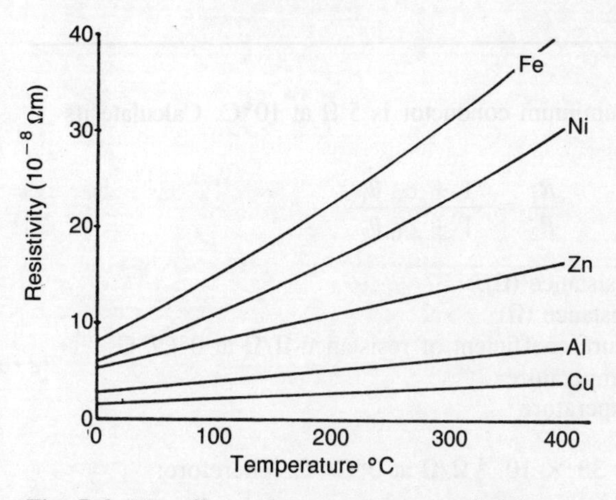

Fig. 7.6 The effects of temperature on resistivity

Composition

Alloys, in general, have a higher resistivity than pure metals and this is largely due to the presence of solute atoms in solid solutions no matter whether such solutions are interstitial or substitutional. The solute atoms interfere with the flow of electrons amongst the positive ions of the metallic crystals. Figure 7.7 shows the effect of composition on the electrical conductivity of cupro-nickel alloys.

Table 7.4 Temperature coefficient of resistance

Material	Temperature coefficient (α_0) Ω/Ω at 0°C/°C
Aluminium	38×10^{-4}
Brass	10×10^{-4}
Carbon (graphite)	-5×10^{-4}
Constantan or Eureka	$+0.1 \times 10^{-4}$ to -0.4×10^{-4}
Copper	43×10^{-4}
German silver	27×10^{-4}
Iron	63×10^{-4}
Manganin	0.25×10^{-4}
Mercury	9.8×10^{-4}
Nickel	62×10^{-4}
Tin	44×10^{-4}
Tungsten	51×10^{-4}
Zinc	37×10^{-4}

Example 7.3

The resistance of an aluminium conductor is 5 Ω at 10°C. Calculate its resistance at 25°C

$$\frac{R_1}{R_2} = \frac{1 + \alpha_0\,\theta_1}{1 + \alpha_0\,\theta_2}$$

where: R_1 = initial resistance (Ω)
R_2 = final resistance (Ω)
α_0 = temperature coefficient of resistance Ω/Ω at 0°C/°C
θ_1 = initial temperature
θ_2 = final temperature

From Table 7.4, $\alpha_0 = 38 \times 10^{-4}$ Ω/Ω at 0°C/°C. Therefore:

$$\frac{5}{R_2} = \frac{1+(38 \times 10^{-4} \times 10)}{1+(38 \times 10^{-4} \times 25)}$$

$$\frac{5}{R_2} = \frac{1.038}{1.095}$$

$$R_2 = \frac{5 \times 1.095}{1.038}$$

$$\underline{\underline{R_2 = 5.275\ \Omega}}$$

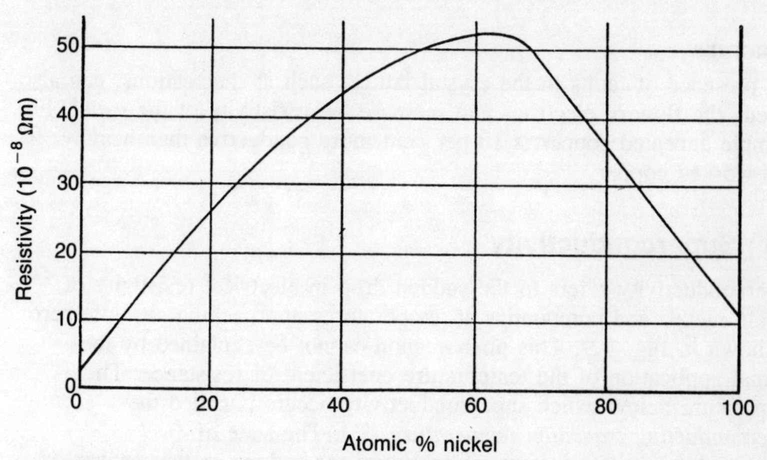

Fig. 7.7 The resistivity of copper-nickel alloys. Alloys have lower conductivities than pure metals

Impurities

Even small amounts of impurities can cause large increases in resistivity. Hence the high purity of 'high conductivity copper'. For example, the presence of traces of phosphorus, silicon, arsenic or even the metal ion can substantially lower the conductivity of copper. It only requires 0.05 per cent phosphorus to be present to reduce the conductivity of copper by some 40 per cent. Figure 7.8 shows the effect of impurities on the electrical conductivity of metals.

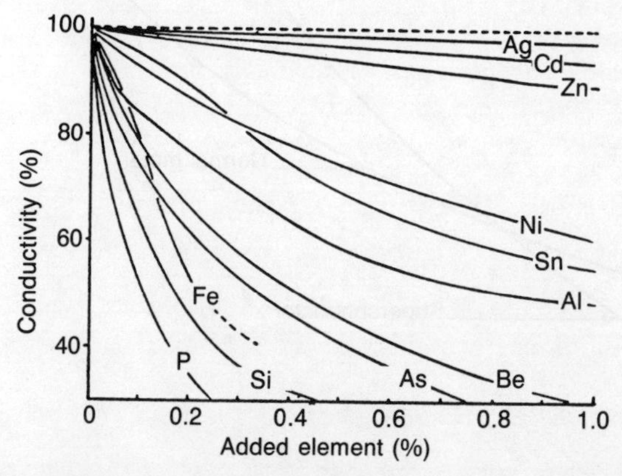

Fig. 7.8 The effects of impurities on the electrical conductivity of copper. Silver and cadmium have little effect and can be used to strengthen overhead conductors (e.g. telephone lines), whilst arsenic and phosphorus must not be present in copper destined for electrical purposes

Structure

The presence of faults in the crystal lattice, such as dislocations, can also impede the flow of electrons and increase the resistivity of the metal. For example annealed copper is 10 per cent more conductive than heavily cold-worked copper.

7.5 Superconductivity

Superconductivity refers to the sudden drop in electrical resistivity of certain metals and compounds at temperatures approaching absolute zero as shown in Fig. 7.9. This phenomenon cannot be explained by the normal application of the temperature coefficient of resistance. The temperature below which superconductivity occurs is called the *superconducting transition temperature* (T_c). The state of superconductivity for a material continues for as long as the material is maintained below its transition temperature. The state of superconductivity is destroyed by strong magnetic fields, either applied from an external source or caused by allowing an excessively heavy current to flow through the conductor.

Some materials which are relatively poor conductors at room temperature become superconducting at very low temperatures. Some

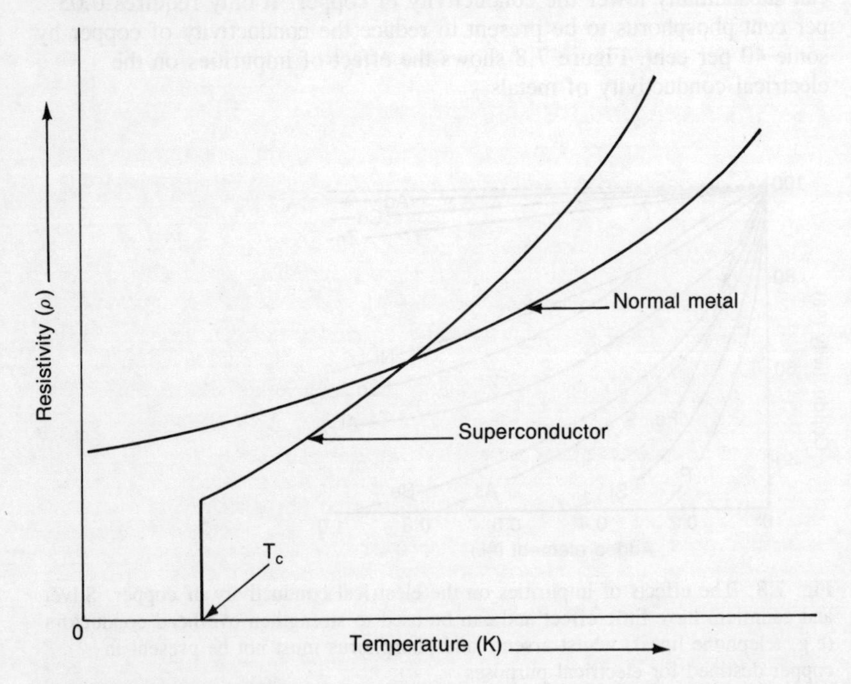

Fig. 7.9 Superconductivity

Table 7.5 Typical superconductors

Material	Transiton temperature (T_c) K
Aluminium	1.19
Lead	7.18
Niobium	9.46
Tantalum	4.48
Technetium	11.2
Vanadium	5.30
Nb_3 Sn	18.1
V_3 Ga	16.8
Mo – Re alloy	10.0

superconducting compounds are listed in Table 7.5 together with their superconducting transition temperatures. The intermetallic compound Nb_3Sn is of particular interest since its superconducting transition temperature of 18 K can be relatively easily achieved using liquid hydrogen or cold helium gas. Current research is directed towards developing materials which become superconducting at higher and more easily achieved temperatures so that they can be used on a commercial basis.

7.6 Insulating materials (dielectrics)

Devices for isolating electric currents and charges are called *insulators* and they are made from materials called *dielectrics*. Dielectric materials must be capable of separating electrical conductors without conducting an electrical charge between them.

An insulating material, or dielectric, is a material in which the valency shell is completely filled and there is a very wide energy gap between its valency band and the next energy band (conduction band). Thus a very large input of energy is required in order for an electron to cross the gap. A very small number of electrons may acquire sufficient energy to cross the gap if the material is subjected to a very high potential difference. Therefore, there is no such thing as a perfect insulator (non-conductor) only materials with very high values of resistivity which, for all practical purposes can be used for insulating purposes.

Dielectric materials may be gases, liquids or solids. With the exception of air, which is the dielectric between the bare conductors of the overhead electrical grid system, gaseous dielectrics are not used commercially. Liquid dielectrics are used mainly as impregnants for high-voltage paper-insulated cables and capacitors, and as filling and cooling media for transformers and circuit-breakers. Such dielectrics may be

Table 7.6 Properties of typical electrical insulators

Material	Resistivity Ωm (20°C)	Dielectric strength V/mm*
Ceramics		
Soda-lime glass	10^{13}	10 000
Pyrex glass	10^{14}	14 000
Vitreous silica	10^{17}	10 000
Mica	10^{11}	40 000
Steatite porcelain	10^{13}	12 000
Mullite porcelain	10^{11}	12 000
Polymerics		
Natural rubber	—	16 000 – 24 000
Phenol formaldehyde	10^{10}	12 000
Polybutadiene	—	16 000 – 24 000
Polyethylene	10^{13} to 10^{16}	20 000
Polystyrene	10^{16}	20 000
Polyvinyl chloride (PVC)	10^{10}	12 000

* Not constant with thickness, (see text)

petroleum oils or, alternatively, silicon oils and fluorinated hydrocarbons where the operating temperature is sufficiently high to cause the oxidation of petroleum oils. Some solid dielectric materials are listed in Table 7.6 together with typical values for their resistivity and dielectric strengths.

Dielectric strength

Dielectric strength is the maximum intensity of electric field which can be placed across an insulating material of unit thickness without breakdown occurring. It should be noted that the breakdown voltage does not increase in direct relationship to an increase in dielectric thickness. This variation is due to the presence of imperfections in the dielectric which may allow local leakage currents to flow resulting in premature failure. In fact it is common practice to use several thin layers of dielectric material in capacitors, rather than a single thick layer, as it is improbable that all the imperfections would coincide. Moisture, contamination, elevated service temperatures, ageing and mechanical stress, all tend to decrease the dielectric strength of insulating materials.

Relative dielectric constant

This is also called the relative permitivity of an insulating material. It is a measure of the displacement or charging effect of a dielectric, and is expressed as a ratio of the capacitance of a capacitor containing the dielectric material to the capacitance of the same capacitor using a vacuum as the dielectric.

$$\epsilon_r = C/C_0$$

where: ϵ_r is the relative dielectric constant;
C is the capacitance of a capacitor using the dielectric;
C_0 is the capacitance of the same capacitor with a vacuum between the plates instead of the dielectric.

Table 7.7 lists some dielectric materials and their relative dielectric constants. In many cases the relative dielectric constant changes with frequency.

Table 7.7 Relative dielectric constant (ϵ_r)

Material	Relative dielectric constant (ϵ_r)
Air	1.000 59
Barium titanate*	6000
Glass	6
Insulating oil	3
Mica	6
Paper	2.5
Polythene	2.3
Vitreous silica	3.5

* The dielectric film in miniature electrolytic capacitors

Note (1) The above values are relative to the absolute dielectric constant ϵ or ϵ_0, for a vacuum, which is unity.
(2) Because of the similarity of their ϵ_r values, C_0 is often taken (for practical purposes) as the capacitance using dry air as a dielectric.

Temperature

It has been shown that, in the case of metallic conductors, the resistivity of the conductor material increases when its temperature increases. It has also been shown that in the case of graphitic carbon the resistivity of the material falls when its temperature rises. This latter phenomenon is common to all non-metals, and the resistivity of insulating materials becomes less as the temperature rises. For example glass is a good insulating material at room temperature yet becomes a conductor when it becomes red-hot. Thus metals become poorer conductors as their temperatures rise and non-metals become poorer insulators as their temperatures rise. Further, as has already been mentioned, the dielectric strength of an insulating material becomes less as the temperature rises and this restricts the operating temperature of electrical devices if an insulation breakdown is to be avoided. For this reason manufacturers of such components as motors, transformers and capacitors usually state the maximum safe operating temperature.

7.7 Semiconductors (intrinsic)

Like the diamond allotrope of carbon, silicon and germanium crytallise into structures in which each atom is covalently bonded to four similar atoms. This is shown in Figure 7.10 for a silicon atom. By sharing their valency atoms in this way each outermost or valency shell of the atom has, effectively, eight electrons and is filled. Both silicon and germanium differ from diamond in that the energy gaps between their valency bands and their conduction bands are much smaller and it is easier for an electron to move between the bands.

When thermal agitation causes an electron to leave its position in a lattice there is a resulting deficiency in the lattice referred to as a 'hole'. A valency electron from an adjacent atom can then move into the hole

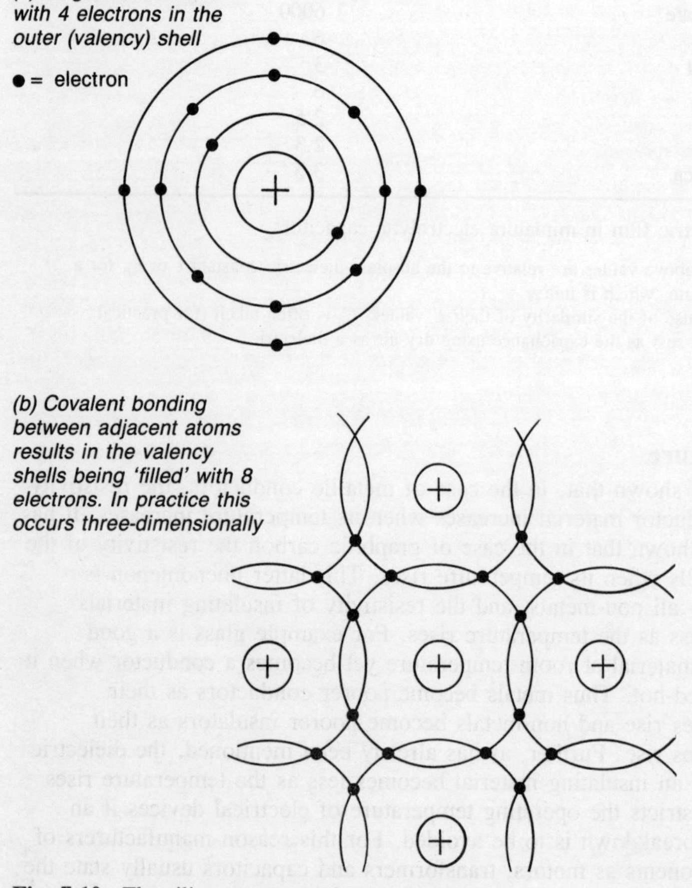

Fig. 7.10 The silicon atom and its covalent bond

leaving yet another hole. This process is repeated in a random manner throughout the material. However, if a potential difference is applied across the material, there is a migration of electrons in one direction and a migration of 'holes' in the opposite direction. Thus, since the migration of the 'holes' is opposite to the migration of electrons, the migration of holes can be considered as the migration of positive charges. In semiconductor parlance, electrons, being negatively charged, are referred to as *n-type charge carriers* and 'holes', being considered to behave as positive charges, are referred to as *p-type charge carriers*. Semiconductor materials which are free from impurities are referred to as being *intrinsic*.

Thermal conduction

In intrinsic semiconductor materials the energy required to free electrons from the valency bond and raise them through the energy gap into the conduction band can be thermal energy, and the electrical conduction resulting from temperature rise is referred to as thermal conduction. Thus, for any intrinsic semiconductor material, the conductivity rises as the temperature rises.

Photo conduction

Light as well as heat energy can raise electrons from the valency band of intrinsic silicon to the conduction band as shown in Fig. 7.11(*a*). For example, a photon of red light has an energy of 1.9 eV which is more than sufficient to cause an electron to migrate across the 1.1 eV energy gap of silicon. Since all materials are more stable when they reduce their energies, the electrons drop back from the conduction band to refill their holes in the valence band when the external light source is removed. This is referred to as *recombination*. For many applications intrinsic semiconductors have to be shielded from external light by ecapsulation.

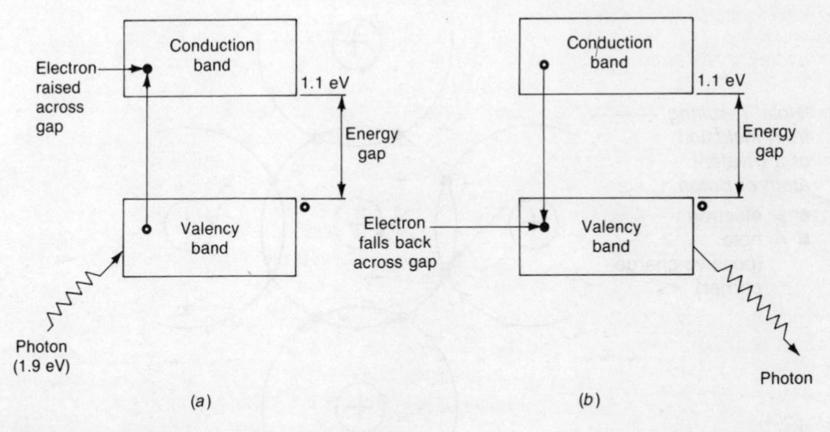

Fig. 7.11 Photoconduction luminescence

Luminescence

When recombination takes place as described above, the energy absorbed initially from the light source has to be given up either as heat energy or as luminescence as shown in Fig. 7.11(*b*).

7.8 Semiconductors (extrinsic)

Unlike intrinsic semiconductor materials which are of very high purity, *extrinsic* semiconductor materials are deliberately 'doped' with impurities during manufacture to give them either *n*-type or *p*-type characteristics. Since silicon and germanium are Group IV elements (that is they have four electrons in their valency shells) they are doped with either Group III or Group V elements which have either three electrons or five electrons respectively in their valency shells.

Figure 7.12(*a*) shows the effect of adding a Group V element such as

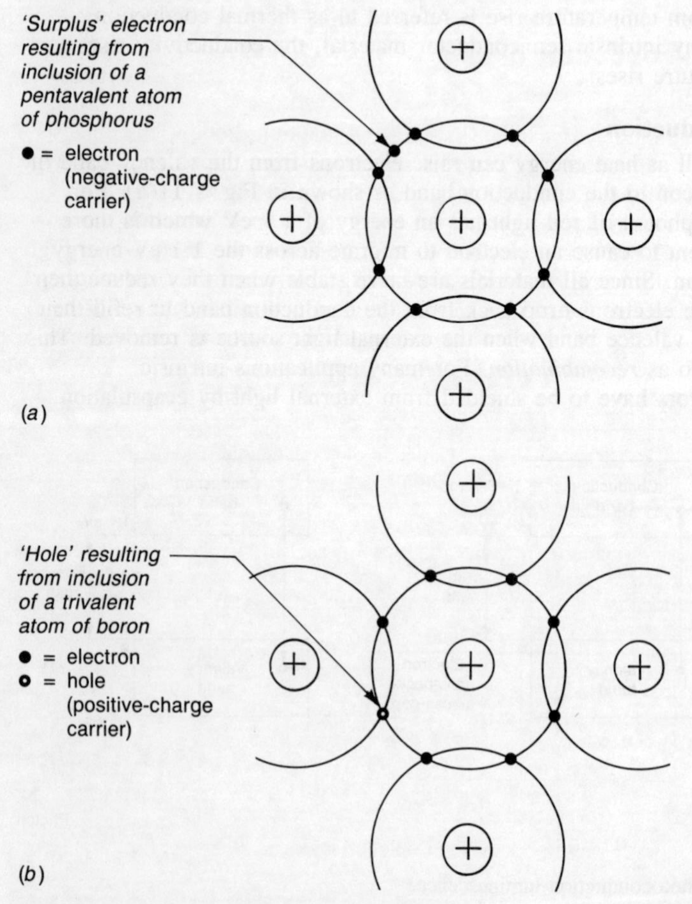

Fig. 7.12 Extrinsic semiconductor material

phosphorus to a semiconductor material such as silicon. It can be seen that the phosphorus atom has a spare electron bonded into its valency shell. Very little energy is required for this electron to break free and rise to the conduction band, and the thermal energy available at room temperature is more than sufficient. An extrinsic semiconductor material in which there are free negatively charged electrons is said to be an n-type material. The element which provides the free electrons is called a *donor*.

Figure 7.12(*b*) shows the effect of adding a group III element such as boron to a semiconductor material such as silicon. It can be seen that as the boron atom has only three electrons in its valence shell it will leave a 'hole' in the bond pattern. As previously explained, 'holes' in semiconductor technology are considered to be positive charge carriers. Therefore an extrinsic semiconductor material in which there are positive charge carriers is said to be a *p-type* material. The element which provides the positive charge carriers ('holes') is called an *acceptor*.

7.9 Semiconductor diodes

Junction diodes

The most commonly used diode is the p–n junction diode. This consists of a piece of semiconductor material which has been doped to give p-type characteristics for half its thickness and n-type characteristics for the other half of its thickness. Where the two regions meet is called the *junction*, hence the name 'junction diode'. Such a diode is shown in Fig. 7.13. Where the positive and negative charge carriers face each other across the junction, the charge carriers tend to neutralise each other by recombination. P-type, acceptor carriers, capture electrons and become

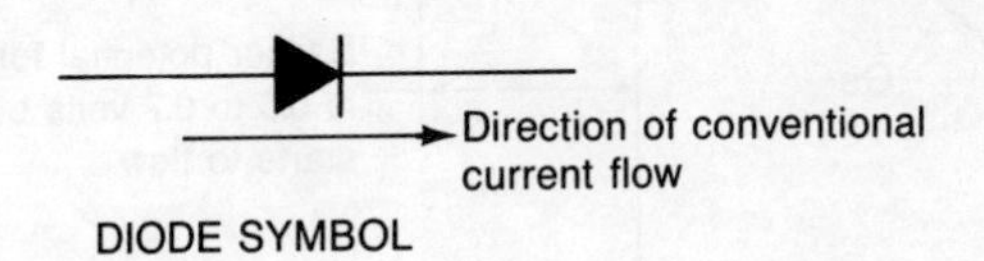

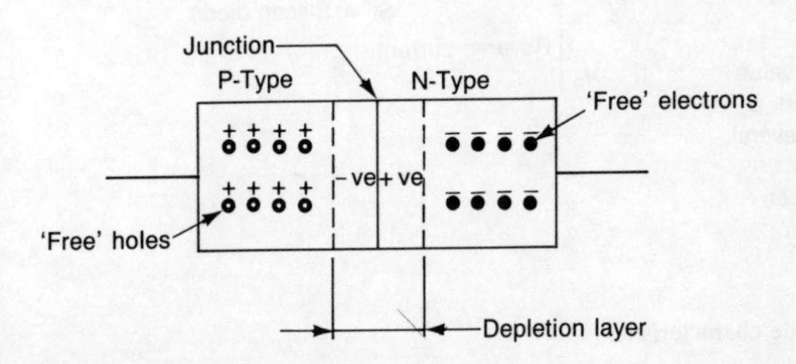

Fig. 7.13 Junction diode

negatively charged ions, whilst n-type, donor carriers, lose free electrons and become positively charged ions. Thus the *depletion layer* develops a negative charge on the 'p' side and a positive charge on the 'n' side of its interface. This results in an electrical potential, called the *barrier potential* appearing across the depletion layer, which has to be overcome before the diode will conduct.

The characteristics for silicon and germanium junction diodes are shown in Fig. 7.14 and their respective barrier potentials can be clearly seen. Although the silicon (Si) diode has a higher barrier potential, its steeper V_F/I_F curve and its ability to withstand a greater peak inverse voltage (p.i.v.) makes it superior to germanium for power handling applications. Further, silicon can sustain higher operating temperatures without destruction than germanium. On the other hand, the lower barrier potential of the germanium (Ge) diode (about 200 mV) makes it more responsive and suitable for small signal radio frequency applications.

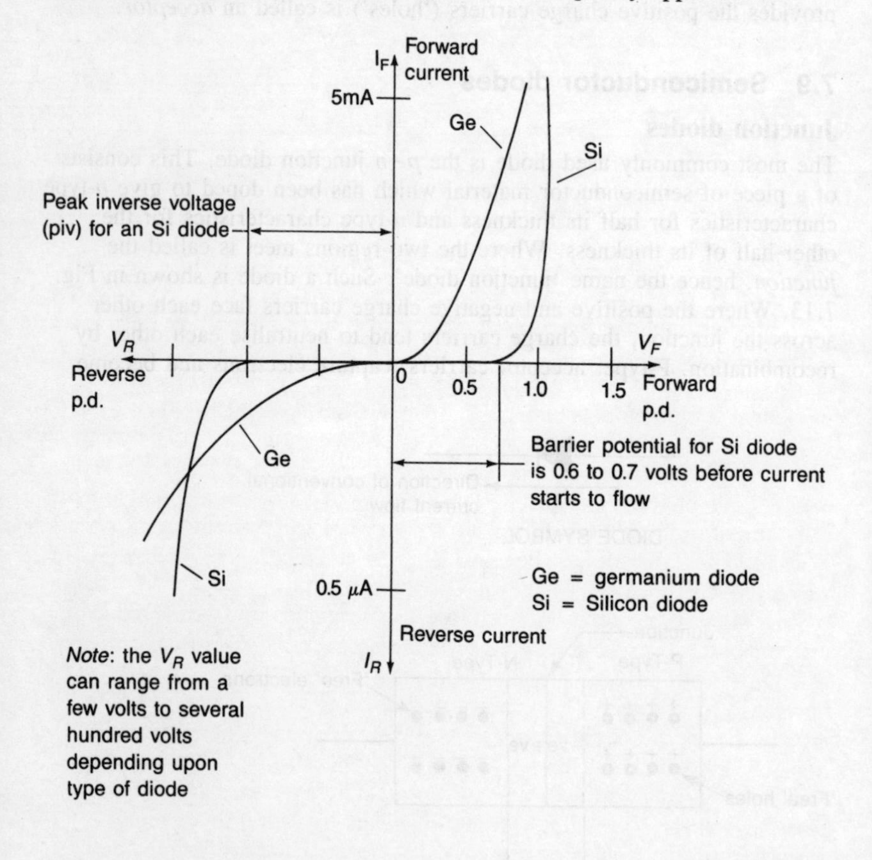

Fig. 7.14 Diode characteristics

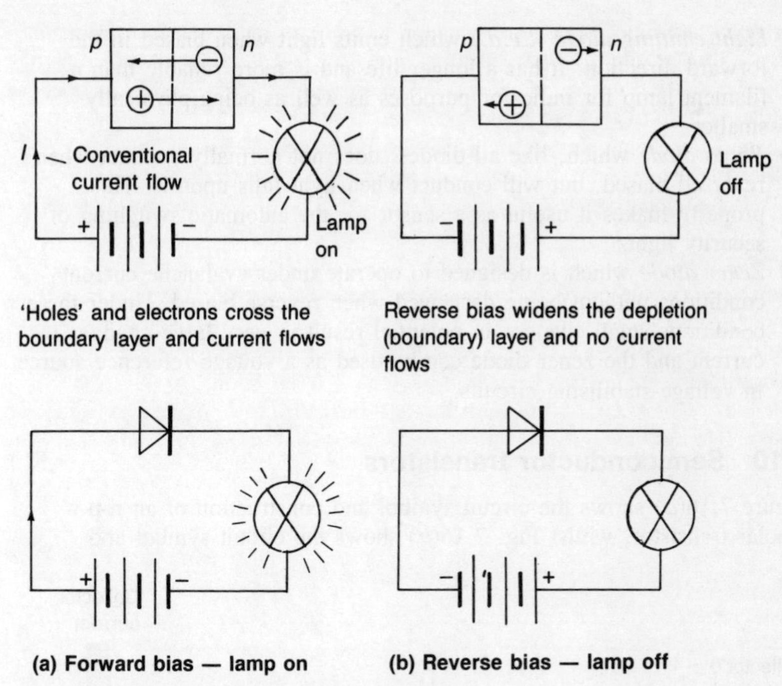

Fig. 7.15 Diode — forward and reverse bias

Figure 7.15 shows two ways of connecting such a diode. In Fig. 7.15(*a*) the lamp will light. This is because the diode has not only been *biased in the forward direction*, but because this bias is sufficient to overcome the 'barrier potential' of the depletion layer. This causes the charge carriers to cross the junction and an electric current to flow. In Fig. 7.15(*b*) the lamp will not light because *reverse bias* has been applied to the diode and this has resulted in the depletion layer widening so as to provide an insulating gap in the circuit preventing a current from flowing. Thus a diode can be considered as an electronic switch which will allow current to flow only in one direction.

Since there is no such thing as a perfect insulator, and semiconductor materials are no exception, a very small current of a few micro-amperes will flow in the reverse direction when the diode is reverse biased. Reference back to Fig. 7.14 shows that if the peak inverse voltage reaches a certain critical value the very small reverse leakage current suddenly increases and there is a heavy flow of current. This is called *avalanche current* and it can be sufficient to destroy the diode. It is caused when the 'insulating' properties of the depletion layer breaks down because its 'dielectric strength' has been exceeded. Some other types of junction diodes are as follows.

(*a*) *Light emitting diode (l.e.d.)* which emits light when biased in the forward direction. It has a longer life and is more reliable than a filament lamp for indicator purposes as well as being physically smaller.
(*b*) *Photo diode* which, like all diodes, does not normally conduct when reversed biased, but will conduct when light falls upon it. This property makes it useful as a sensor for the automatic switching of security lights.
(*c*) *Zener diode* which is designed to operate under avalanche current conditions without being destroyed when reverse biased. Under these conditions small changes in potential result in very large changes in current and the zener diode can be used as a voltage reference source in voltage-stabilising circuits.

7.10 Semiconductor transistors

Figure 7.16(*a*) shows the circuit symbol and construction of an n-p-n bipolar transistor, whilst Fig. 7.16(*b*) shows the circuit symbol and

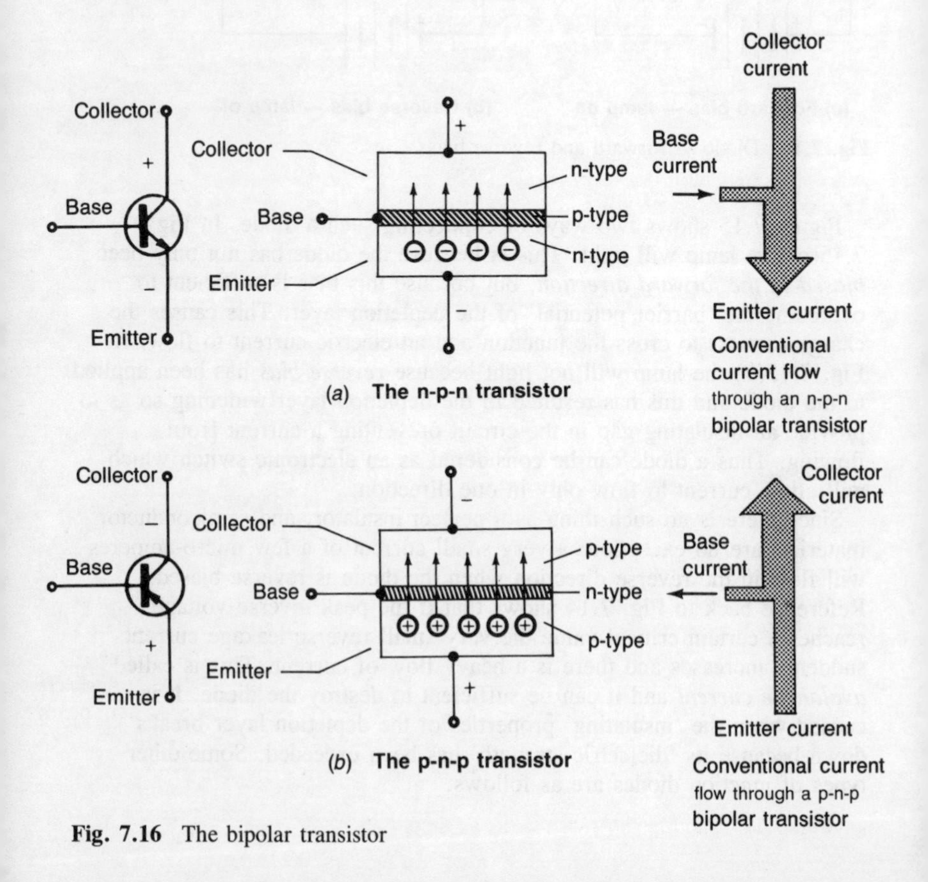

Fig. 7.16 The bipolar transistor

construction of a p-n-p transistor. In both cases the transistor consists of layers of n-type and p-type semiconductor material. Wires are connected to each layer and these layers are called the *emitter, base*, and *collector*. The 'emitter' emits (sends) charge carriers through the thin 'base' layer to be collected by the 'collector' layer.

In an n-p-n type transistor the emitter sends electrons through the base to the collector, whilst in a p-n-p type transistor the emitter sends positively charged 'holes' through the base to the collector. In both cases the arrowhead on the emitter symbol shows the direction of conventional current flow.

One of the uses of a transistor is as a switching device (electronic relay) with a small current in the base circuit controlling a larger current in the emitter-collector circuit. Figure 7.17(*a*) shows such a simple

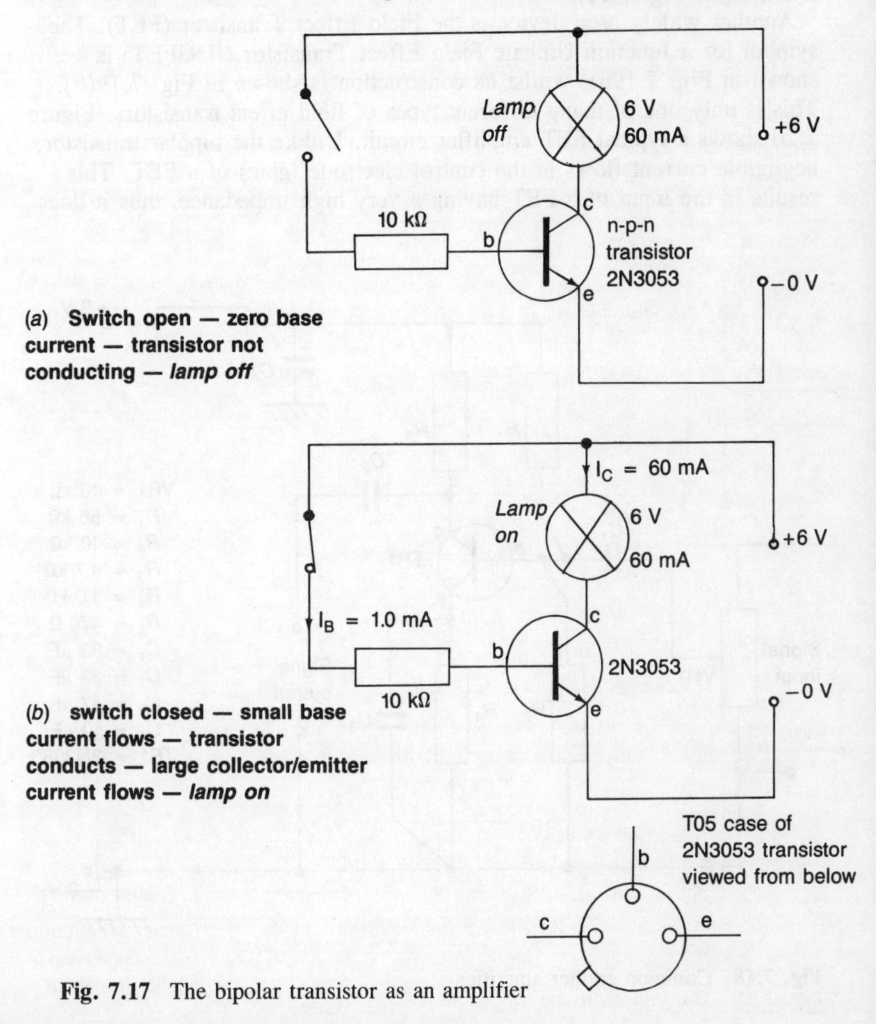

Fig. 7.17 The bipolar transistor as an amplifier

switching circuit. When the switch is closed, the base of the transistor is connected to the 6 V supply via the 10 kΩ current limiting resistor, the emitter-collector circuit conducts and the lamp lights. When the switch is opened, and the base current ceases to flow, the lamp goes out. Since in this circuit a base current (I_B) of 1 mA is controlling an emitter-collector current (I_C) of 60 mA, the current gain (amplification) is 60.

$$\text{Current amplification } (\beta) = I_C/I_B$$

The transistor can also be used for amplifying the fluctuating or alternating base currents found in audio frequency and radio frequency amplifiers and also in oscillator circuits. A typical common emitter, small-signal audio-frequency amplifier using a bipolar junction transistor is shown in Fig. 7.18.

Another widely used device is the Field Effect Transistor (FET). The symbol for a Junction Unigate Field Effect Transistor (JUGFET) is shown in Fig. 7.19(*a*) whilst its construction is shown in Fig. 7.19(*b*). This is only one of many different types of field effect transistors. Figure 7.20 shows a typical FET amplifier circuit. Unlike the bipolar transistor, negligible current flows in the control electrode (gate) of a FET. This results in the input of a FET having a very high impedance, thus it does

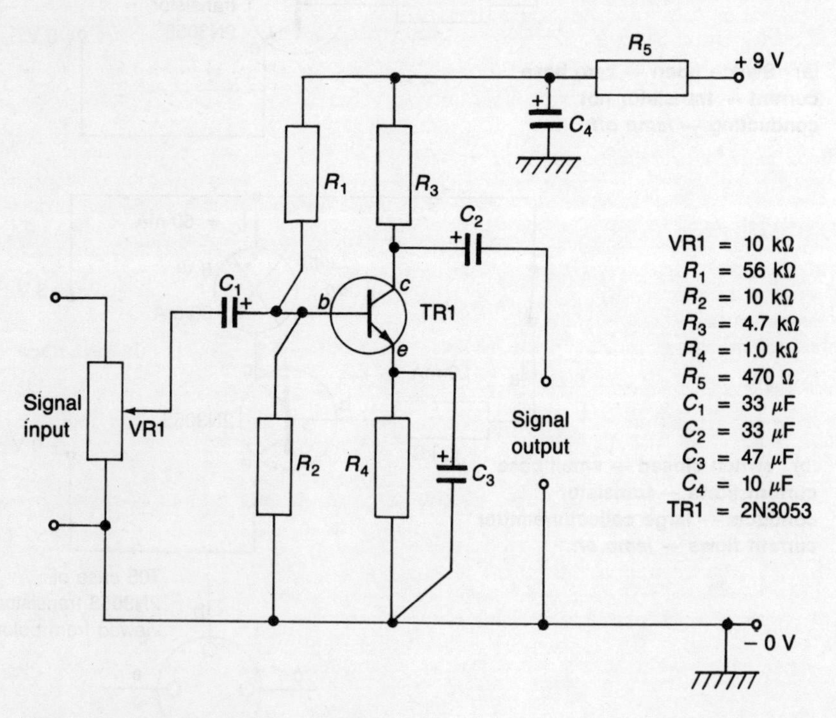

Fig. 7.18 Common emitter amplifier

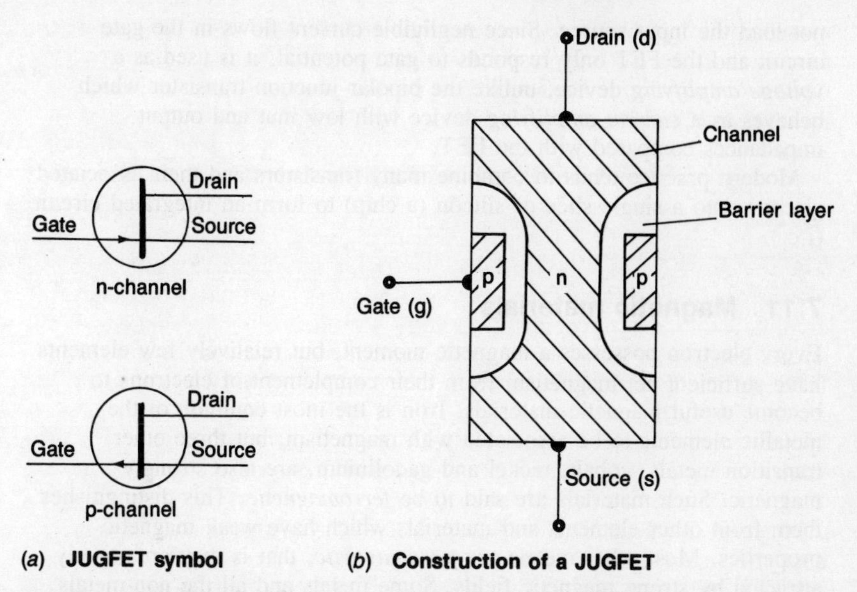

(*a*) **JUGFET symbol** (*b*) **Construction of a JUGFET**

Note: there are many other types of field effect transistors (FET) for special applications.

Fig. 7.19 Field effect transistor

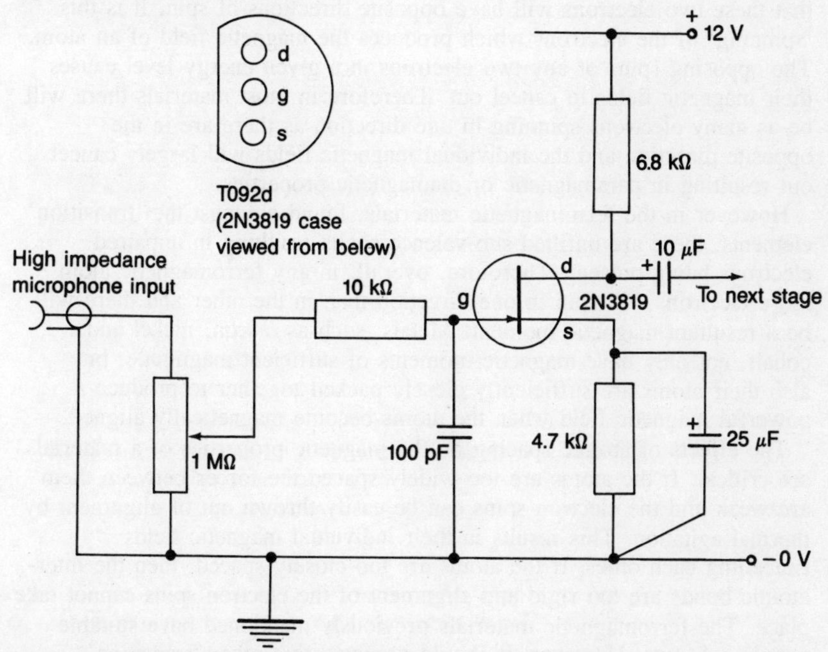

Fig. 7.20 FET as a small signal amplifier

not load the input source. Since negligible current flows in the gate circuit and the FET only responds to gate potential, it is used as a *voltage amplifying* device, unlike the bipolar junction transistor which behaves as a *current amplifying* device with low inut and output impedances compared with the FET.

Modern practice tends to combine many transistors and their associated circuits onto a single slice of silicon (a chip) to form an integrated circuit (i.c.).

7.11 Magnetic materials

Every electron possesses a magnetic moment, but relatively few elements have sufficient net magnetism from their complement of electrons to become useful magnetic materials. Iron is the most common of the metallic elements to be associated with magnetism, but three other transition metals, cobalt, nickel and gadolinium, are also strongly magnetic. Such materials are said to be *ferromagnetic*. This distinguishes them from other elements and materials which have weak magnetic properties. Most other metals are *paramagnetic*, that is they are weakly attracted by strong magnetic fields. Some metals and all the non-metals are *diamagnetic*, that is they are repelled by strong magnetic fields.

Reference to Pauli's exclusion principle has shown that in a stable atom not more than two electrons can occupy the same energy level and that these two electrons will have opposite directions of spin. It is this 'spinning' of the electrons which produces the magnetic field of an atom. The opposing spins of any two electrons in a given energy level causes their magnetic fields to cancel out. Therefore in most materials there will be as many electrons spinning in one direction as there are in the opposite direction and the individual magnetic fields will largely cancel out resulting in paramagnetic or diamagnetic properties.

However in the ferromagnetic materials, found amongst the 'transition' elements, there are unfilled sub-valence shells resulting in unpaired electrons being present. Therefore, overall, in any ferromagnetic atom more electrons will spin in one direction than in the other and there will be a resultant magnetic moment. Metals, such as α-iron, nickel and cobalt, not only have magnetic moments of sufficient magnitude, but also their atoms are sufficiently closely packed together to produce a powerful magnetic field when the atoms become magnetically aligned.

The effects of atomic spacing on the magnetic properties of a material are critical. If the atoms are too widely spaced the forces between them are weak and the electron spins can be easily thrown out of alignment by thermal agitation. This results in their individual magnetic fields cancelling each other. If the atoms are too closely spaced, then the inter-atomic bonds are too rigid and alignment of the electron spins cannot take place. The ferromagnetic materials previously mentioned have suitable atomic spacings. However, it should be noted that other 'transition' elements such as chromium, manganese and titanium have atomic spacings only just outside the ideal. In fact if manganese contains interstitial nitrogen atoms, the resulting modification to its atomic spacing results in it becoming ferromagnetic.

Within the crystals of ferromagnetic materials are small regions referred to as magnetic *domains*. These are regions in which groups of atoms' magnetic fields are aligned in the same direction. When the material is not magnetised, the magnetic fields in the domains are arranged in a random manner so that there is no resultant field (Fig. 7.21(*a*)). However, when the material becomes magnetised the fields of the domains become orientated in the same direction and the material has a resultant magnetic field (Fig. 7.21(*b*)).

Some ceramic materials can also show magnetic properties. For example, the mineral magnetite (Fe_3O_4), formerly known as 'lodestone', is an example of a naturally occurring magnetic ceramic material. A commercial ceramic magnet material is 'Ferroxdur' ($BaFe_{12}O_{19}$). Such materials are said to be *ferrimagnetic*.

Magnetic materials are classified as *hard* magnetic materials or *soft* magnetic materials. Hard magnetic materials are used for 'permanent' magnets since they retain their magnetism when the magnetising field is removed. Such materials are also physically hard and can only be machined by grinding (for example quench-hardened high carbon steel). The magnetism of permanent magnets can be destroyed by subjecting them to:

(*a*) an alternating magnetic field;
(*b*) temperatures above the *Curie* point;
(*c*) mechanical vibration (hammering).

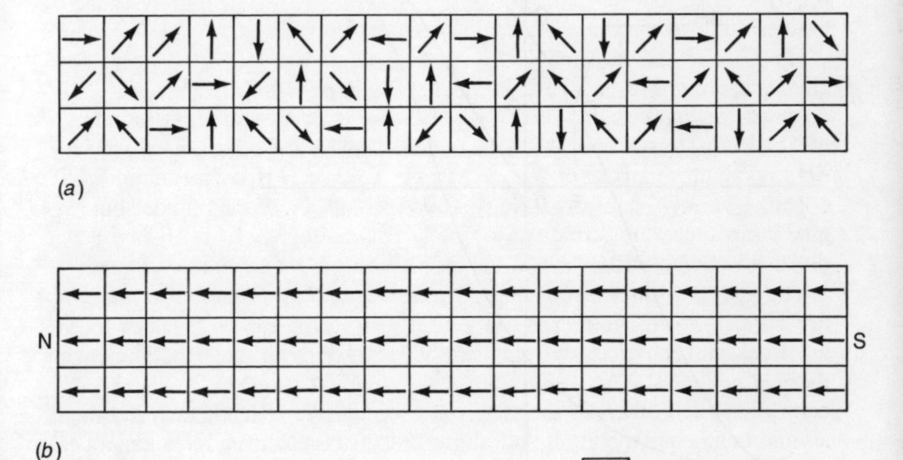

Fig. 7.21 Alignment of domains in a bar magnet

All the above provide sufficient agitation to allow randomisation of the domain alignment.

Soft magnetic materials retain their magnetism only as long as they are energised by an external magnetic field. Once the magnetising field is removed, soft magnetic materials immediately lose most of their magnetism. Such materials are used for such applications as transformer cores, motor rotor and stator laminations and electromagnet cores.

7.12 Magnetic properties (hard magnetic materials)

Table 7.8 lists some typical 'hard' magnetic materials together with their magnetic properties. The most important characteristics of a permanent magnet are:

(*a*) *Coercive force*. This is the resistance of the material to demagnetisation by electromagnetic techniques.
(*b*) *Remanence*. This the intensity of the residual magnetism after the magnetising field has been removed.
(*c*) *Energy product value*. This is an index of energy required to demagnetise and to reverse the polarity of a permanent magnet and thus a measure of the amount of magnetic energy stored in a magnet after the magnetising field is removed.

Figure 7.22 shows a typical *magnetic hysteresis loop* for a hard

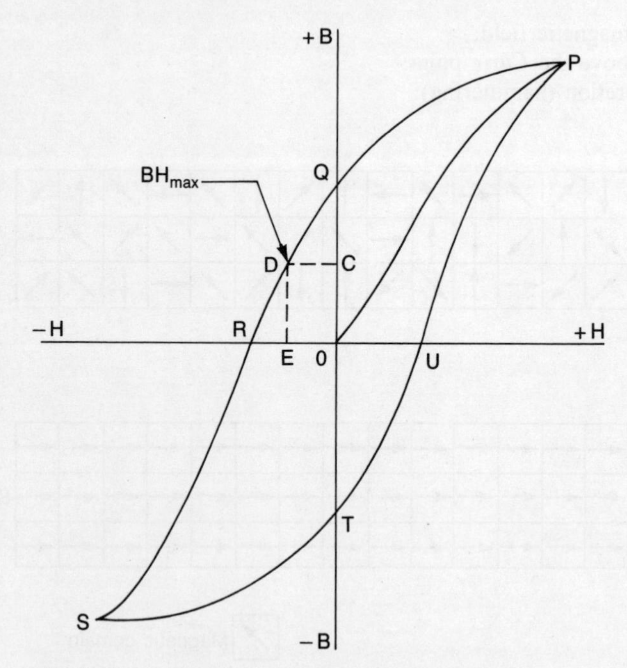

Fig. 7.22 Hysteresis loop for a magnetic material

Table 7.8 Permanent magnet materials

Name	Composition (%)										Magnetic properties		
	C	Cr	W	Co	Al	Ni	Cu	Nb	Ti	Fe	B_{rem} (T)	H_c (A/m)	BH_{max} J/m^2
Quench hardened high-carbon steel	1.0	—	—	—	—	—	—	—	—	Rem	0.9	4 400	1 560
35% cobalt steel	0.9	6.0	5.0	35.0	—	—	—	—	—	Rem	0.9	20 000	7 800
Alnico				12.0	9.5	17.0	5.0	—	—	Rem	0.73	44 500	13 500
Alcomax III*				24.5	8.0	13.5	3.0	0.6	—	Rem	1.26	51 700	38 000
Hycomax III*				34.0	7.0	15.0	4.0	—	5.0	Rem	0.88	115 400	35 200
Columax**				24.5	8.0	13.5	3.0	0.6	—	Rem	1.35	58 800	52 800
Ferroxdur ($BaFe_{12}O_{19}$)	—	—	—	—	—	—	—	—	—	—	0.4	150 000	20 000

* Anisotropic alloys whose magnetic properties are measured along the preferred axis
** This alloy derives its very high BH value from the way it is cooled during casting which orientates its columnar crystals parallel to the preferred axis of magnetisation

C = carbon Cr = chromium W = tungsten Co = cobalt Al = aluminium Ni = nickel
Cu = copper Nb = Niobium Ti = titanium Fe = iron

magnetic material. The induced magnetic flux density B (teslas) is plotted against the magnetising field H (ampere-turns per metre). The starting point O on the curve indicates zero magnetic field strength. The magnetising field strength H is gradually increased, and corresponding values of B and H are plotted, until magnetic saturation occurs at point P and the curve 'levels off'. The magnetising field strength is then reduced to zero and the corresponding values of B and H are once more plotted to give the curve PQ. Thus the residual magnetism is represented by OQ, that is the *remanence* (B_{rem}).

To determine the coercive force the material must be demagnetised by reversing the magnetising field and gradually increasing its field strength from zero at point Q on the curve until point R is reached. Thus OR represents the field strength (force) to demagnetise the material completely and it is called the *coercive force* (H_c).

The strength of the reverse magnetising field is increased until 'negative saturation' is reached at point S, whereupon the magnetising field is again reversed and the curve SUP is plotted to complete the loop. The hysteresis loop represents the amount by which the induced magnetic flux lags behind the magnetising field. For a permanent (hard) magnetic material the loop must be as large as possible. The ultimate requirement of a magnetic material is that the product of B and H (the energy product value BH_{max}) must be as large as possible as this is the maximum energy that the magnet can provide external to itself. The BH_{max} value occurs when the product of CD and DE on the demagnetisation curve is at a maximum.

The selection of a permanent magnetic material depends upon whether a high value of remanence is required or whether a high value of coercive force is required. For example, 'Columax' has a high remanence whilst Hycomax III and the ceramic 'Ferroxdur' have high coercive forces.

Some permanent magnetic materials are said to be *anisotropic*, that is, their properties along a preferred axis are much enhanced. To achieve this effect, the magnetic alloy material is raised to a high temperature and then cooled in a powerful magnetic field. The thermal agitation decreases as the temperature falls resulting in groups of atoms becoming aligned along the direction of the external field, so that this alignment becomes 'frozen in' by the time the magnet material has reached room temperature.

7.13 Magnetic properties (soft magnetic materials)

A major requirement of soft magnetic materials is that they should have a high *permeability* (μ). This is the ratio of the flux density (magnetic induction) B, to the total magnetic field H.

$$\mu = B/H$$

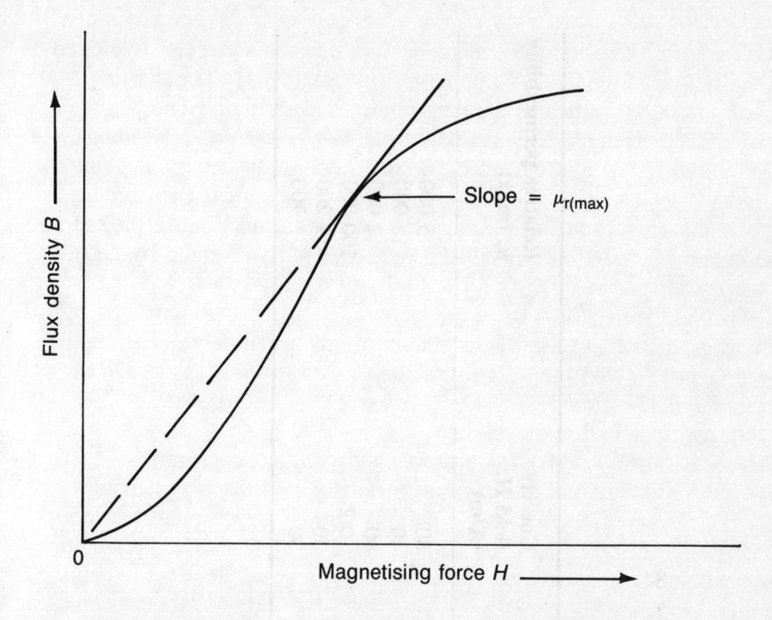

Fig. 7.23 Magnetising curve for 'soft' magnetic materials

The permeability value, for a given material, is not constant but alters with the magnetising field as shown in Fig. 7.23.

Since soft magnetic materials often operate under the influence of magnetic fields generated by alternating currents they are magnetised, demagnetised, then remagnetised and demagnetised with reverse polarity, this cycle being repeated many times each second. Such an alternating cycle produces a closed hysteresis loop and the area of the loop represents the energy wasted in overcoming the remanence for the material. This wasted energy causes heating of the magnetic material. Therefore soft magnetic materials are formulated to have very narrow hysteresis loops and Fig. 7.24 compares typical hysteresis loops for hard and soft magnetic materials. Table 7.9 lists a range of soft magnetic materials together with their properties.

Permeability is not the only property to be considered when selecting a soft magnetic material, otherwise the ceramic magnetic materials (Ferroxcube A and B) would not be used. However at high frequencies (over 1 MHz) a compromise between permeability and electrical conductivity is required. Metallic materials are unsuitable at such high frequencies because of their high conductivity which results in excessive eddy-current heating. Eddy-current heating not only represents wasted energy, but very high temperatures are attained. In fact, it is possible to melt metals using high-frequency eddy current heating.

Table 7.9 'Soft' magnetic materials

Material	Saturation induction B_s (T)	Coercive field H_c (A/m)	Relative permeability μ_r (max)
α-iron	2.2	80	5 000
Silicon-ferrite transformer sheet	2.0	40	15 000
Permalloy Ni-Fe	1.6	10	2 000
Superpermalloy Ni-Fe-Mo	0.2	0.2	100 000
Ferrox cube A (Mn Zn) Fe_2O_4	0.4	30	1 200
Ferrox cube B (Ni Zn) Fe_2O_4	0.3	30	700

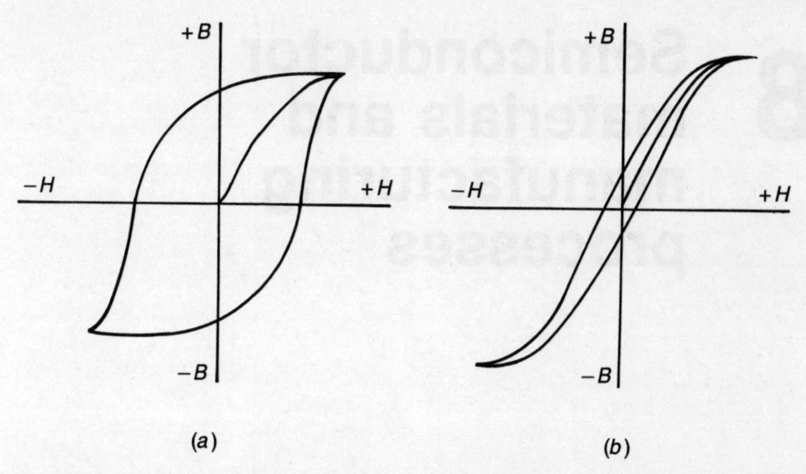

Fig. 7.24 Hysteresis loops for 'hard' and 'soft' magnetic materials

The silicon-ferrite sheet is used for transformer core lamination and motor and generator rotor and stator stampings. This is a relatively low cost material and therefore suitable for such large-scale production. The more expensive permalloy alloys are used for more specialised applications such as screening cans and the cores for audio-frequency coupling and matching transformers in telecommunications equipment.

8 Semiconductor materials and manufacturing processes

8.1 Silicon as a semiconductor material

Solid-state semiconductor devices were introduced in Chapter 7. Such devices are manufactured from high-purity silicon or germanium monocrystals which are free from crystallographic defects such as dislocations and which have been given the requisite electrical characteristics, that is n-type or p-type characteristics and the required resistivity.

Since silicon is by far the most widely used of the semiconductor materials, this chapter will be restricted to its manufacture and use. *Silicon* must not be confused with the *silicones* which are inorganic silicon-oxygen structures whose many uses include the production of silicone polymers which are useful heat-resistant, flexible, insulating materials.

Pure elemental silicon is a silvery-grey material with a density of 2300 kg/m^3, a melting point of 1680 K, and a boiling point of 2628 K (760 mm Hg). Pure silicon has a resistivity of the order of 2.3×10^3 ohm metre (at 20°C), and thus lies part way between the values expected for conductors and for insulators. However, the resistivity of commercially hyperpure (*intrinsic*) silicon rarely exceeds 2×10^2 ohm metre, and it will have residual n-type or p-type characteristics. The conductivity of silicon not only increases as its temperature increases, as explained in Section 7.7, but the conductivity is also dependent upon its purity.

Most silicon is supplied to manufacturers of semiconductor devices as an *extrinsic* semiconductor material. Traces of trivalent or pentavalent impurities (dopants) having been added during manufacture to give the material specific p-type or n-type characteristics and levels of resistivity to suit customer requirements. The presence of the dopants increases the conductivity of the material significantly depending upon the amount and type of dopant present. The level of dopants present is only of the order of parts per million (p.p.m.) and close quality control is essential.

8.2 Purification of silicon

Silicon is the second most abundant element occuring in nature where it is nearly always in association with oxygen (silicon dioxide) and other elements to form mineral silicates as in quartz, sand and clays. It is from these raw materials that commercially pure silicon is nearly always produced by chemical decomposition (reduction).

The high-purity silicon used for the manufacture of semiconductor devices is nearly always produced by the decomposition or reduction of a volatile silicon compound such as *trichlorosilane* ($SiHCl_3$). This is produced by combining hydrogen chloride with silicon in a fluid-bed reactor at approximately 300°C.

$$Si + 3HCl \rightarrow SiHCl_3 + H_2$$

A simplified diagram of the plant is shown in Fig. 8.1. In practice the plant is much more complicated. For example, the unused gases from the vapour deposition equipment are regenerated and returned to the trichlorosilane plant for re-use. Trichlorosilane has the advantage of a low boiling point (31.8°C) which allows it to be purified by fractional distillation.

Trichlorosilane is the most widely used gaseous compound for the manufacture of hyperpure silicon because of its low price, high volume availability, ease of handling, low toxicity, low flammability, and

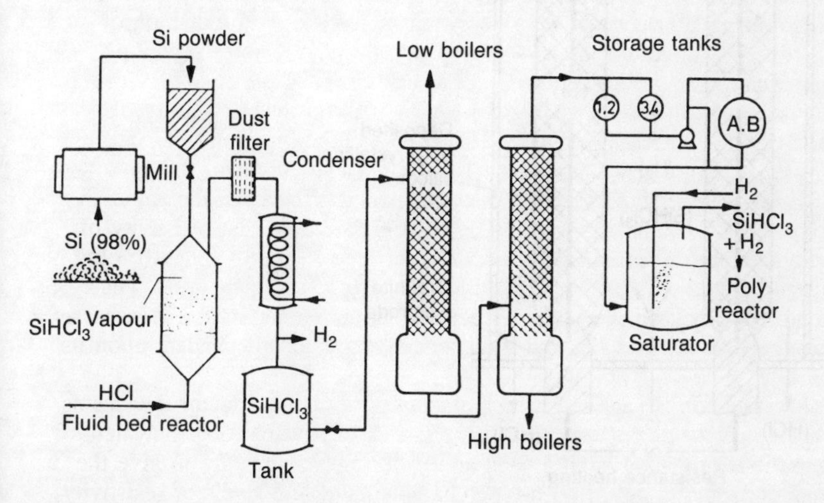

Fig. 8.1 Preparation and refining of trichlorosilane: flowchart

commercially viable convertibility into silicon by a high-temperature reduction process.

Polycrystalline silicon, of high purity, is produced by *chemical vapour deposition (CVD)*. In this process trichlorosilane is reacted with hydrogen gas in the presence of a thin, high-purity silicon rod as shown, diagrammatically, in Fig. 8.2. The rod is heated by the passage of an electric current and silicon is deposited coherently upon it in polycrystalline form. Hydrogen chloride gas is generated as a by-product.

Provided that the rod is at the correct temperature and the ratio of trichlorosilane to hydrogen lies within critical limits, the reaction proceeds quite readily. The deposition rate and the evenness of the deposit depend upon a number of parameters including the reaction vessel size, the temperature of the 'thin rod', and the ratio of the reaction gases.

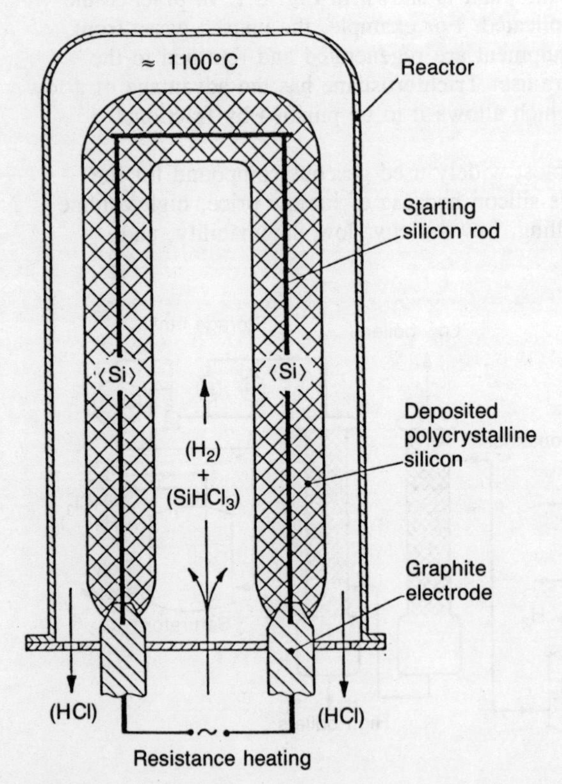

Fig. 8.2 Production of high-purity polycrystalline silicon by chemical vapour deposition

By careful control, polycrystalline rods having a good coherent structure and high density can be formed.

To ensure silicon of the highest purity, care must be taken in the choice of raw materials and in the design and operation of the plant. The following precautions must be taken.

(*a*) Only a limited range of materials can be used for the reaction vessel and for those parts of the plant coming into contact with the reaction gases to prevent contamination.
(*b*) The trichlorosilane must be purified by careful distillation so that the level of impurities (particularly phosphorus and boron — which are electrically active) are reduced to a fraction of one part per million (p.p.m.).
(*c*) The hydrogen gas must be free from traces of oxygen, dried to a dew point of $-100\,^{\circ}\mathrm{C}$ and dust free. Hydrocarbons and ammonia must be rigorously excluded.

The point in the process at which the dopants are added depends upon the quantity and type of dopant, and the manufacturing processes used. In practice, the electrical characteristics required by the customer may be given to the polycrystalline silicon by adding the dopants under very close control during the refinement and/or the polycrystallisation process, or it may be added to the crucible prior to crystal pulling. The amount of intrinsic silicon produced is very small compared with the amount of extrinsic p-type or n-type silicon produced.

Before the high purity polycrystalline silicon can be used in the manufacture of electrical semiconductor devices, it has to be converted into monocrystalline rods of sufficient size that they can be sliced into wafers. Such monocrystalline rods must have a dislocation-free atomic structure. The two processes most widely used for growing single crystals of silicon are the *Czochralski (CZ)* process and the *float-zone (FZ)* process. These will now be considered in detail.
considered in detail.

8.3 Czochralski (CZ) process

Worldwide, 80 per cent of all solid state devices are manufactured from monocrystalline silicon produced by the Czochralski process. Crystals grown by this process are particularly suitable for low-power devices such as large-scale integrated (LSI) circuits and very-large-scale integrated (VLSI) circuits which are used in logic and control electronics.

The principle of the Czochralski process is shown in Fig. 8.3. Polycrystalline silicon, produced as described in Section 8.2, is broken into small pieces and then cleaned. It is loaded into a crucible under clean room conditions and the dopant is added, except where volatile dopants are used and these are added directly to the melt. The most common dopants are boron, phosphorus, antimony and arsenic.

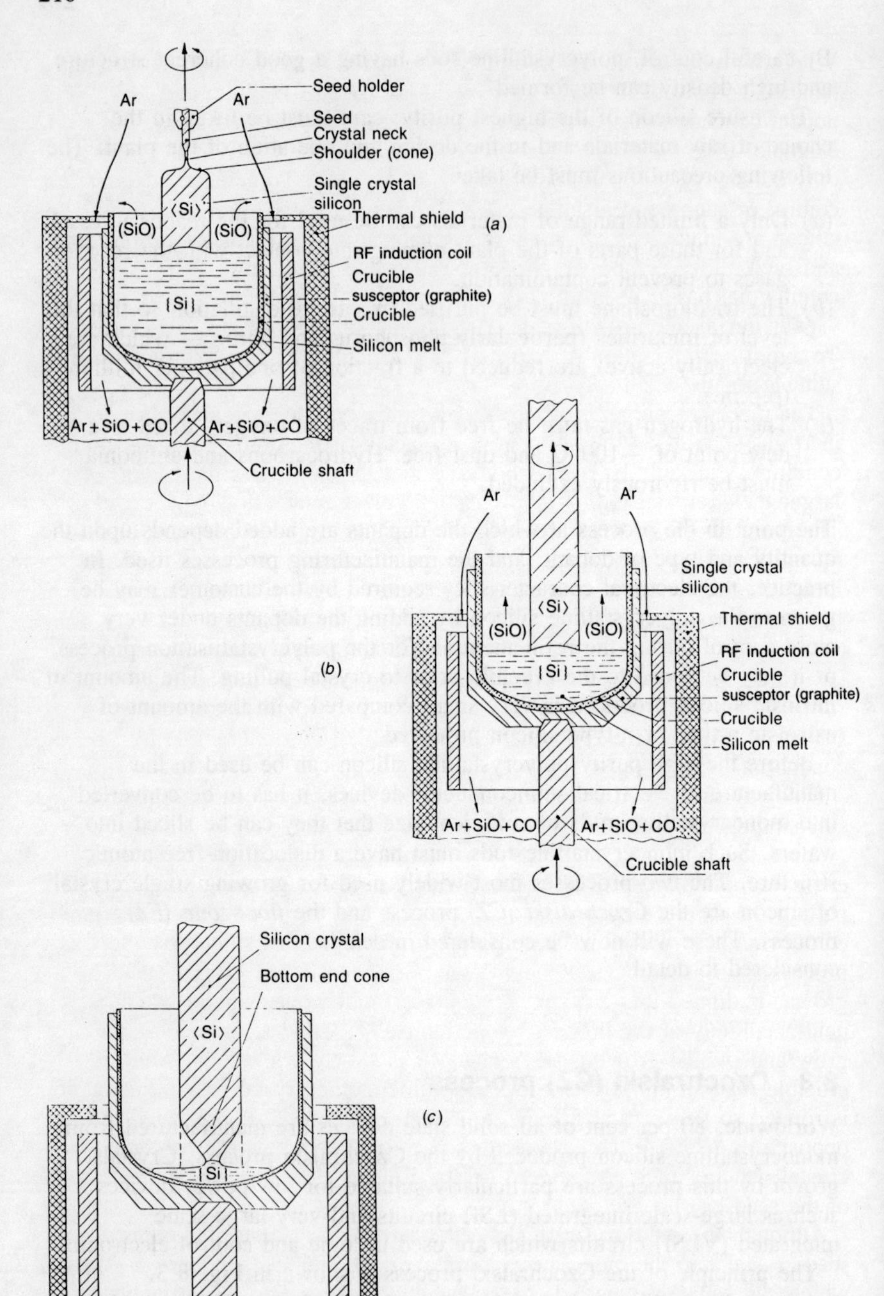

Fig. 8.3 Czochralski single crystal pulling: (*a*) initial stage of process, (*b*) advanced stage of process, (*c*) final stage of process

Where high dopant concentrations are required these can be weighed and handled without difficulty. However in the case of medium and low dopant concentrations this is not possible on a commercial basis. For example, to grow p-type silicon crystals with a resistivity of 0.1 ohm m, only 0.1 mg of boron is require for a 10 kg melt. To achieve adequate control of such concentrations on a commercial basis, dilute alloys of silicon are made which contain the dopant in a higher concentration range and such alloys can be weighed out and added to the melt on a commercial basis. Such alloys are prepared by adding the dopant as a vapour during the growth of polycrystalline silicon by the CVD process. To avoid contamination of the melt only crucibles of pure quartz are suitable for the CZ process.

The pieces of polycrystalline silicon in the crucible are melted under an inert gas such as argon or under high vacuum conditions. High vacuum conditions can only be used where the melt is small (less than 1 kg); for larger melts crystal pulling can only take place under flowing inert gas. There are two main pressure ranges, either atmospheric pressure or 5 to 50 millibar. The inert gas has to flow permanently downwards through the pulling chamber, as shown in Fig. 8.3, to carry off the reaction products which are evaporated in considerable amounts.

To give close temperature control, modern furnaces use radio frequency (RF) induction heating. The crucible is surrounded by a graphite *susceptor* in which eddy currents are induced causing heating of the susceptor. This heat energy is transmitted to the quartz crucible by conduction.

Older furnaces used a cylindrical graphite resistance heating element surrounding the crucible and heating it by radiation.

When the silicon in the crucible is completely molten, the electrical energy supplied to the RF induction coil is reduced so that during crystal pulling the whole furnace including crucible and melt are kept in thermal equilibrium. To produce a high-purity single crystal of silicon by the CZ process a tiny seed crystal, having the perfection of the crystal to be grown, is dipped into a quartz crucible of molten pure silicon which is held accurately at the process temperature. The crystal commences to grow and, whilst growing takes place, it is slowly withdrawn from the crucible (*crystal-pulling*). At the same time as the crystal is withdrawn, it is rotated so that crystal growth is uniform. The crucible is rotated in the opposite direction. Under correctly controlled conditions the nascent crystal which grows on the seed crystal follows the structural perfection of the seed crystal.

Although the seed crystal is dislocation free, the action of dipping the seed crystal into the melt causes thermal shock and surface tension effects which can cause dislocations to appear which move into the growing crystal, particularly in crystals of large diameter. Because of the high strains and temperatures in the crystals, the dislocations are not confined just to their own glide planes, but can spread to adjacent glide planes by cross-slip, multiplication processes and by climb (see Chapter 10).

To ensure dislocation-free crystals the crystal pulling technique developed by W.C. Dash has to be used. This technique generates a 'neck' between the seed crystal and the crystal being pulled. This neck reduces the cooling strain to a very low value and, as a result, the remaining strain energy may not be sufficient to move the existing dislocations or generate new ones. In any case, necking will result in any residual dislocation motion being slower than the rate of crystal growth.

(*a*) The crystal diameter is gradually reduced to about 2 to 4 mm.
(*b*) The growth velocity is raised to a maximum of 6 mm/minute, depending upon orientation.

A (111)-orientation crystal does not require a very small crystal neck but it does need a high growth velocity. However, a (100)-orientation crystal requires both a thin neck and a low growth velocity. With a suitable combination of neck diameter and growth velocity the crystal becomes dislocation free after a few centimetres of growth.

The dislocation-free state of the grown crystal shows itself in the development of 'ridges' on the crystal surface. The transition region from the seed node to the cylindrical part of the crystal is referred to as the seed-cone and can vary between the almost flat to a very pronounced taper. Shortly before the desired crystal diameter is reached, the pulling velocity is raised to the specific velocity at which the crystal grows. Rotation of the seed crystal results in the monocrystalline silicon being almost circular in cross-section.

The pulling velocity is reduced towards the bottom end of the crystal. This is because the heat loss from the walls of the crucible increases as the level of the melt sinks; the heat transference required for crystallisation becomes more difficult, and more time is required to grow a given length of crystal. To complete a crystal free from dislocation an end cone has to be produced so that the crystal diameter is gradually reduced. For this reason the pulling speed is raised and the crystal diameter decreases and, if the diameter becomes small enough, the crystal can be separated from the melt without dislocations forming in the cylindrical part of the crystal from which the wafers are made. Care must be taken so that the increase in pulling velocity is not too great otherwise thermal shock will occur resulting in plastic deformation (slip) in the lower part of the crystal. Conventionally, the residual melt left in the crucible is discarded together with the crucible since there are many practical difficulties not only in using up the whole of the melt but also in emptying the crucible.

Since several hours of machine time is lost in cooling the furnace, replacing the crucible and charge, reheating the furnace, evacuating the furnace if a vacuum is used or purging and flooding the furnace if an inert atmosphere is used, solitary charging is costly and inefficient. Therefore current practice is moving towards techniques for recharging the hot crucible whilst pulling is in progress so that continuous or semi-continuous pulling can be achieved.

However there are severe limitations to the number of times the crucible may be recharged. The more important of these limitations are:

(*a*) a build up of impurities in the melt;
(*b*) contamination from the crucible itself;
(*c*) lack of mechanical strength in the crystal itself which results in handling problems and limits its physical size.

An important aspect of the CZ process is the possibility for *in situ* remelting of faulty crystals.

(*a*) The entire crystal can be remelted and growth started again.
(*b*) The crystal is only partially remelted until the defect is reached, whereupon the Dash technique is applied to create a neck and a new dislocation-free crystal is grown from the new neck and shoulder.

The possibility of remelting is an important economic advantage of the CZ crystal pulling process. It is not possible in either the float-zone or the pedestal-pulling processes. Table 8.1 lists typical data for modern pulling apparatus used in Czochralski silicon growth, whilst Table 8.2 lists the electrical and mechanical characteristics of wafers made from monocrystalline silicon produced by the CZ process.

Table 8.1 Typical data for modern pulling apparatuses used in Czochralski Si growth

Crucible:	diameter	from 180 to	350 mm
	height	from 160 to	280 mm
	capacity	from 6 to	50 kg Si
Crystals:	diameter	from 50 to	150 mm
	length	from 500 to	2200 mm
	weight	from 5 to	48 kg Si
Seed shaft travel		up to	2500 mm
Seed shaft rotation		from ≈0 to	50 RPM (reversible)
Seed shaft speed		from ≈0 to	10 mm/min (slow),
		from 20 to	800 mm/min (fast)
Crucible shaft travel		up to	500 mm
Crucible shaft rotation		from ≈0 to	20 RPM (reversible)
Crucible shaft speed		from ≈0 to	1 mm/min (slow),
		from ≈0 to	200 mm/min (fast)
Power control		up to 150 kW	
Overall height		up to 9 m	
Overall weight		up to 9500 kg	
Pressure range		10^{-5} to 2 bar	
Gas flow (argon)		0.4 to 3 m^3/h	
Vacuum pumps (mechanical)		up to 200 m^3/h	

Table 8.2 Czochralski silicon wafers — polished
Resistivity

Type, dopant	Range ohm-cm (target)	Minimum tolerance %	Maximum radial variation % (6 mm edge exclusion) (111)	(100)
p, Boron	0.005–25	±10	8	8
	>25–60	±20	12	12
n, Phosphorus	0.030–15	±15	20	12
	>15–40	±20	25	15
n, Antimony	<0.020 for diameter up to 125 mm		20	12
	<0.025 for diameters up to 125 mm		22	8
	>0.025–0.050		22	9

Geometry

Diameter mm		**76.2 ± 0.3**	**100 ± 0.3**	**125 ± 0.1**	**150 ± 0.1**
Thickness μm	Standard	381	525	625	675
	Minimum	290	350	400	500
Thickness tolerance μm	Minimum	±10	±10	±10	±10
Global flatness μm Front side reference (TIR)	Typical	< 2.0	< 2.0	< 2.5	< 3.0
	Maximum	3.0	4.0	5.0	5.5
Local Thickness Variation (LTV) μm Site $15 \cdot 15$ mm^2	Typical	—	< 1.0	< 1.0	< 1.0
	Maximum	—	2.0	2.0	2.0
Total Thickness Variation (TTV) μm (ASTM F 533)	Typical	< 3.0	< 5.0	< 5.0	< 5.0
	Maximum	7.0	8.5	8.5	10.0
Warp μm (ASTM F 657) (damage-free etched backside)	Typical	<10	<15	<20	<25
	Maximum	20	25	30	40

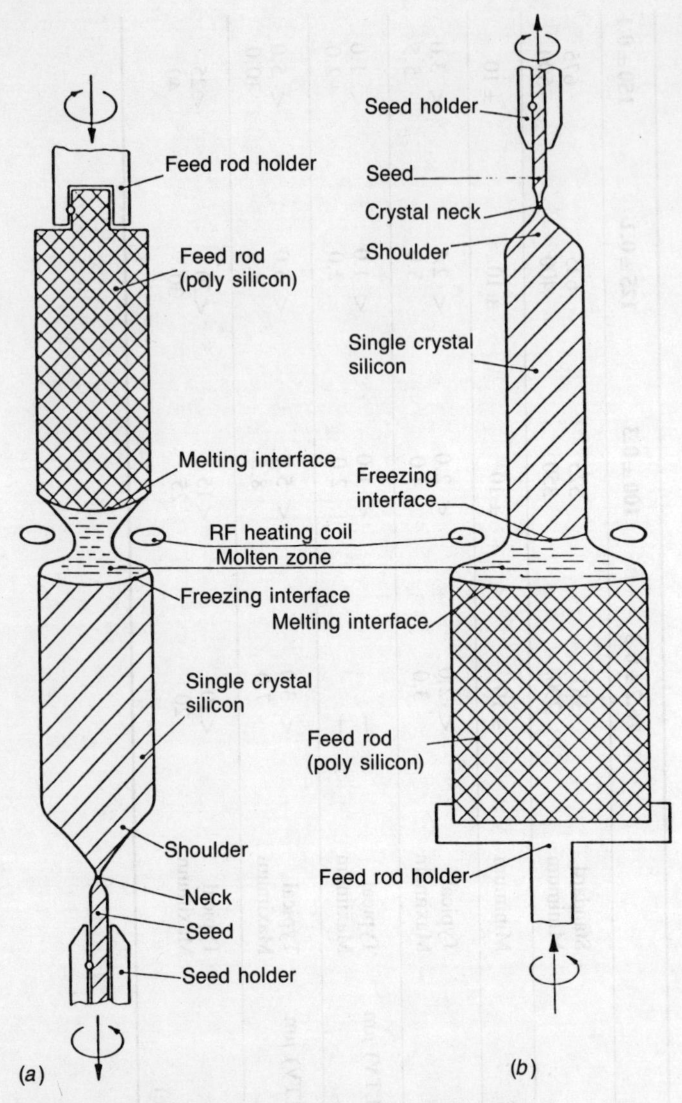

Fig. 8.4 Principles of float-zone and pedestal pulling

8.4 Float-zone (FZ) process

The principle of this process is shown in Fig. 8.4. A rod of polycrystalline silicon is mounted vertically and there is no containing vessel. A narrow zone of silicon is melted using radio-frequency induction heating, and the molten zone is made to traverse the rod vertically by relative movement of the rod and the induction coil. Surface tension and levitation forces from the electric field prevent the molten zone from falling out even in large-diameter rods of 100 mm to 150 mm diameter. The process may be carried out in a vacuum or in an inert atmosphere such as argon. For the highest purities a vacuum is preferable. If a seed crystal is introduced at one end of the polycrystalline rod and the molten zone is made to traverse the rod from this end, then a single crystal will be produced which reflects the structural perfection of the seed crystal. To ensure complete freedom from dislocation, the Dash technique of introducing a crystal 'neck' is also used in the float-zone process.

In the Czochralski (CZ) process some reaction occurs between the polycrystalline charge and the quartz crucible at the high melting temperatures involved. For the very highest purity and crystallographic integrity, the more expensive float-zone (FZ) process is used. In this process there is no contamination from the crucible since none is used. Further, refinement takes place as well as crystal growth, and any residual impurities are slowly swept along the rod from one end to the other and then discarded.

A variation on the float-zone process is *pedestal pulling* and this is also shown in Fig. 8.4. Again there is no crucible to cause contamination and crystals of the highest purity and crystallographic integrity can be produced by this process. The Dash technique of 'necking' is also used when pedestal pulling to ensure freedom from dislocation.

Monocrystalline silicon produced by the float-zone and pedestal pulling processes are used for high power and very high power devices, as well as for detector devices. Table 8.3 compares some of the properties of crystals produced by the CZ and the FZ processes.

8.5 Production of wafers

Modern silicon crystal production equipment is highly automated including automatic sizing. However, although variations in ingot size and

Table 8.3 Typical crystal data (special materials excluded)

Parameter		Crucible pulling (CZ)	Float-zone (FZ)
Crystal quality		Dislocation-free	
Max. diameter		150 mm	150 mm
Resistivity	p-type	0.005–50 ohm cm	0.1–3000 ohm cm
Range	n-type	0.005–50 ohm cm	0.1– 800 ohm cm
Dopants		B, P, Sb, As	B, P
Orientations		[111] [110] [100]	[111] [100] [511]
Lifetime		10–50 μsec	100–3000 μsec
Oxygen content		10^{16}–10^{18} cm^{-3}	below detection limit
Carbon content		10^{17} cm^{-3}	below detection limit

roundness are very small, the ingot is still not dimensionally accurate enough for modern device-processing equipment. Therefore the monocrystalline ingot is centreless ground to an accuracy of better than ±0.2 mm. Flats are then surface ground along the ingot for identification and location purposes as shown in Fig. 8.5.

Unfortunately, all grinding processes result in some structural damage to the crystal surface and, if this is not rectified slipping of the crystal lattice and stacking faults will develop during subsequent high-temperature treatment. The usual treatment is to chemically etch the ground surfaces to remove the damaged layer when all the mechanical cutting processes have been completed.

The next process is to cut the ingot into thin wafers ready for device-processing. Depending upon the application for which the wafers are required, the ingot is cut perpendicular to the crystal axis or with a well defined misorientation of several degrees. Off-axis wafers are usually used, after polishing, as the substrates for epitaxial processes used in the manufacture of discrete devices or bipolar integrated circuits. Cutting techniques are now becoming sufficiently sophisticated that, after cleaning and etching, the wafers can be used without further processing. However for the majority of applications, the cut wafers are cleaned and lapped to meet the thickness and flatness tolerances required by some discrete devices and by most modern mask printing techniques. The majority of wafer diameters used are in the 100 mm to 150 mm range with a typical thickness of 0.3 to 0.7 mm for the smaller diameters and 0.5 to 1.0 mm thickness for the larger diameters to provide adequate handling strength. Since the electrical performance of semiconductor materials is affected by changes in temperature, chips produced from silicon wafers have to be kept as thin as possible to allow adequate heat dissipation. For this reason, wafers for low voltage, heavy current diodes may only be 0.2 mm thick with the diameter restricted to 75 mm. The production processes for silicon wafers are summarised in Fig. 8.6. Note that unlike most circular saws, those used for cutting silicon are in the form of an annulus with the teeth on the inside diameter.

Wafer yield during device processing has been increased considerably by 'edge-rounding'. This offers the following advantages,

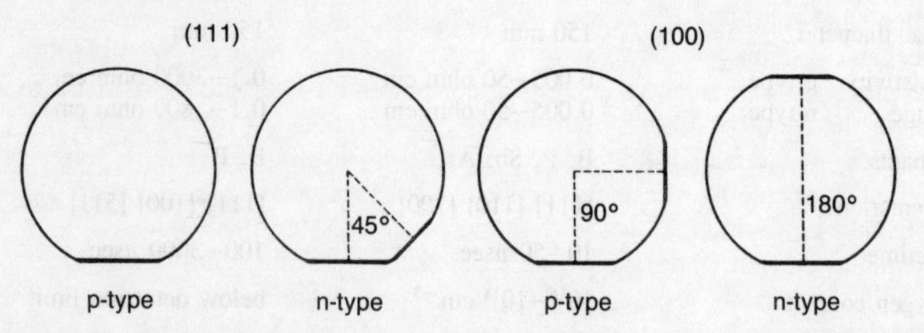

Fig. 8.5 Wafer characterisation by different flats

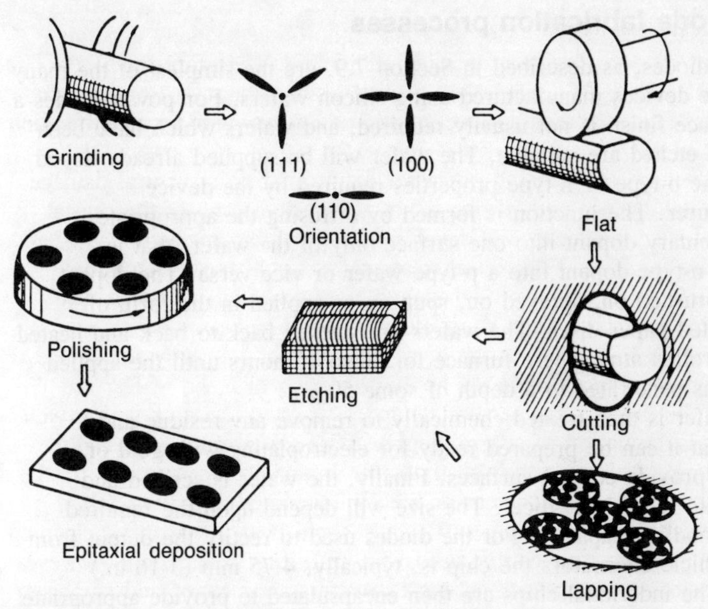

Fig. 8.6 Crystal machining and polishing: flowchart

(*a*) Rounded edges prevent the formation of an 'epi-crown' during epitaxial deposition processes. An 'epi-crown' is caused by the build-up of epitaxial layers on the sharp edges of untreated wafers at a greater rate than the build up of the epitaxial film on the flat surface of the wafer.

(*b*) Similarly, the photoresist film is smoother on an edge-rounded wafer and there is no increase in thickness adjacent to the wafer edge.

(*c*) Edge-rounding also improves the mechanical properties of the wafers with less tendency to chipping and cracking at the edges during handling. This results in less chance of process-induced crystallographic faults developing.

After edge-rounding the wafers are etched in order to remove any mechanical damage. Care has to be taken during etching not to disturb the plane-parallel surfaces of the wafer, otherwise it would be difficult to obtain the required flatness during subsequent polishing of the wafer surfaces. Modern photolithographic processing techniques require very high surface quality. To avoid damage of the wafer surface, abrasive polishing is impractical and a combined chemical and mechanical technique is used. After polishing a carefully controlled, multi-step cleaning process is used so that the wafers are completely free from particles and residues.

8.6 Diode fabrication processes

Junction diodes, as described in Section 7.9, are the simplest of the many solid state devices manufactured using silicon wafers. For power diodes a high surface finish is not usually required, and wafers which have been sawn and etched are suitable. The wafer will be supplied already doped to give the p-type or n-type properties required by the device manufacturer. The junction is formed by diffusing the appropriate complementary dopant into one surface only of the wafer, that is diffusing n-type dopant into a p-type wafer or vice versa. The dopant may be brushed on, sprayed on, spun on or applied in the form of impregnated paper discs. The wafers are stacked back to back and heated in a controlled atmosphere furnace for 30 to 40 hours until the applied dopant has penetrated to a depth of some 50 μm.

The wafer is then treated chemically to remove any residue and to ensure that it can be prepared ready for electroplating with gold or nickel to provide contact surfaces. Finally, the wafer is scribed and broken into 'chips' or 'dice'. The size will depend upon the required current handling capacity. For the diodes used to rectify the output from motor vehicle alternators the chip is, typically, 4.75 mm (3/16 in.) square. The individual chips are then encapsulated to provide appropriate protection and heat dissipation, and provided with wire 'tails', solder tags, or terminals for connection to the external circuit.

8.7 Planar fabrication

Devices with a *planar* configuration are manufactured so that the preparation of the various p-type and n-type layers and the metallised contacts are on one flat surface of the chip and not on its sides or ends. A section through a typical bipolar junction transistor is shown in Fig. 8.7. The substrate is made from a relatively thick wafer of n-type silicon which has been doped to give it a low resistivity for each conduction. The layers which are built up on this substrate are referred to as *epitaxial* layers. These are layers which grow onto the substrate surface as a continuation of the underlying crystal. The growth comes from the gaseous mixtures in which the wafer is heated to give p-type or n-type conducting layers or silicon dioxide insulating layers. The fabrication of planar devices requires wafers with one surface lapped and polished to a high degree of flatness and surface finish.

Figure 8.7 shows only a single transistor. However since the chip on which it is fabricated is only some 4 mm or 5 mm square, very many such devices can be made at the same time on wafers whose diameters lie between 100 mm diameter and 150 mm diameter. High-precision photographic processes are used to produce the masks used during the fabrication of the device. Only one such device is drawn out and the camera takes a succession of photographs of the device on the same negative, moving by an increment equal to one chip spacing between each

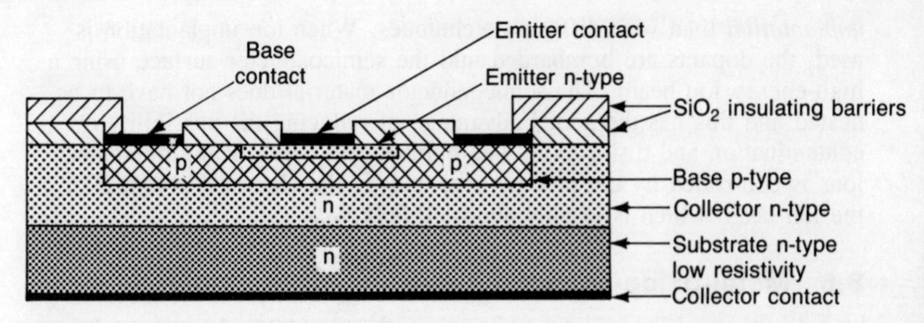

Fig. 8.7 Bipolar transistor manufactured using planar technology

exposure. Thus the negative is covered in a pattern of chip masks suitable for printing photographically onto the wafer which is coated with a photo-sensitive emulsion. The resolution required to reproduce the fine detail and intricacy of modern solid state devices precludes the use of white light, and ultraviolet light and laser light sources are used for making the exposures. For simplicity, the fabrication of only one device on one chip will be considered.

Since the photo-resist cannot withstand the processing temperature, silicon oxide is first grown on the epitaxial layer. Then the photo-resist is added and exposed, photographically, to obtain the pattern required. Next, the wafer is etched in hydrofluoric acid to remove those areas of silicon oxide which are not protected by the exposed photo-resist. The residual photo-resist is then washed away and the silicon oxide pattern acts as a mask for the infusion of the next doping process.

The overlying (epitaxial) layer of n-type silicon is grown on the substrate and it has a higher, but controlled, resistivity. Following a photographic masking procedure, boron is diffused from the surrounding gas into the epitaxial silicon. This changes the n-type material into p-type material in the zone unprotected by the photo-resist mask and forms the base of the transistor. A second masking and diffusion process using phosphorus establishes the n-type emitter layer. Final masking and processing allows the formation of silicon dioxide (SiO_2) insulating barriers. Metallisation and electroplating produces the contact surfaces on those surfaces not protected by the silicon dioxide.

The thickness of the diffused base layer is limited to only 0.5 μm in order to prevent any recombination. The thickness of the layers and the dopant concentrations are controlled by the diffusion time, the furnace temperature and by the furnace atmosphere gas composition. However, although the diffusion technique has been developed into a highly reproducible production process, it is difficult to achieve a well defined wall (abrupt profile) by conventional techniques. The need for an abrupt profile becomes increasingly important with the development of large scale integration (LSI) where very large numbers of components are built up on a single chip with component separation measured on a molecular scale. Such an abrupt profile can be more easily attained using *ion implantation* than with diffusion techniques. When ion implantation is used, the dopants are bombarded into the semiconductor surface using a high-energy ion beam. The semiconductor material does not have to be heated and this has the added advantage of reducing the possibility of contamination and dislocation. The number and depth of the implanted ions is controlled by the process time and by the electrical potential of the ion beam which is usually about 10 000 eV.

8.8 Metal-oxide-semiconductor (MOS) technology

The development of metal-oxide-semiconductor (MOS) devices established the superiority of monocrystalline silicon as a semiconductor material. One of the main reasons for this was the ease with which an ideally isolating silicon dioxide film can be grown on silicon wafers by simple heat treatment in an oxygen rich atmosphere. Silicon dioxide layers serve many useful functions in device manufacture. For example, such oxide layers are not easily penetrated by dopant elements and thus form useful diffusion masks. Further, such films form the gate oxide in MOS devices and determine the electrical characteristics of MOS transistors. Silicon dioxide films are also used during the processing and finishing of solid state devices to protect p-n junctions (passivation) and, in multi-level integrated circuits, such films serve as electrical insulators for the various metal or polysilicon layers.

The field effect transistor (FET) concept only became a reality with the perfection of MOS technology. MOS devices are essentially planar surface devices and the drain and source regions, even in large MOS integrated circuits, have a thickness equal to or less than 1 μm. The several insulation and interconnection layers are situated directly on the wafer surface. Figure 8.8 shows, schematically, a cross-section through a typical MOS-RAM device. The main high-temperature processes, which govern the performance of such a device, can be outlined as follows:

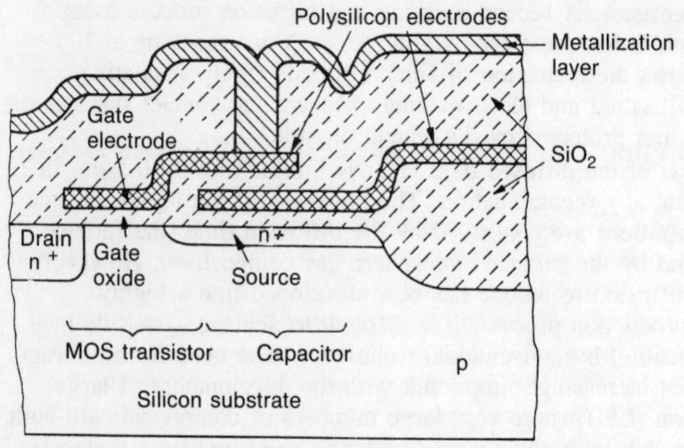

Fig. 8.8 Cross-section of a storage cell of a typical MOS-RAM

(*a*) gate oxidation;
(*b*) field oxidation;
(*c*) drain and source implantation;
(*d*) annealing of implantation damage and drive in diffusion;
(*e*) chemical vapour deposition of polysilicon, nitrides and (phosphorus) oxides.

The chemical vapour deposition (CVD) processes may take place at relatively low temperature, depending upon the system used and the nature of the layer to be deposited. Some annealing of ion-implantation damage may be coupled with subsequent processes. To obtain a high accuracy of alignment, modern devices are processed by the so-called self-alignment technique. This requires that the field and gate oxidaton have to take place during the first part of the process. To keep the gate oxide, which is the most sensitive oxide, free from impurities, it is covered by a nitride layer, and the field oxide is formed at temperatures below 1100°C. Since the drain and source are rather shallow, they are realized by brief and/or low temperature diffusion at relatively high push-pull rates, or by ion-implantation to obtain well-defined diffusion walls.

Another way of realising a densely packed device, e.g. a memory, starts with the deposition of a phosphorus-doped oxide which, after a masking step, serves as a diffusion source. Such an approach does not allow a subsequent long heat treatment at relatively high temperatures as needed, for example, for the field-oxide formation, since the diffused regions would be washed out. Here other ways have to be found to insulate the different diffused islands. One way is to pre-bias a polysilicon layer electrically.

The p-channel MOS technology, which was first used to manufacture MOS devices has largely lost its importance. However, medium-scale consumer devices are still produced by this technology. The gate oxides are thick and aluminium layers are used for the interconnections. Since the dimensions need not be controlled with a high precision, the drain and source diffusion can be deeper and the process temperatures higher, in order to obtain a higher throughput.

8.9 Complementary metal-oxide-semiconductor (CMOS) technology

Complementary metal-oxide-semiconductor (CMOS) technology combines p-channel and n-channel transistors on one chip. CMOS devices are more economical in power consumption and are superior in reliability to normal MOS devices and, therefore, are gaining increasing attention. However, integrating two field-effect transistors on one chip needs more process steps. A distinct feature is the so-called p-well into which the n-channel FET is built. In order to obtain the correct surface concentration of dopants, a long 'drive-in' has to be performed at relatively high temperatures.

8.10 Bipolar technology

Standard bipolar technologies include a process step not used in MOS technologies: the epitaxial deposition of a monocrystalline layer of silicon on the polished surface of the wafer. The electrically active parts of the devices are confined to this epilayer; thus stringent demands are made on its crystalline perfection.

Figure 8.9 shows a section through a basic element of a standard bipolar integrated circuit. The main process steps to realize such a device structure are:

(*a*) buried layer diffusion;
(*b*) deposition of epitaxial layer;
(*c*) isolation diffusion;
(*d*) collector diffusion;
(*e*) base diffusion;
(*f*) emitter diffusion.

A thermal oxidation plus a photo-lithography step is needed before every diffusion in order to establish the diffusion masks.

Recently, the isolation diffusion has been replaced by an oxidation step, and the silicon dioxide produced is used to isolate the single transistors dielectrically as shown in Fig. 8.10. Since silicon dioxide has a higher specific volume than silicon, part of the silicon in the isolation well has to be removed before oxidation. Oxide isolation techniques are also widely used in MOS and CMOS devices.

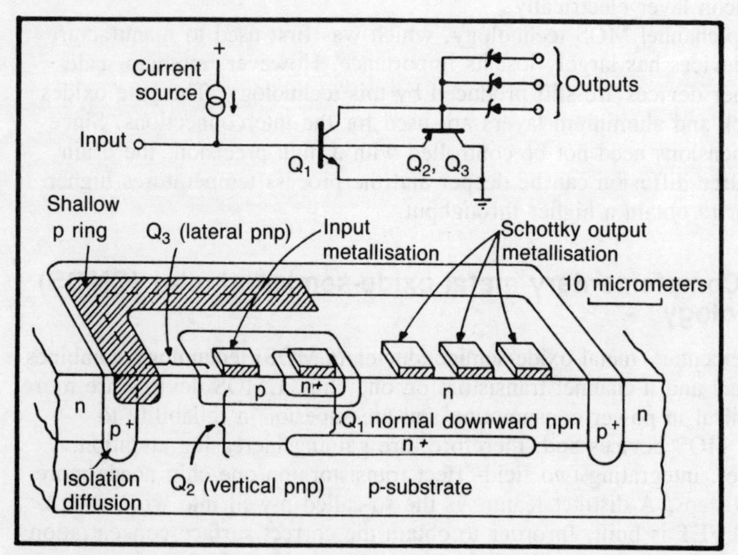

Fig. 8.9 Bipolar device with diffused isolation wells

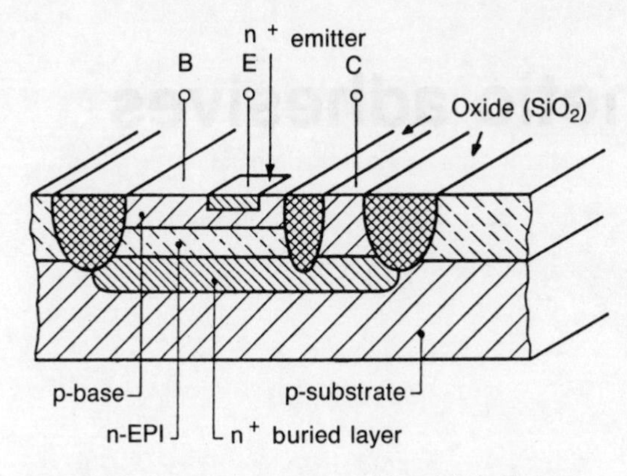

Fig. 8.10 Bipolar structure with oxide isolation

With particular respect to the performance of CZ-silicon wafers, the main differences between the device technologies can be described as follows.

(*a*) MOS devices are predominantly surface devices with shallow junctions; they are extremely sensitive to surface inhomogeneities and impurities.
(*b*) CMOS devices, although basically still surface devices, extend much more into the substrate because of the relatively deep p-well. The formation of the p-well adds a long-lasting, high-temperature process step to normal MOS technology.
(*c*) Bipolar devices have a more pronounced three-dimensional character. The electrically active parts are built into the epilayer. Compared to MOS technology, bipolar technology uses deeper junctions and high-temperature processes have to be used. This, in turn, requires greater perfection and integrity in the crystalline structure of the wafer and the epilayer if process-induced discontinuities and other defects are to be avoided.

The production of monocrystalline silicon and the production of solid state devices from monocrystalline silicon wafers is extremely complex and the basic principles discussed in this chapter can only be but a very brief introduction to this subject. The author is indebted to Wacker-Chemitronic GmbH, Postfach 1140, D-8263 Burghausen, West Germany, for most of the data upon which this chapter is based.

9 Synthetic adhesives

9.1 Introduction

Polymeric materials for moulding and extrusion were introduced in volume 1. Many of these materials are capable of being used directly for, or as the bases for, adhesives. Adhesives can be divided into two main categories:

(*a*) natural adhesives;
(*b*) synthetic adhesives.

The natural adhesives are vegetable and animal derivatives which may be used directly as adhesives with little modification. Compared with the majority of synthetic adhesives they are relatively weak but have the advantage of not being toxic. Unfortunately, they are adversely affected by damp and soften when raised much above room temperature. Natural adhesives may be subdivided further into *gums* and *glues*.

Gums are made from vegetable matter, resins and rubbers being extracted from the sap of trees, and starch derivatives being extracted from the byproducts of flour milling.

Glues are derived from the horns, hooves and bones of animals and the bones of fish. Derivatives of milk and blood are also used. Such glues soften at the boiling point of water and were largely used for joining wood in the furniture and toy-making industries.

Natural glues and gums are still widely used for low-strength applications, but they are being increasingly supplanted by high-strength synthetic adhesives and the range of applications of adhesive bonding is ever increasing. Table 9.1 lists some of the main groups of adhesives, whilst Table 9.2 lists some of the more important advantages and limitations of adhesive bonding as compared with mechanical and thermal

Table 9.1 Main groups of adhesives

Origin and basic type		Adhesive material	
Natural	Animal	Albumen, animal glue (inc. fish), casein, shellac, beeswax	
	Vegetable	Natural resins	(gum arabic, tragacanth, colophony, Canada balsam, etc.); oils and waxes (carnauba wax, linseed oils); proteins (soyabean); carbohydrates (starch, dextrines)
	Mineral	Inorganic materials	(silicates, magnesia, phosphates, litharge, sulphur, etc.); mineral waxes (paraffin); mineral resins (copal, amber); bitumen (inc. asphalt)
	Elastomers	Natural rubber	(and derivatives, chlorinated rubber, cyclised rubber, rubber hydrochloride)
Synthetic		Synthetic rubbers and derivatives	(butyl, polyisobutylene, polybutadiene blends (inc. styrene and acrylonitrile), polyisoprenes, polychloroprene, polyurethane, silicone, polysulphide, polyolefins (ethylene vinyl chloride, ethylene polypropylene))
		Reclaim rubbers	
	Thermoplastic	Cellulose derivatives	(acetate, acetate-butyrate, caprate, nitrate, methyl cellulose, hydroxy ethyl cellulose, ethyl cellulose, carboxy methyl cellulose)
		Vinyl polymers and copolymers	(polyvinyl-acetate, alcohol, acetal, chloride, polyvinylidene chloride, polyvinyl alkyl ethers)
		Polyesters (saturated)	(Polystyrene, polyamides (nylons and modifications))
		Polyacrylates	(methacrylate and acrylate polymers, cyano-acrylates, acrylamide)
		Polyethers	(polyhydroxy ether, polyphenolic ethers)
		Polysulphones	
	Thermosetting	Amino plastics	(urea and melamine formaldehydes and modifications)
		Epoxides and modifications	(epoxy polyamide, epoxy bitumen, epoxy polysulphide, epoxy nylon)
		Phenolic resins and modifications	(phenol and resorcinol formaldehydes, phenolic-nitrile, phenolic-neoprene, phenolic-epoxy)
		Polyesters (unsaturated)	
		Polyaromatics	(polyimide, polybenzimidazole, polybenzothiazole, polyphenylene)
		Furanes	(phenol furfural)

Table 9.2 Advantages and limitations of bonded joints

Advantages

1. The ability to join dissimilar materials, and materials of widely different thicknesses
2. The ability to join components of difficult shape that would restrict the application of welding or riveting equipment
3. Smooth finish to the joint which will be free from voids and protrusions such as weld beads, rivet and bolt heads, etc.
4. Uniform distribution of stress over entire area of joint. This reduces the chances of the joint failing in fatigue
5. Elastic properties of many adhesives allow for flexibility in the joint and give it vibration-damping characteristics
6. The ability to electrically insulate the adherends and prevent corrosion due to galvanic action between dissimilar metals
7. The joint will be sealed against moisture and gases
8. Heat-sensitive materials can be joined

Limitations

1. The bonding process is more complex than mechanical and thermal processes, i.e. the need for surface preparation, temperature and humidity control of the working atmosphere, ventilation and health problems caused by the adhesives and their solvents. The length of time that the assembly must be jigged up whilst setting (curing) takes place
2. Inspection of the joint is difficult
3. Joint design is more critical than for many mechanical and thermal processes
4. Incompatibility with the adherends. The adhesive itself may corrode the materials it is joining
5. Degradation of the joint when subject to high and low temperatures, chemical atmospheres, etc.
6. Creep under sustained loads

jointing processes. This chapter will be concerned with the characteristics and applications of synthetic adhesives.

9.2 The adhesive bond

Figure 9.1(*a*) shows a typical bonded joint and explains the terminology used for the various features of the joint. The strength of the bond depends upon two factors:

(*a*) adhesion;
(*b*) cohesion.

Adhesion is the ability of the bonding material (adhesive) to stick (adhere) to the materials being joined (adherends). There are two ways in which the bond can occur and these are shown in Fig. 9.1(*b*).

Cohesion is the ability of the adhesive and/or the adherend to resist the applied forces within itself. Figure 9.1(*c*) shows three ways in which a

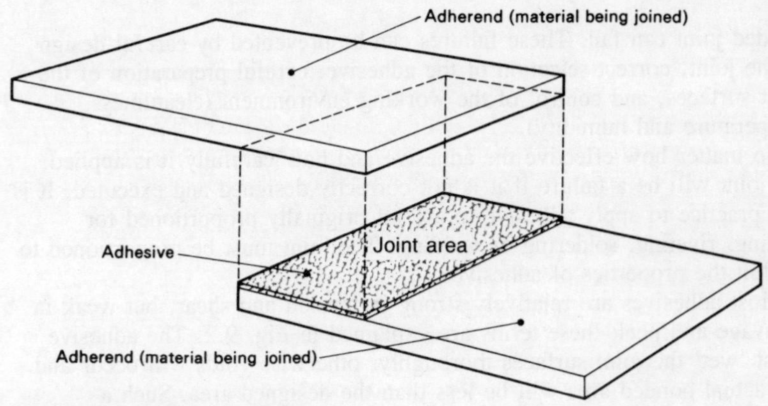

(a) **Elements of the bonded joint**

A simple cemented joint in which the adhesive penetrates the pores of the adherends to form the bond. This occurs with rough or porous surfaces.

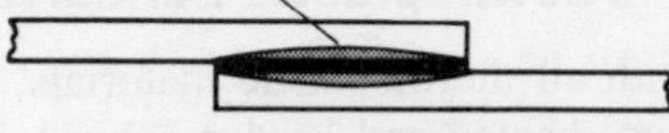

The adhesive and the adherends react together chemically so that an intermolecular bond is formed.

(b) **Types of bond**

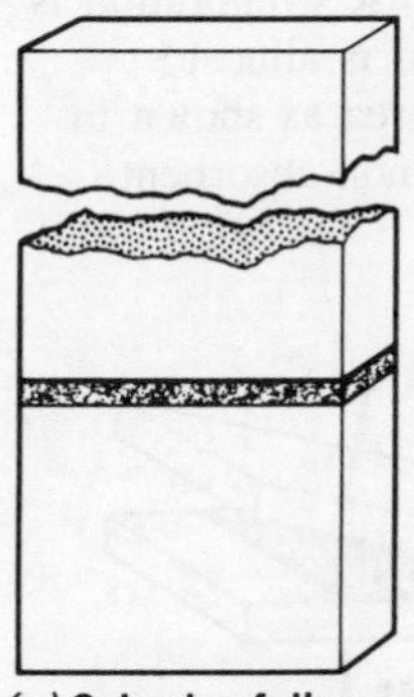

(a) **Cohesive failure of the adherend**

(over-strong adhesive)

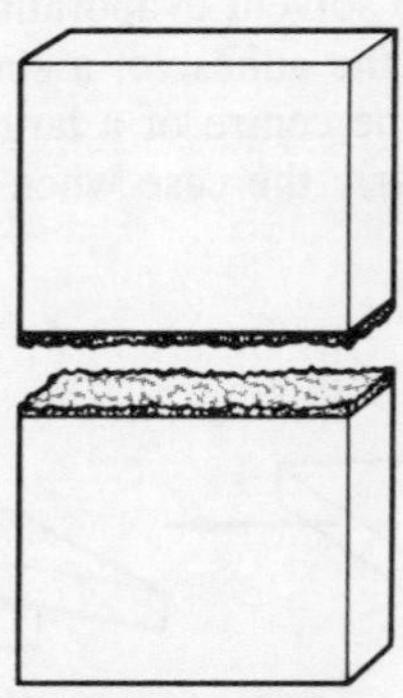

(b) **Cohesive failure of the adhesive**

(weak adhesive)

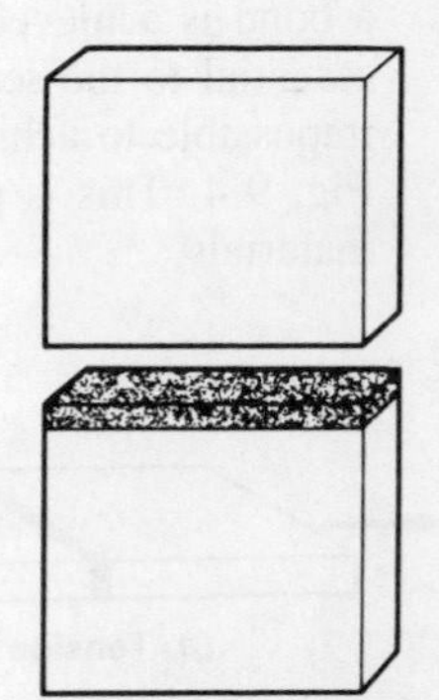

(c) **Adhesive failure**

(inadequate preparation of the joint faces resulted in a poor bond)

Fig. 9.1 Bonded joint

bonded joint can fail. These failures can be prevented by careful design of the joint, correct selection of the adhesive, careful preparation of the joint surfaces, and control of the working environment (cleanliness, temperature and humidity).

No matter how effective the adhesive and how carefully it is applied, the joint will be a failure if it is not correctly designed and executed. It is bad practice to apply adhesive to a joint originally proportioned for bolting, riveting, soldering or welding. The joint must be proportioned to exploit the properties of adhesives.

Most adhesives are relatively strong in tension and shear, but weak in cleavage and peel; these terms are explained in Fig. 9.2. The adhesive must 'wet' the joint surfaces thoroughly, otherwise voids will occur and the actual bonded area will be less than the designed area. Such a reduction in area will seriously weaken the strength of the joint. Figure 9.3 shows the effects of 'wetting' on the formation of the joint.

9.3 Thermoplastic adhesives

As with all thermoplastic materials, thermoplastic adhesives soften when they are heated and harden again when they are cooled. They may be classified into three categories.

(*a*) *Heat activated*. The adhesive is softened by heating until it is fluid enough to spread freely over the whole joint surface. Upon cooling to room temperature the adhesive adheres to the materials being joined and a bond is achieved.

(*b*) *Solvent activated*. The adhesive is softened by a suitable solvent and a bond is achieved by the solvent evaporating. Because evaporation is essential to the setting of the adhesive, a sound bond is almost impossible to achieve at the centre of a large joint area as shown in Fig. 9.4. This is particularly the case when joining non-absorbent materials.

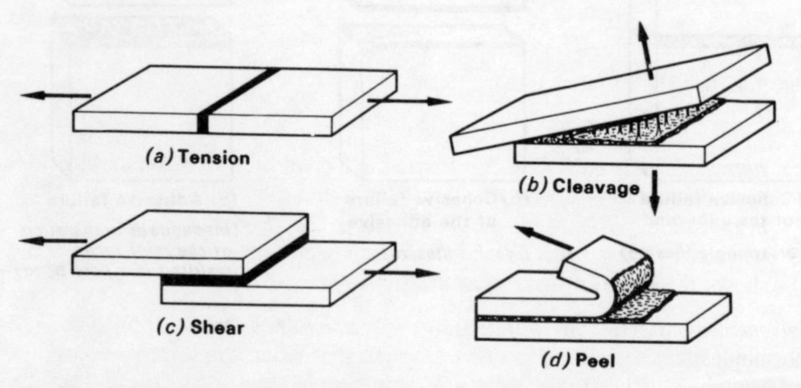

Fig. 9.2 The stressing of bonded joints

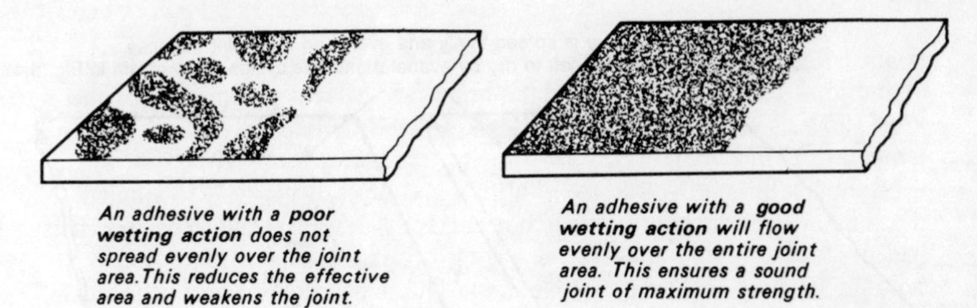

Fig. 9.3 Wetting capacity of an adhesive

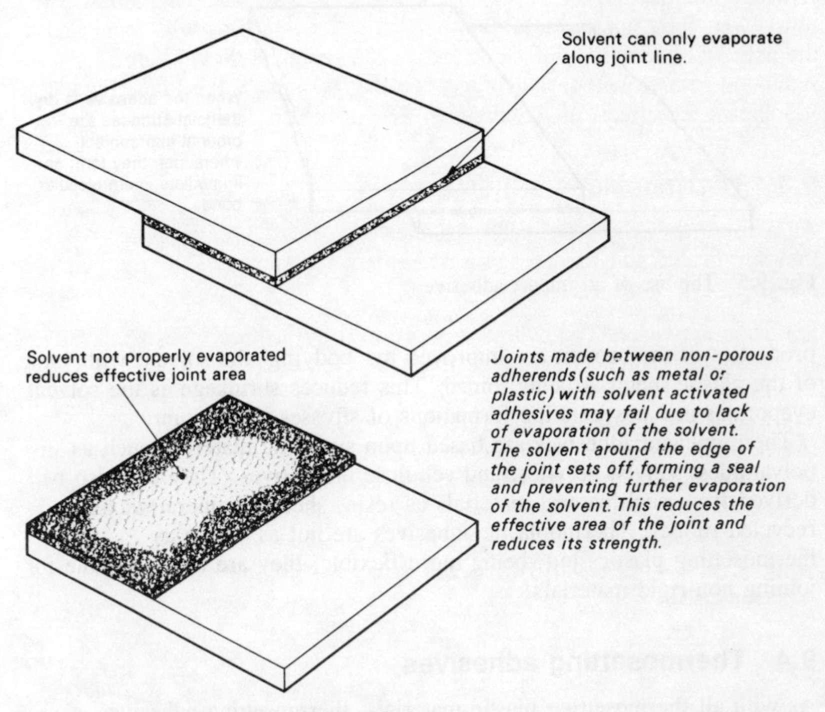

Fig. 9.4 Solvent activated adhesive fault

(*c*) *Impact adhesives*. These are solvent activated adhesives which are spread separately on the two joint faces and then left to dry by evaporation. When dry, the treated faces are brought together whereupon they form a bond by intermolecular attraction. Figure 9.5 shows the steps in making an impact joint.

Solvent cements (solvent welding) are solvents which, when applied to thermoplastics, soften the joint surfaces so that when the joint surfaces are brought together under pressure they form a bond. Gap-filling

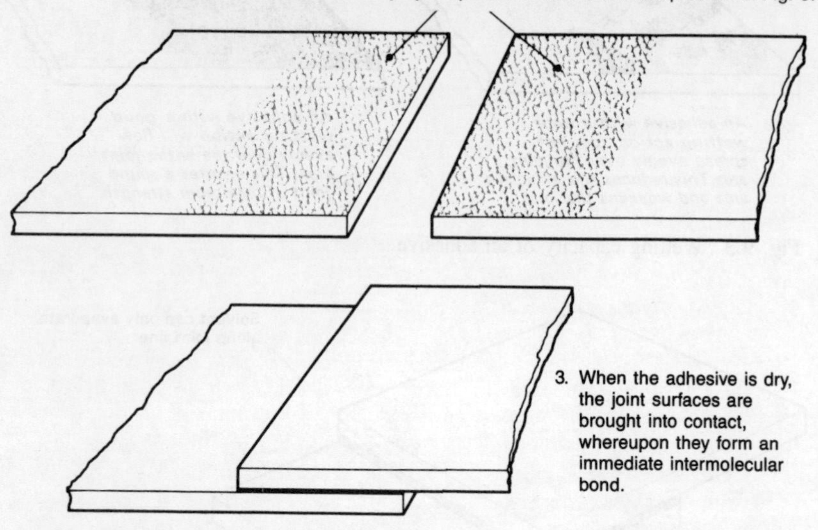

Fig. 9.5 The use of an impact adhesive

properties of the cement are improved by 'bodying' the solvent with some of the plastic material being joined. This reduces shrinkage as the solvent evaporates and prevents the formations of stresses in the joint.

Thermoplastic adhesives are based upon synthetic materials such as polyamides, vinyls, acrylics, and cellulose derivatives. They can also be derived from such natural materials as resin, shellac, mineral waxes and recycled rubber. Thermoplastic adhesives are not as strong as thermosetting plastics but, being more flexible, they are more suitable for joining non-rigid materials.

9.4 Thermosetting adhesives

As with all thermosetting plastic materials, thermosetting adhesives require heat to make them set. The setting (curing) process causes chemical changes to take place within the adhesive. Once set (cured) they cannot be softened again by the reapplication of heat. This makes them less temperature sensitive than thermoplastic adhesives.

The heat necessary to cure the adhesive can be applied externally by means of an oven (*autoclave*) or by radiant heat (for example, when phenolic resins are used), or internally by adding a chemical hardener (for example, when epoxy resins are used). The hardener is a chemical which reacts with the adhesive to generate heat internally (an exothermic reaction). Since the setting process is a chemical reaction and not dependent upon solvent evaporation, the area of the joint can be made as large as is necessary to achieve the required joint strength.

Thermosetting adhesives are much stronger than thermoplastic adhesives and can be used for making structural joints between high-strength materials such as metals. The body shells of motor cars and stressed members in aircraft are increasingly dependent upon adhesives for their joints in place of spot welding and riveting. The stresses are more uniformly distributed, and the joints are sealed against corrosion. Further, the relatively low process temperatures involved in adhesive bonding do not affect the crystallographic structure of the metal. Thermosetting adhesives tend to lack flexibility when cured and, therefore, they are not suitable for joining flexible (non-rigid) materials.

9.5 Joint design

It has already been shown that adhesive bonded joints are strong in tension and shear but weak in cleavage and peel. Therefore, all joints designed for use with adhesives should subject the bond only to tension or shear. Mechanical interlocking is also desirable for highly stressed joints particularly if it places the bond in compression.

Figure 9.6 shows a simple lap joint. It can be seen that the applied tensile forces are not axially aligned and that the joint is subjected to a distorting couple. If the distortion becomes sufficiently severe, it results in the adhesive bond no longer being in pure shear but being subjected to some cleavage and peel as well. Figure 9.7 shows some alternative lap joints in which the tensile forces are aligned so that the adhesive bond is only subjected to shear forces.

As has just been described, the forces acting upon bonded joints are not always as simple as they may seem. Figure 9.8 shows correct and incorrect designs for some further joints.

9.6 Surface preparation

For optimum adhesion of the adhesive to the adherend, the correct preparation of the joint faces is supremely important. The composition of the surface layer of any adherend depends upon its previous history and

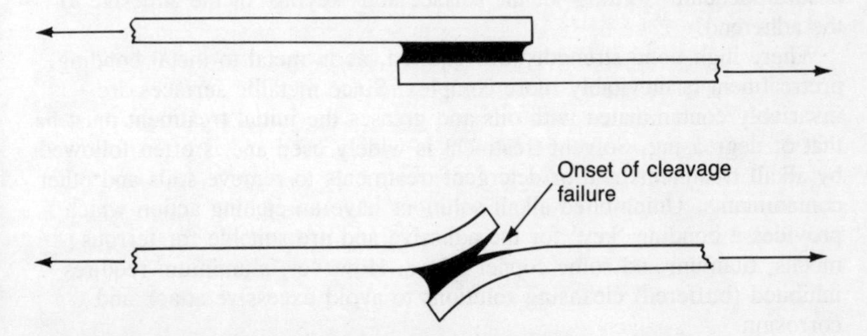

Fig. 9.6 Distortion and failure of a simple lap joint

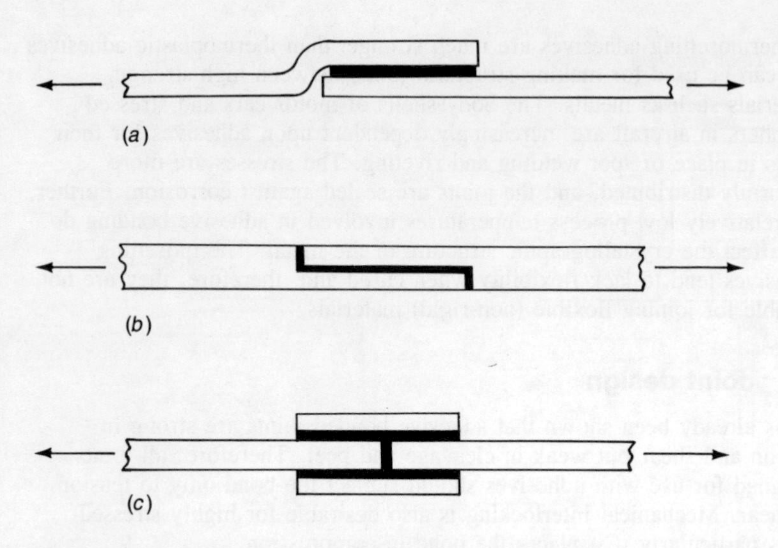

Fig. 9.7 Alternatives to the simple lap joint

is largely an unknown quantity. Therefore, the surface treatments discussed in this section can only be of a general nature. The choice of treatment largely depends upon such factors as the adherend material and its surface condition, the adhesive type selected, joint loading requirements, service conditions and service life, processing costs and resources available.

Pretreatments are similar to those described in Sections 6.16 to 6.20 inclusive, and may range from the simple to the complex, but in all cases the adhesive must be applied immediately after treatment to avoid the surface becoming contaminated again. If this is not possible then the surface must be protected immediately after treatment with an easily removable coating which, itself, will not affect the adhesion of the adhesive. The simplest treatments may rely simply upon the cleansing action of solvents and/or abrasives to remove surface contaminants to ensure adequate 'wetting' of the surface and 'keying' of the adhesive to the adherend.

Where high bond strengths are required, as in metal to metal bonding, pretreatment is inevitably more complex. Since metallic surfaces are invariably contaminated with oils and greases the initial treatment must be that of degreasing. Solvent treatment is widely used and is often followed by alkali treatments and/or detergent treatments to remove soils and other contaminants. Uninhibited alkali solutions have an etching action which provides a bonding 'key' for the adhesive and are suitable for ferrous metals, titanium and some copper alloys. However, aluminium requires inhibited (buffered) cleansing solutions to avoid excessive attack and corrosion.

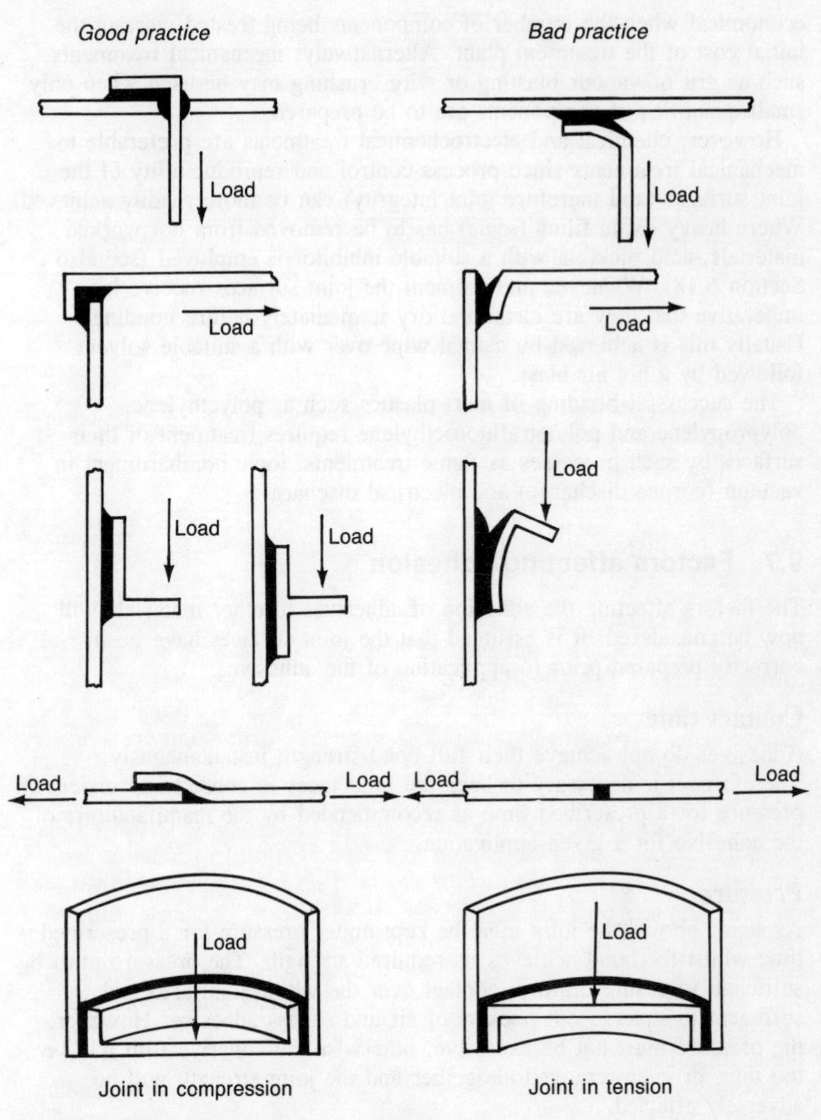

Fig. 9.8 Joint design for adhesive bonding

Structural adhesives invariably require specific pretreatments if they are to achieve permanent bonds of maximum strength, durability and integrity. For example, solvent-cleaned aluminium joints show a marked decrease in strength when weathering occurs, whilst acid-etched aluminium joints retain their bonded strength over many years when subjected to the same environmental conditions. However, such chemical treatments are only economical when the number of components being treated warrant the initial cost of the treatment plant. Alternatively, mechanical treatments such as grit or vapour blasting or wire brushing may be used when only small quantities of components are to be prepared.

However, chemical and electrochemical treatments are preferable to mechanical treatments since process control and reproducibility of the joint surfaces (and therefore joint integrity) can be more readily achieved. Where heavy oxide films (scale) has to be removed from hot worked materials, acid pickling with a suitable inhibitor is employed (see also Section 6.18). Whatever pretreatment the joint surfaces receive it is imperative that they are clean and dry immediately before bonding. Usually this is achieved by a final wipe over with a suitable solvent followed by a hot air blast.

The successful bonding of inert plastics such as polyethylene, polypropylene and polytetrafluoroethylene requires treatment of their surfaces by such processes as flame treatments, ionic bombardment in vacuum (corona discharge) and electrical discharge.

9.7 Factors affecting adhesion

The factors affecting the adhesion of adhesives to other materials will now be considered. It is assumed that the joint surfaces have been correctly prepared prior to application of the adhesive.

Contact time

Adhesives do not achieve their full bond strength instantaneously. Therefore, it is necessary to keep the joint faces in contact and under pressure for a prescribed time as recommended by the manufacturers of the adhesive for a given application.

Pressure

As stated above, the joint must be kept under pressure for a prescribed time whilst the bond achieves its required strength. The pressure must be sufficient to ensure uniform contact over the whole joint area and sufficient to squeeze out pockets of air and excess adhesive. However, the pressure must not be excessive, otherwise the adhesive film will be too thin, or even removed altogether and the joint strength will be adversely affected.

Temperature

This depends upon the type of adhesive used. In the case of thermosetting adhesives the temperature of the joint must be sufficient to ensure complete curing of the adhesive. The joint is usually heated in an autoclave or, if the assembly is too large, radiant heaters are used. Even 'cold setting' adhesives, which use the exothermic reaction of a hardener to ensure curing, must be kept at a sufficiently high ambient temperature to ensure that the exothermic reaction occurs in a satisfactory manner. In the case of thermoplastic adhesives the ambient temperature must be sufficiently high to ensure complete evaporation of the solvent. Usually a warm, dry, working environment around 20°C is suitable for both 'cold setting' thermosetting adhesives and thermoplastic adhesives.

Thickness of adhesive film

With correctly prepared surfaces, the adhesion at the interface is usually greater than the strength of the adhesive itself and, under most conditions, failure occurs within the adhesive film. Failure of the adhesive film is usually caused by the propagation of cracks which are accelerated by the presence of discontinuities and flaws. Therefore, thin layers of adhesive, providing the joint faces are adequately covered, provide the strongest joints (see joint pressure).

Molecular weight

The bonding together of two materials by means of an adhesive is dependent upon *mechanical adhesion* in which the surface roughness or absorption properties of the adherends provides a 'key' for the adhesive to grip, and upon *specific adhesion* which depends principally upon the Van der Waals forces between the molecules of the adhesive and the surface molecules of the adherends.

Mechanical adhesion is more likely to occur when bonding porous materials such as wood, but for hard, smooth materials such as metals, plastics and ceramics, specific adhesion is the more important. Since the Van der Waals forces are greater for large molecules than for small molecules, it follows that adhesives should have large molecules. Thus most adhesives are organic compounds composed of very large and complex molecules having a high density of polar groups.

9.8 Selection of adhesives

The selection of an adhesive for a particular application is a complex exercise dependent upon a large number of factors. There are a number of specialist directories and handbooks which can be consulted in order to draw up a 'short-list' of possible adhesives. The final choice invariably involves consultation with the manufacturers of the adhesives under consideration. This section merely reviews the more general factors involved in the selection of an adhesive and reviews a small number of adhesive/adherend combinations.

Joint requirements

The type and method of assembly of components using bonded joints are amongst the most important factors influencing the selection of an adhesive. Primarily, the adhesive is selected to provide a bond of adequate strength under service conditions and for the duration of the service life of the assembly. The joint characteristics must be reproducible and satisfy any quality control requirements. In addition, a number of secondary factors influencing the choice of adhesive also have to be considered. These include:

(*a*) its suitability for the application process demanded by the economics and method of assembly, size of components and batch quantity;
(*b*) its ability to act as a sealant against liquids and gases;
(*c*) its ability to act as a thermal or electrical insulator;
(*d*) its ability to resist vibration and fatigue;
(*e*) its ability to prevent corrosion in joints involving metal components.

Materials to be bonded

The mechanical and physical properties of the materials being bonded, and the economic and practical restraints upon joint surface pretreatment are important factors in the selection of an adhesive. Reference back to Fig. 9.1 shows that failure can occur in the adherend, or at the adherend/adhesive interface, or cohesively within the adhesive. Usually it is preferable for cohesive failure to occur. There is nothing to be gained from using a high-strength adhesive when bonding low-strength adherends. Further, a rigid adhesive is unsuitable when bonding flexible or semi-rigid materials. Adhesion must also be considered. For example, epoxy adhesives have a much higher cohesive strength than a solvent cement such as polystyrene cement. However, the polystyrene cement will provide the stronger bond when joining polystyrene mouldings since the epoxy resin will have difficulty in adhering to the joint surfaces.

Compatibility

Adhesives and adherends must be mutually compatible.

(*a*) They must have similar thermal expansion properties to prevent stresses developing in the joint with changes of temperature.
(*b*) The adhesive must not cause corrosion of the adherend. Some acidic adhesives will attack metallic adherends.
(*c*) Any solvents or volatiles in the adhesive must not affect the adherends.
(*d*) The adhesives must not be adversely affected by plasticisers in flexible adherends. They must resist the migration of such plasticisers.

Bond stress

Usually, the adhesive selected should have similar strength characteristics to the adherends being bonded together. An exception would be where bonding is only temporary pending some other joining processes being used. In this case, bond strength need only be sufficient to withstand handling during the final joining process and it should not interfere with that process.

As well as the magnitude of the stresses created in the bond by the application of external forces, the direction of loading and the conditions under which the load is applied must also be considered. The joint may be subjected to tension, compression, shear, peel or cleavage, or a combination of these loads depending upon the design of the joint. The bond loading may be constant, intermittent or vibratory. Most adhesives show optimum strength characteristics when in tension or compression closely followed by shear. A small joint area requires an adhesive which is strong in tension or shear. However, such an adhesive may be weak in peel or cleavage. A lower strength adhesive may have better cleavage and peel characteristics. In this case, its lower tension and shear strength characteristics can be accommodated by increasing the joint area.

Often the high strength, thermosetting adhesives form brittle bonds which are adversely affected by vibration and impact loading which causes the bond to crack or shatter. Under such conditions a slightly weaker but more resilient adhesive may perform more satisfactorily. Other adhesives may show satisfactory strength characteristics under test conditions but will tend to 'creep' under sustained loads in service.

Joint thickness has already been considered earlier. With high moduli adhesives, optimum tensile and shear strengths are obtained when the adhesive film thickness lies between 0.06 mm and 0.12 mm. Thinner films lead to adhesive starvation and catastrophic failure of the joint. Thicker films may lead to early cohesive failure. Adhesives based upon the elastomers generally achieve their optimum performance characteristics with a film thickness in excess of 0.12 mm as this enhances their peel strengths by allowing more 'give' in the joint. Such elastomer adhesives and joint proportions are usually superior under vibratory loads for the same reason.

Processing

There is often a considerable difference between the requirements for an adhesive which is suitable for manual application in the case of prototype and small quantity production compared with an adhesive which is to be applied and spread by machine under quantity production conditions. Other factors include the environmental conditions in the workshop, working life of the adhesive once it has been mixed, drying time, curing temperature and the method of achieving that temperature, the effect of the curing temperature on the adherend material, and such safety considerations as flammability, toxicity, and odour. Again, pretreatment processes for the joint surfaces which are suitable under experimental prototype and small quantity production conditions may be quite impractical under quantity production conditions.

Cost

The cost of the adhesive is often only a small part of the total cost of achieving a satisfactorily bonded joint. Therefore, the process costs should include such items as the cost of pretreatment, cost of processing equipment, the cost of ensuring a correct working environment, the cost of safety requirements, labour cost, and the setting time. In the case of a lap joint the cost of adherend material overlap required to give an

Table 9.3 Bonding of plastics with solvent cements

This table indicates the solvents (and mixtures) which may be used to join thermoplastic materials by solvent-welding. The surfaces to be joined are softened, by the application of a suitable solvent and then pressed together to effect a bond. Gap-filling properties of the cement are improved by bodying the solvent with some of the thermoplastic: this reduces shrinkage at the joint which would otherwise result in stress formation.

	Solvent cements																													
Plastics	Acetic acid (glacial)	Acetone	Acetone : ethyl acetate : cellulose acetate butyrate (40 : 40 : 20)	Acetone : ethyl lactate (90 : 10)	Acetone : methoxyethyl acetate (80 : 20)	Acetone : methyl acetate (70 : 30)	Butyl acetate : acetone : methyl acetate (50 : 30 : 20)	Butyl acetate : methyl methacrylate monomer (40 : 60)	Ethyl acetate	Ethyl acetate : ethyl alcohol (80 : 20)	Ethylene dichloride	Ethylene dichloride : methylene chloride (50 : 50)	Glycerine : water (15 : 85)	Methyl acetate	Methylene chloride	Methylene chloride : methyl methacrylate monomer (60 : 40)	Methylene chloride : methyl methacrylate monomer (50 : 50)	Methylethyl ketone	Methyl isobutyl ketone	Methyl methacrylate monomer	Tetrachloroethylene	Tetrachloroethane	Tetrahydrofuran : cyclohexanone (80 : 20)	Toluene	Toluene : ethyl alcohol (90 : 10)	Toluene : methylethyl ketone (50 : 50)	(1.1.2) Trichloroethane	Trichloroethylene	Xylene	Xylene : methyl isobutyl ketone (25 : 75)
Acrylonitrile butadiene styrene																		X	X							X				X
Cellulose acetate film		X		X	X	X	X		X					X																
Cellulose acetate butyrate		X	X	X	X	X	X		X					X																
Cellulose propionate						X																								
Cellulose nitrate		X							X					X																
Ethyl cellulose										X															X					
Polyamide (nylon)	X																													
Polymethyl methacrylate											X					X	X			X										
Polycarbonate											X	X				X						X					X			
Polystyrene									X						X			X			X			X				X		
Polyvinyl chloride and copolymers (acetate)																		X	X				X						X	
Styrene acrylonitrile								X	X									X												
Styrene butadiene		X																X	X											
Polyvinyl alcohol													X																	
Polyphenylene oxide																														X

adequate joint area must also be considered when comparing strong, and potentially more expensive adhesives, with weaker and cheaper adhesives which require a larger joint area.

Service conditions

Adhesives which prove satisfactory under short-term test conditions may fail in service. Such failure may be due to changes in temperature and humidity, degradation of the adhesive due to ultraviolet radiation, load characteristics such as unexpected vibrations and shock loads, and environmental pollutants. The service life of the adhesive must match the service life of the assembly as a whole. Therefore, the rate and pattern of degradation of the adhesive must match the rate and pattern of degradation of the adherend material.

The above factors, which should be considered when selecting an

Table 9.4 Chart showing complementary adhesives and adherends. (From Shields, J., *Adhesive Bonding*, The Design Council.)

ADHERENDS \ ADHESIVES	**Natural**	Animal glues	Starch	Dextrine	Casein	**Elastomers**	Acrylonitrile butadiene	Polychloroprene	Polyurethane	Silicone rubber	Polybutadiene	Natural rubber	Butyl	**Thermoplastics**	Cellulose nitrate	Polyvinyl alcohol	Polyvinyl acetate	Polyacrylate	Silicone resin	Cyanoacrylate	**Thermosets**	Phenolic formaldehyde	Urea formaldehyde	Resorcinol formaldehyde	Melamine formaldehyde	Polyesters (unsaturated)	Epoxy resins	Polyimides	Phenolic-vinyl formal	Phenolic-polyvinylacetal	Phenolic nitrile	Phenolic epoxy	**Inorganic**	Sodium silicate
Metals							X	X				X					X			X							X	X	X		X	X		
Glass, Ceramics		X						X							X		X			X							X			X		X		X
Wood					X							X			X		X					X	X	X	X		X							
Paper		X	X	X	X										X	X	X																	X
Leather		X					X	X				X			X																			
Textiles, felt		X					X					X			X		X																	
Elastomers																																		
Polychloroprene (Neoprene)								X																										
Nitrile							X													X														
Natural							X					X								X									X					
Silicone										X																								
Butyl							X						X																					
Polyurethane							X	X	X																									
Thermoplastics																																		
Polyvinyl chloride (flexible)							X	X	X																									
Polyvinyl chloride (rigid)							X	X	X																		X							
Cellulose acetate									X						X					X														
Cellulose nitrate									X						X					X														
Ethyl cellulose															X					X							X				X			
Polyethylene (film)								X			X							X																
Polyethylene (rigid)																											X				X			
Polypropylene (film)								X			X							X																
Polypropylene (rigid)																											X				X			
Polycarbonate									X																		X							
Fluorocarbons												X							X			X					X							
Polystyrene									X											X							X							
Polyamides (nylon)								X														X		X			X				X			
Polyformaldehyde (acetals)									X											X						X	X							
Methyl pentene								X												X														
Thermosets																																		
Epoxy																				X		X		X			X							
Phenolic									X											X			X				X				X			
Polyester																										X	X							
Melamine								X	X																		X							
Polyethylene terephthalate							X	X											X							X								
Diallyl phthalate							X																			X	X							
Polyimide																											X	X						

Note: In general, any two adherends may be bonded together if the chart shows that they are compatible with the same adhesive.

adhesive for a particular application, are in no way exhaustive but are intended merely to give an indication of the problems to be considered in arriving at an initial 'short list' suitable for deeper consideration.

Table 9.3 lists some typical solvent cement adhesives and the adherend materials for which they are appropriate. Table 9.4 lists some thermoplastic and elastomer adhesives and the adherend materials for which they are appropriate, together with some thermosetting adhesives and the adherend materials for which they are appropriate.

10 Materials in service

10.1 Structure of materials

An understanding of the structure of materials is essential to the understanding of the behaviour of such materials in service. The crystalline structure of metals and crystallinity in some polymeric materials were discussed in volume 1, together with the solidification of molten metal and casting defects. These matters will now be developed further.

From time to time in Chapter 8, crystal orientation was referred to using the notation [100] or [111] for example. These numbers are derived from the *Miller indices*. The Miller indices are used to describe the planes and directions of atoms in crystals. Miller indices are proportional to the reciprocals of the intercepts which the planes make with the three principal crystal axes, namely the x, y and z axes.

Figure 10.1 shows how these indices are derived. The plane PQR is shown shaded and it can be seen that the intercepts which it makes with the three axes are: $x = 2$ unit lengths; $y = 4$ unit lengths; $z = 4$ unit lengths. The reciprocals of these intercept values are: ½; ¼; ¼, respectively. These reciprocals are then converted to the smallest integers which are in the same ratio, that is: 2; 1; 1. Thus the Miller indices for the plane PQR in Fig. 10.1 is (211). A plane on the opposite side of the origin would have negative intercepts on the axes and its indices would be denoted as $(\bar{2}\bar{1}\bar{1})$.

Directions through a crystal can be specified in a similar manner by the use of indices which give the integral coordinates of a point on a line drawn between the point and the origin. Direction indices are enclosed in square brackets [ ] to distinguish them from Miller indices which are enclosed in parentheses ( ). Figure 10.2 shows the Miller indices for

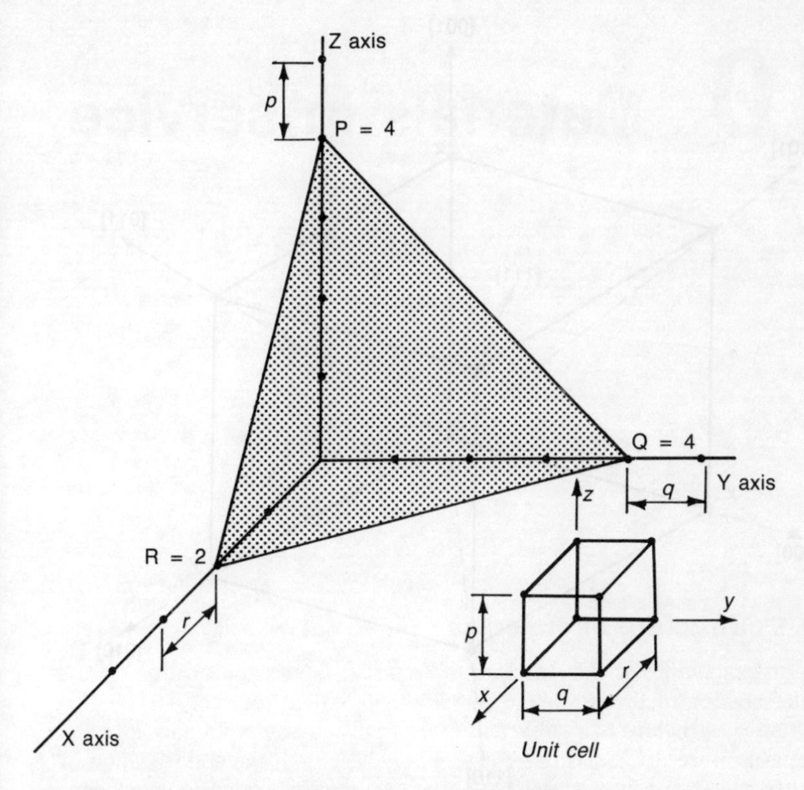

Fig. 10.1 Derivation of Miller indices for plane PQR

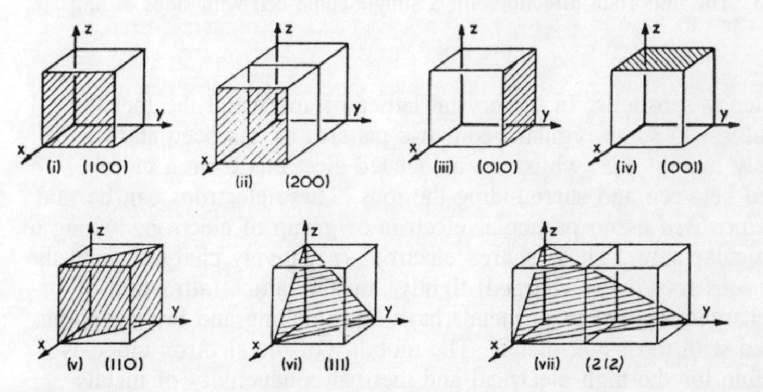

Fig. 10.2 Miller indices for some typical planes in a simple cubic system

some typical planes in a simple cubic system, whilst Fig. 10.3 shows some important directions in a simple cubic cell.

The properties of metals are also dependent upon the *metallic bond* which results from the fact that metallic atoms have only few electrons in

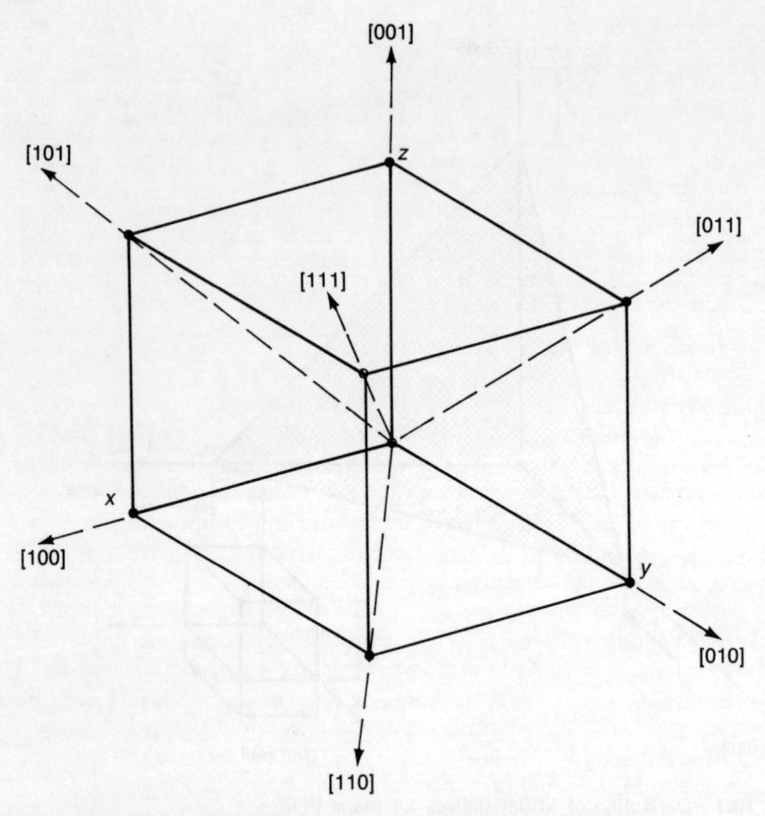

Fig. 10.3 The important directions for a simple cubic cell with sides of unit length

their valency subshells. In the crystal lattice of any metal the metallic ions conform to some regular geometric pattern, as has been shown previously in Fig. 3.3, whilst the associated electrons form a cloud dispersed between and surrounding the ions. These electrons can be said to be *delocalised* as no particular electron or group of electrons belong to any particular atom. These shared electrons (negatively charged) bind the metallic ions (positively charged) tightly into the lattice (attraction of unlike charges) so that most metals have high melting and boiling points compared with most non-metals. The mobility of the electron cloud is responsible for the high electrical and thermal conductivity of metals compared with non-metals.

Non-metals also form crystals, for example ice crystals when water freezes, and crystals of sodium chloride (common table salt). These crystals are much weaker than metallic crystals and the reason for this was shown in Fig. 3.3.

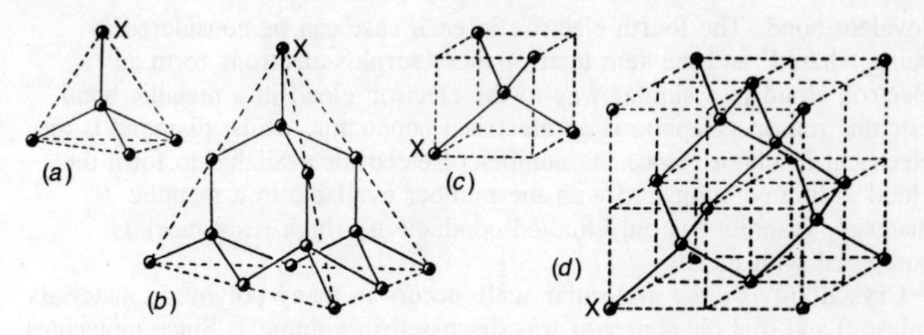

Fig. 10.4 The crystal structure of diamond

Some materials are allotropic, that is, they can exist in more than one form, with each form having its own special properties. Consider the non-metallic element carbon; it can exist in the form of diamond or in the form of graphite. The unit cell for a diamond crystal is a tetrahedron. As the crystal grows, not all the 'fringe' atoms are linked up so that the diamond can be considered as having a *cubic* structure. Figure 10.4 shows the stages in the crystal growth for a diamond.

On the other hand, carbon can assume a *layer structure*. The carbon atoms within the layers are covalently bonded together and are thus relatively strong, whilst the layers themselves are only held together by relatively weak Van der Waals forces. For this reason, the layers or plates can easily slide over each other and this results in graphite being an excellent lubricant.

Reference to Fig. 10.5 shows that only three of the four atoms available in the valency sub-shell of graphite are used in forming the

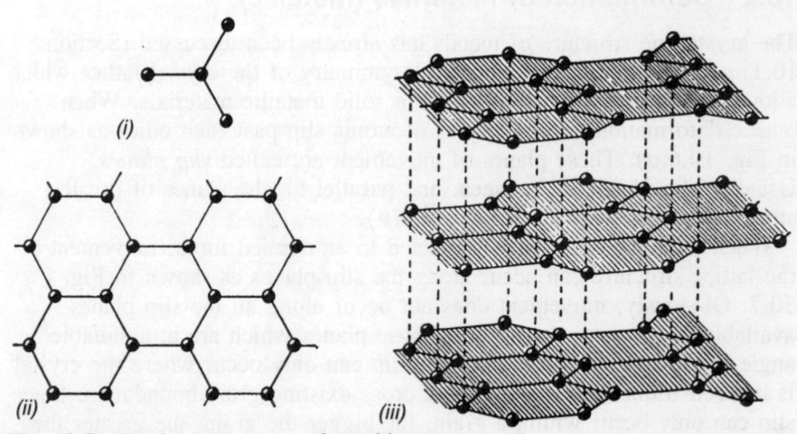

Fig. 10.5 The layer structure of graphite

covalent bond. The fourth electron in each case can be considered as being 'shared' and the sum total of these surplus electrons form an electron cloud in a similar way to the electron cloud in a metallic bond. For this reason, graphite is an electrical conductor, whilst diamond is an electrical insulator. Since the number of electrons available to form the cloud is limited compared with the number available in a metallic material, graphite has only limited conductivity (high resistance) compared with metals.

Crystallinity on the molecular scale occurs in many polymeric materials (plastic) and this phenomenon was discussed in volume 1. Since molecules as well as atoms are self-contained units, it is possible for molecules to arrange themselves into crystal lattices. The atoms within the molecules are held together by covalent or ionic bonding but the molecules are held together by the weaker Van der Waals forces. Such materials usually have a mixture of amorphous and crystalline regions (micelle regions). Although usually found in polymeric materials, some metals (tellurium, for example) also form long chain molecules with areas of crystallinity on the molecular scale.

Amorphous (without shape) substances are not true solids since all true solids must, by definition, be crystalline. Some amorphous substances such as glass and pitch appear to be solid and will shatter when subjected to sudden impact. However, when lightly loaded they will 'flow' over a period of time. Pitch will settle to the shape of its container and a vertical pane of glass in a window will eventually become thicker at the lower end than at the top. Further, such amorphous materials do not show a clearly defined melting point but become increasingly softer as their termperature is raised, and eventually resemble high viscosity liquids. Many polymeric materials behave in this manner.

10.2 Deformation of materials (metallic)

The crystalline structure of metals has already been discussed (Section 10.1) and it is the strict geometrical symmetry of the crystal lattice which allows plastic deformation to occur in solid metallic materials. When plastic deformation occurs, planes of atoms slip past each other as shown in Fig. 10.6(a). These planes of movement are called *slip planes*. Usually, slip planes lie between, and parallel to, the planes of greatest atomic density as shown in Fig. 10.6(b).

When a ductile material is subjected to an applied force, movement of the lattice structure can occur along the slip planes as shown in Fig. 10.7. Obviously, movement does not occur along all the slip planes available in a crystal, but only in those planes which are at a suitable angle to the applied force. Further, slip can only occur where the crystal is not constrained, thus slip cannot cross existing grain boundaries. Since slip can only occur within a grain, the bigger the grain, the greater the amount of slip which can take place. This is borne out in practice, since fine grain materials are generally less ductile and malleable than when the same material has been processed to enlarge its grain structure. The

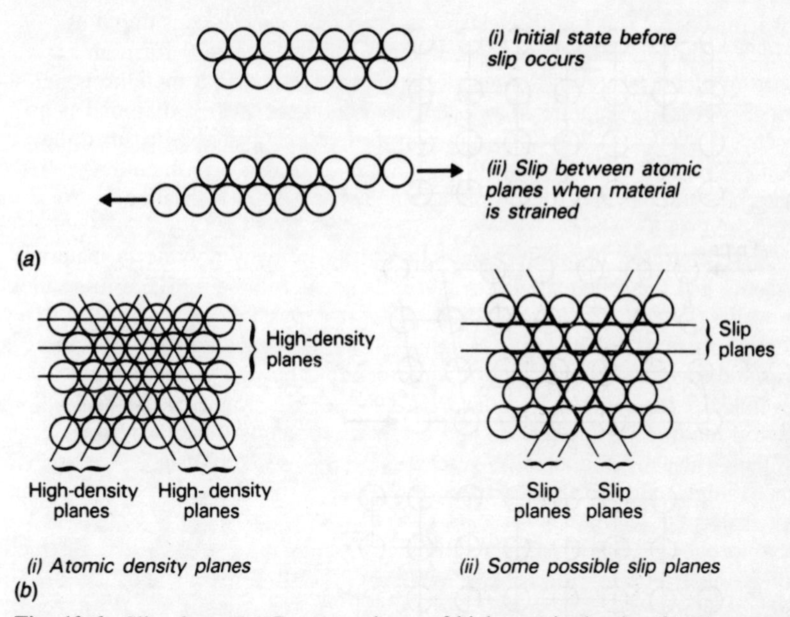

Fig. 10.6 Slip planes (*a*) Between planes of high atomic density (*b*) The orientation of slip planes

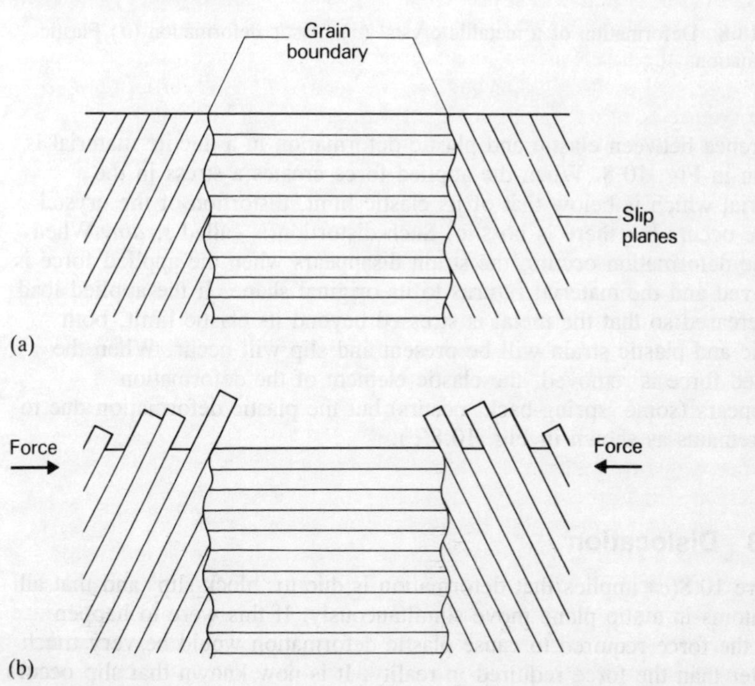

Fig. 10.7 Formation of slip bands during plastic deformation of a metal (*a*) Slip planes before the application of a force (*b*) Movement of slip planes after the application of a force

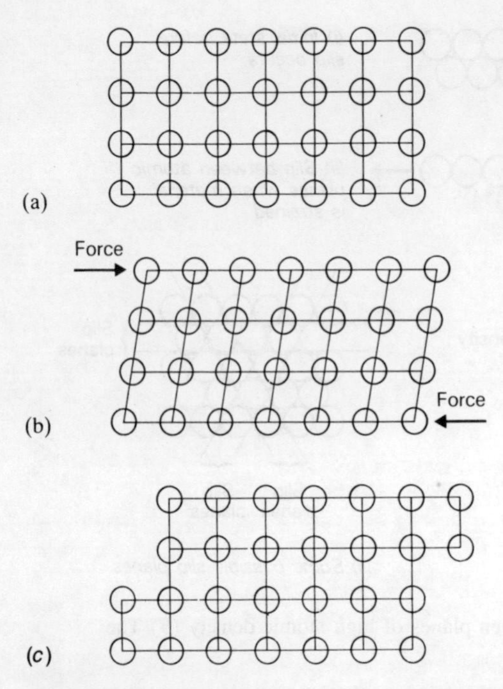

Fig. 10.8 Deformation of a metallic crystal (*a*) Elastic deformation (*b*) Plastic deformation

difference between elastic and plastic deformation in a ductile material is shown in Fig. 10.8. When the applied force creates a stress in the material which is below that of its elastic limit, distortion of the crystal lattice occurs but there is no slip. Such distortion is called *strain*. When elastic deformation occurs, the strain disappears when the applied force is removed and the material returns to its original shape. If the applied load is increased so that the metal is stressed beyond its elastic limit, both elastic and plastic strain will be present and slip will occur. When the applied force is removed, the elastic element of the deformation disappears (some 'spring-back' occurs) but the plastic deformation due to slip remains as shown in Fig. 10.8(*c*).

10.3 Dislocation

Figure 10.8(*c*) implies that deformation is due to 'block-slip' and that all the atoms in a slip plane move simultaneously. If this were to happen then the force required to cause plastic deformation would be very much greater than the force required in reality. It is now known that slip occurs through a system of dislocations rather like moving a carpet along a floor a little at a time by bunching it and moving the ruck along as shown in

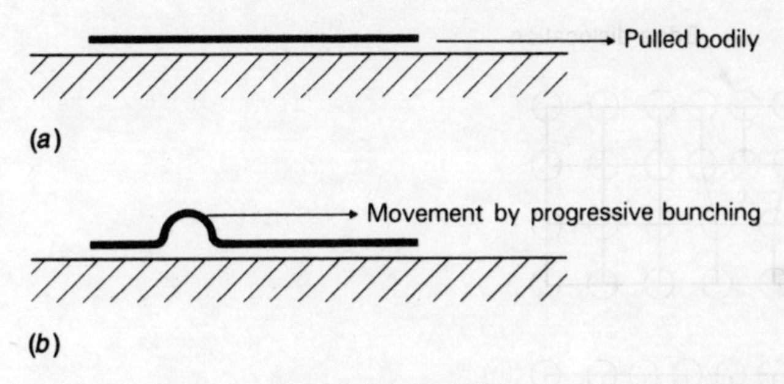

Fig. 10.9 'Carpet analogy' of dislocation (*a*) The equivalent of block slip (*b*) The equivalent of deformation by dislocation

Fig. 10.9. Plastic deformation due to dislocation is possible because crystals are not as perfect as has been implied so far. Whilst the majority of atoms will follow the general lattice pattern, there will also be a variety of faults and deficiencies, referred to as 'point defects', leading to irregularities and distortions in the surrounding crystal lattice. Such point defects may be interstitial or substitutional solute atoms, vacancies, coherent precipitates, impurity atoms, and simple edge dislocations.

Of the above point defects, *edge dislocations* are the most important when considering plastic deformation. Figure 10.10(*a*) shows a typical edge dislocation, whilst Fig. 10.10(*b*) to 10.10(*d*) shows how the dislocation moves progressively through the crystal, under the action of an applied force, until a *slip step* is formed. The presence of a dislocation is indicated diagrammatically by the sign $\perp$. The applied force must be sufficient to stress the material beyond its elastic limit and dislocation will cease when the force is removed. The dislocation may also be halted when it meets some other fault or reaches the crystal boundary.

A *screw dislocation* may also cause slip as shown, in principle, in Fig. 10.11. Slip resulting from screw dislocation is usually the result of the application of offset or 'shear' forces being applied to the material as shown. In practice the situation is often more complex with edge and screw dislocations occurring simultaneously as shown in Fig. 10.12.

Another mechanism by which deformation can take place is *twinning*. The principle of deformation by twinning is shown in Fig. 10.13. Unlike slip, where all the atoms in a block move the same distance, in twinning each successive plane of atoms moves a different distance. When twinning is complete, the deformation of the crystal lattice will result in one half of the twin becoming the mirror image of the other half, as shown in Fig. 10.13(*b*). Like slip, twinning proceeds by a series of dislocations. The force required to produce twinning dislocations is generally greater than that required to produce slip.

Twinning is not as common as block slip and occurs mainly when metals are shock loaded at low temperatures. For example, cold-heading rivets made from α-iron (which has body-centred cubic crystals) causes them to develop thin lamellar twins referred to as Neumann bands.

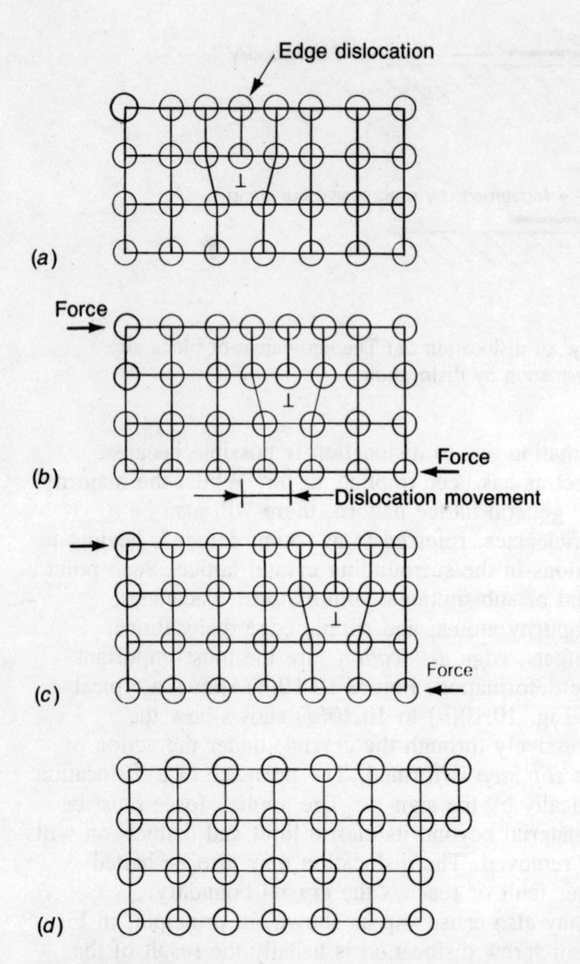

Fig. 10.10 Deformation by dislocation (*a*) Crystal showing an edge dislocation at ⊥ (*b*) & (*c*) Application of a force causes dislocation movement (*d*) Dislocation (slip) complete

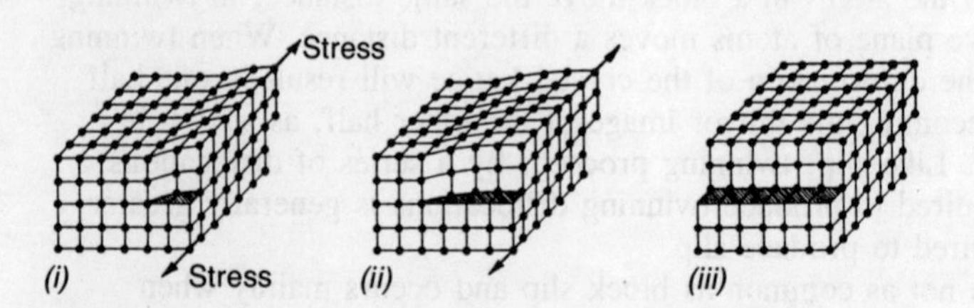

Fig. 10.11 The movement of a screw dislocation

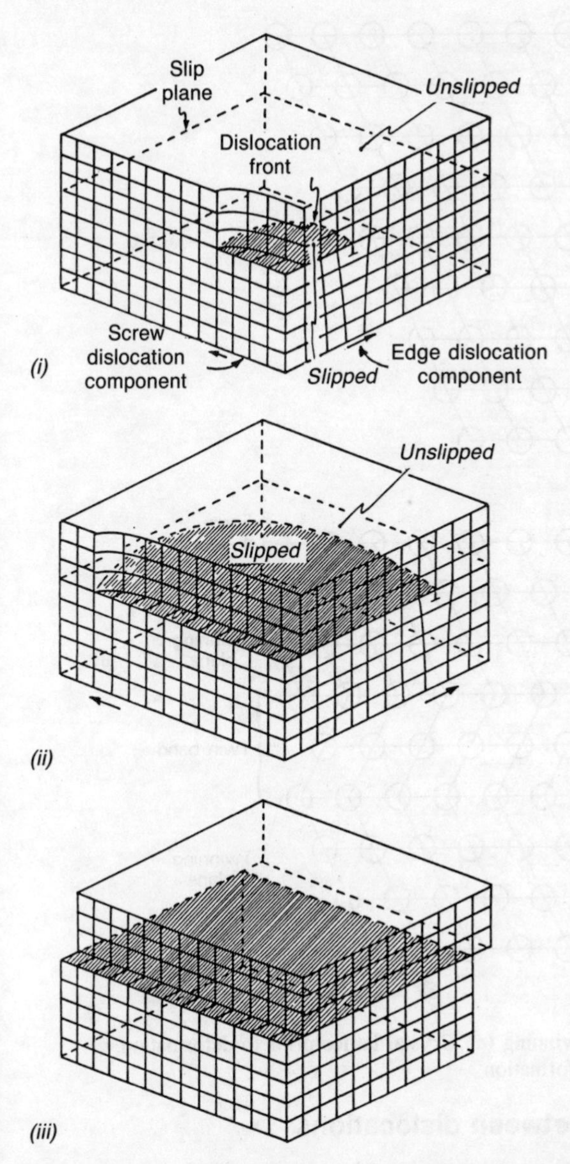

Fig. 10.12 Slip by the movement of a dislocation loop over a slip plane

Twinning can also be caused by heat treatment and is common in copper, brass alloys, bronze alloys and austenitic steel alloys when these metals are annealed. Such twins are the result of recrystallisation and grain growth, rather than the result of mechanical stressing as previously described.

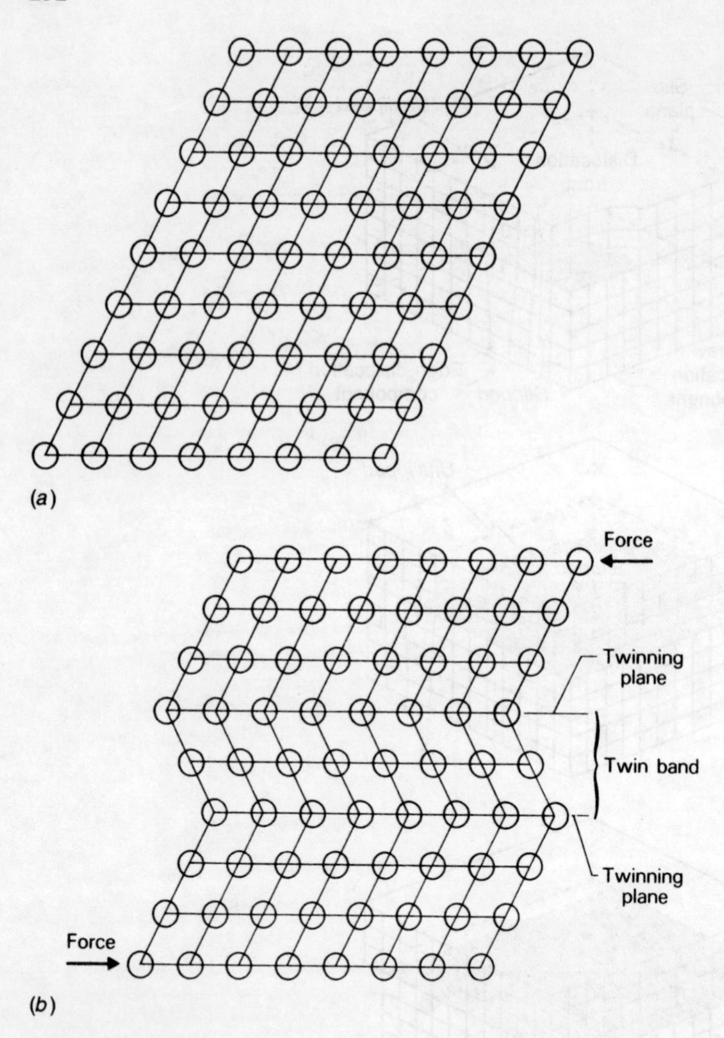

Fig. 10.13 Principle of twinning (*a*) Crystal lattice prior to deformation (*b*) Crystal lattice after deformation

10.4 Interaction between dislocations

The extra half plane of atoms or ions introduced by a simple edge dislocation results in an increase in strain energy in the vicinity of the dislocation. When two or more dislocations occur in the same vicinity they will interact with each other. The additional half plane of an edge dislocation introduces crowding of the atoms resulting in a region of compressive strain energy, whilst the corresponding separation of the atoms beyond the edge dislocation results in a region of tensile strain energy, as shown in Fig. 10.14.

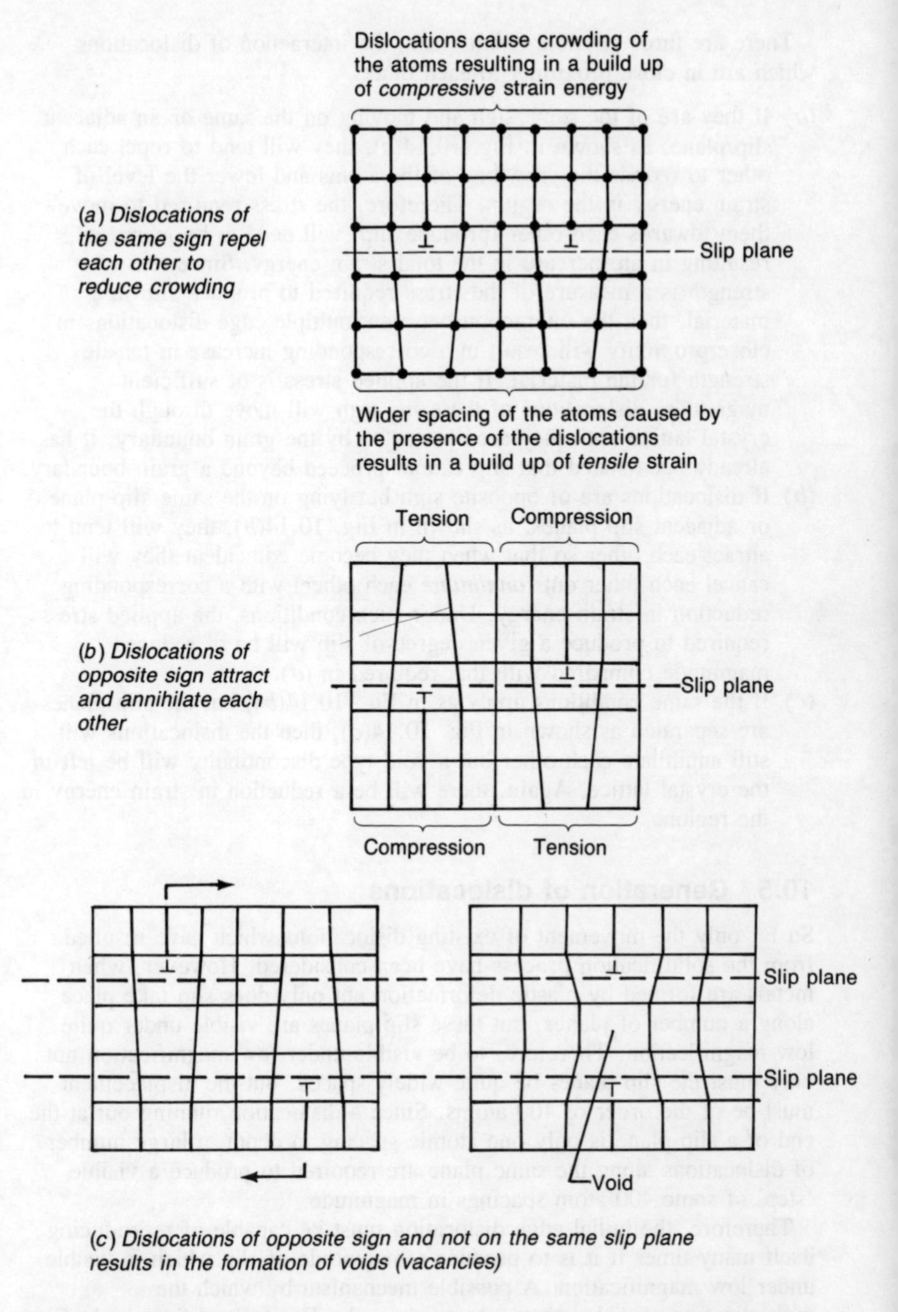

Fig. 10.14 Multiple dislocations (*a*) Dislocations of the same sign (*b*) Dislocations of opposite signs (*c*) Dislocations of opposite signs and on separate slip planes

There are three possible results from the interaction of dislocations which are in close proximity to each other.

(*a*) If they are of the same sign and moving on the same or an adjacent slip plane, as shown in Fig. 10.14(*a*), they will tend to repel each other to reduce the crowding of the atoms and lower the level of strain energy in the region. Therefore, the stress required to move them towards each other (produce slip) will need to be increased, resulting in an increase in the total strain energy. Since tensile strength is a measure of the stress required to produce slip in a material, then the interaction between multiple edge dislocations in close proximity will result in a corresponding increase in tensile strength for the material. If the applied stress is of sufficient magnitude, dislocations of the same sign will move through the crystal lattice until they are obstructed by the grain boundary. It has already been stated that slip cannot proceed beyond a grain boundary.

(*b*) If dislocations are of opposite sign but lying on the same slip plane or adjacent slip planes, as shown in Fig. 10.14(*b*), they will tend to attract each other so that when they become coincident they will cancel each other out (*annihilate* each other) with a corresponding reduction in strain energy. Under such conditions, the applied stress required to produce a given degree of slip will be of reduced magnitude compared with that required in (*a*).

(*c*) If the same conditions apply as in Fig. 10.14(*b*), but the slip planes are separated as shown in Fig. 10.14(*c*), then the dislocations will still annihilate each other but a void type discontinuity will be left in the crystal lattice. Again, there will be a reduction in strain energy in the region.

10.5 Generation of dislocations

So far only the movement of existing dislocations which have resulted from the solidification process have been considered. However, when metals are formed by plastic deformation not only does slip take place along a number of planes, but these slip planes are visible under quite low magnification. Therefore, to be visible under low magnification not only must the slip planes be quite widely spaced, but the displacement must be of the order of 400 atoms. Since a dislocation running out at the end of a slip plane is only one atomic spacing in depth, a large number of dislocations along the same plane are required to produce a visible 'step' of some 400 atom spacings in magnitude.

Therefore, the initial edge dislocation must be capable of reproducing itself many times if it is to produce a magnitude of slip which is visible under low magnification. A possible mechanism by which the multiplication of dislocations can occur is the *Frank-Read Source*. In Fig. 10.15 the initial dislocation is shown restricted at its ends, possibly by other dislocations, grain boundaries, or lattice imperfections, so that when

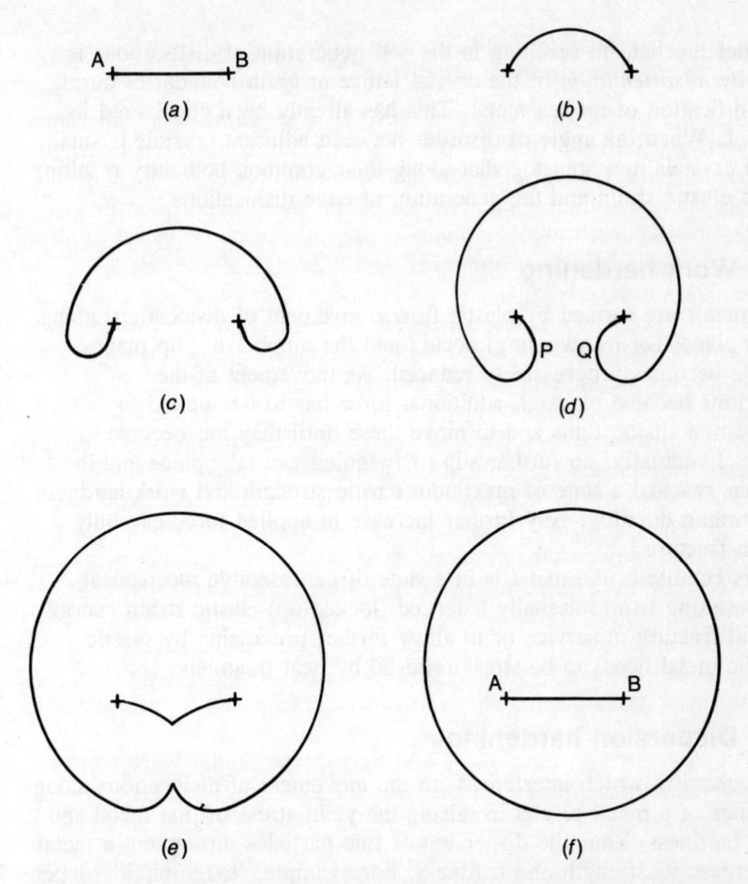

Fig. 10.15 The Frank-Read source (*a*) Dislocation with anchored ends (*b*) Dislocation starts to bow (*c*) Dislocation starts to sweep round anchored ends (*d*) Dislocation loop continues to expand (*e*) Dislocation loop breaks away (*f*) Dislocation loop complete but original dislocation line AB remains to reproduce itself again

a force is applied the ends of the dislocation cannot move. This causes the dislocation to bow outwards. Initially, the radius of the bow will be large but as the applied force is increased the radius is reduced. Once the bow has become a semi-circle the applied force has reached a maximum value, after which the force required to maintain growth of the loop can be progressively reduced. Eventually, proximity of the strain regions *P* and *Q* will result in their annihilation and the dislocation loop will be complete. Since the original dislocation line still remains, the process can keep repeating itself as long as adequate external force is applied, thus allowing the extensive slip necessary for forming by plastic flow to take place.

Another mechanism resulting in the self generation of dislocations is caused by *misorientation* of the crystal lattice at grain boundaries during the solidification of molten metal. This has already been considered in volume 1. When the angle of disorder between adjacent crystals is small, the two crystals may join together along their common boundary resulting in some elastic strain and the generation of edge dislocations.

10.6 Work hardening

When metals are formed by plastic flow, movement of dislocations along the slip planes (or by twinning) occurs and the number of slip planes available becomes progressively reduced. As movement of the dislocations become blocked, additional force has to be applied to generate new dislocations and to move these until they too become blocked. Eventually, no further slip or twinning can take place and the metal has reached a state of maximum tensile strength and work-hardness but minimum ductility. Any further increase in applied force can only result in fracture.

In this condition, the metal is in a state of considerable mechanical stress resulting from internally balanced (locked-up) elastic strain energy. To avoid fracture in service or to allow further processing by plastic flow, the metal needs to be stress relieved by heat treatment.

10.7 Dispersion hardening

Any mechanism which interferes with the movement of dislocations along slip planes of a metal results in raising the yield stress of that metal and also its hardness. Thus the dispersion of fine particles throughout a metal will increase its strength and hardness. For example, 'tough-pitch' copper is strengthened and hardened by the dispersion of particles of copper oxide. Further, heat-treatable aluminium alloys are hardened by the precipitation of intermetallic particles which hinder the movement of dislocations. The alloying of metals can introduce alien atoms into the crystal lattice to form interstitial or substitutional solid solutions. Again, such point defects hinder the movement of dislocations resulting in an increase in the yield stress and hardness of the metal.

10.8 Stress relief and recrystallisation

The stress relief and recrystallisation (annealing) processes for work-hardened metals has already been introduced in volume 1. The processes so described will now be reconsidered in terms of dislocation theory.

Reference back to Fig. 10.14 shows that in the vicinity of an edge dislocation there are regions of tension and compression in the crystal lattice. At the point where the lattice is in tension the region will possess high strain energy. Further, high potential energy is associated with dislocations congregating at the grain boundaries due to misorientation and to 'misfits' between adjacent crystals as previously discussed. These high energy regions initiate seed crystals (nucleation) during annealing processes.

When the temperature of cold-worked metals is raised, the first change to take place in the metal is stress-relief. The increased vibration of the atoms associated with the rising temperature allows them to approach their equilibrium positions. There is no change in the distortion of the grain structure and no reduction in the hardness and tensile strength of the metal. In fact, in the case of α-brass, not only is the tensile strength of the metal slightly increased by stress-relief annealing, but the corrosion resistance of the metal is improved. The intergranular corrosion of severely cold-worked metals due to the congregation of edge dislocations was discussed in Section 6.8.

At the relatively low temperatures associated with stress-relief only 'glide' of the dislocations along the slip planes can occur. However, if the temperature is raised, edge dislocations actually start to move out of their slip planes by a mechanism known as *climb*. This is shown diagrammatically in Fig. 10.16, where it can be seen that the last row of atoms of an edge dislocation moves to a new location. This mechanism is known as *positive climb* and it results in a reduction in strain energy in

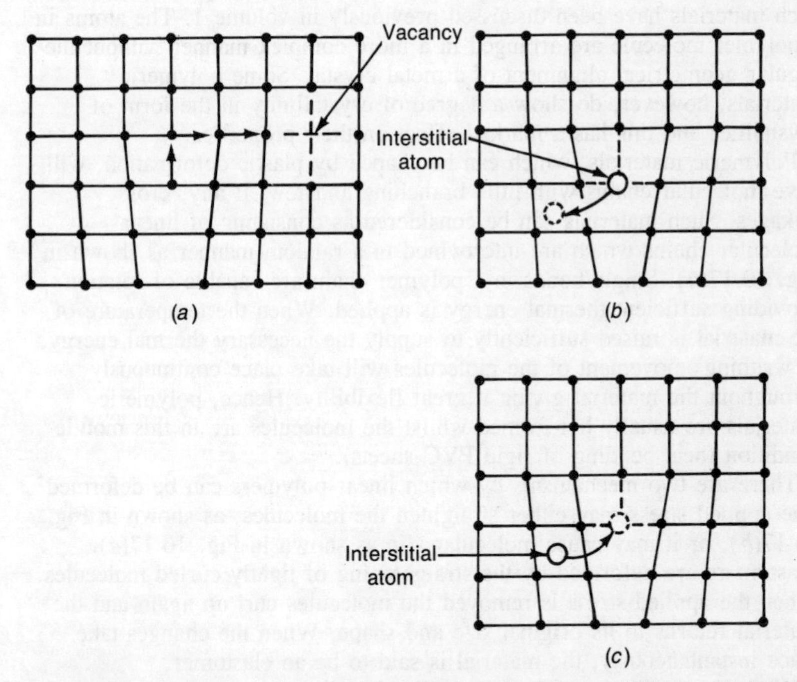

Fig. 10.16 Dislocation climb (*a*) Positive climb by vacancy diffusion (*b*) Positive climb (interstitial atom) (*c*) Negative climb (interstitial atom)

the region adjacent to the dislocation. If the last row of atoms of an edge dislocation is augmented as shown in Fig. 10.16(*b*), then the mechanism is known as *negative climb* and there is an increase in the strain energy in the region adjacent to the dislocation. Whole rows of atoms do not move simultaneously during climb, any more than they move en bloc during slip. It is assumed that the atoms move singly or in small groups in a series of 'jogs' until the total movement is complete. When heated, the movement (diffusion) of interstitial atoms in an alloy can also give rise to positive and negative climb. When heating of a cold-worked metal results in stress relief, it can be assumed that positive climb is taking place since there is a reduction in locked-up internal strain energy.

If the temperature is raised still higher to that required for annealing as described in volume 1, the elastic strains are totally eliminated and recrystallisation takes place, with seed crystals being initiated at the most heavily deformed and stressed areas. If the annealing temperature is maintained, grain growth occurs by the merging of adjacent grains. The annealing reduces the strength and hardness of the metal but improves its ductility and malleability so that it can again be cold-worked.

10.9 Deformation of materials (polymeric)

The molecular structures of polymeric materials and the properties of such materials have been discussed previously in volume 1. The atoms in a polymer molecule are arranged in a more complex manner without the regular geometrical alignment of a metal crystal. Some polymeric materials, however, do show a degree of crystallinity in the form of crystallites and this has a marked effect on their properties.

Polymeric materials, which can be shaped by plastic deformation, will have molecular chains with little branching and few, if any, cross linkages. Such materials can be considered as consisting of linear molecular chains which are intertwined in a random manner as shown in Fig. 10.17(*a*). Single bonds in a polymer chain are capable of rotation providing sufficient thermal energy is applied. When the temperature of the material is raised sufficiently to supply the necessary thermal energy, a 'writhing' movement of the molecules will take place continuously throughout the material giving it great flexibility. Hence, polymeric materials are usually hot-formed whilst the molecules are in this mobile condition (heat bending of rigid PVC sheets).

There are two mechanisms by which linear polymers can be deformed. The applied stress may either straighten the molecules, as shown in Fig. 10.17(*b*), or it may cause molecular slip as shown in Fig. 10.17(*c*). Elastomers are deformed by the straightening of tightly curled molecules. When the applied stress is removed the molecules curl up again and the material returns to its original size and shape. When the changes take place instantaneously, the material is said to be an elastomer.

When the applied stress causes molecular slip, permanent plastic deformation takes place and the process is not reversible since little or no

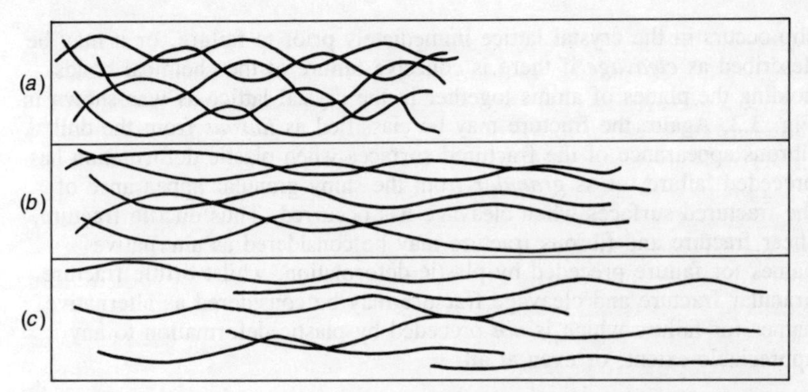

Fig. 10.17 Deformation of polymeric materials (*a*) No stress applied (*b*) Deformation by molecular straightening (*c*) Deformation by molecular slip

elastic strain occurs. In this instance, the applied stress must be sufficient to overcome the Van der Waals forces between the molecules. The introduction of a plasticiser, which separates the molecular chains and weakens the Van der Waals forces, acts as a lubricant, and aids the deformation of plastic materials. The molecular slip does not occur instantaneously, as in the straightening of elastomer molecules, but is time-dependent upon the viscosity of the material and the distortion is said to be *viscoelastic*.

It has been previously stated that for plastic deformation to occur, molecular branching and linking should be on a limited scale or not present at all. Where branching and linking is present, or when thermoplastics are below their glass-transition temperature, all the atoms are held together either by chemical bonds or intermolecular forces. Under such conditions only limited elastic deformation and recovery is possible and, if the applied stress is too great, the material will shatter.

10.10 Failure of metals (fracture)

The failure of metals through fracture is a complex subject which can only be dealt with briefly in this section. Fracture is said to have occurred when there is cohesive failure of a metal resulting in the metal separating into two or more portions.

When an external force (load) is applied to a metal component, the metal is put into a state of stress, and a corresponding elastic and/or plastic strain occurs. Eventually, if the stress is increased sufficiently the metal will fracture. Fracture is described in various ways depending upon the behaviour of material under stress or upon the mechanism of fracture, or even its appearance. For example, the fracture may be classified either as *ductile* or as *brittle* depending upon whether or not plastic deformation precedes cohesive failure. The fracture may also be classified as *shear* if slip occurs in the crystal lattice immediately prior to failure, or it may be described as *cleavage* if there is cohesive failure of the chemical bonds holding the planes of atoms together in the crystal lattice as was shown in Fig. 3.3. Again, the fracture may be classified as *fibrous* from the dull fibrous appearance of the fractured surfaces when plastic deformation has preceded failure, or as *granular* from the shiny granular appearance of the fractured surfaces when cleavage has occurred. Thus ductile fracture, shear fracture and fibrous fracture may be considered as alternative names for failure preceded by plastic deformation, whilst brittle fracture, granular fracture and cleavage fracture may be considered as alternative names for failure which is *not* preceded by plastic deformation to any appreciable extent, or even at all.

10.11 Ductile fracture

Figure 10.18 shows a typical, ductile tensile specimen during and after testing. The 'necking' of the specimen which occurs before fracture and the typical 'cup and cone' of this type of fracture are clearly shown. Initially, only elastic deformation occurs but, as the applied load is increased, plastic deformation occurs and the 'neck' begins to form. The increased load coupled with the reduction in cross-sectional area in the region of the neck results in the specimen being subjected to increased stress. This stress rapidly reaches a magnitude where small internal cavities start to appear. These cavities 'nucleate' (form) more easily if impurities or a second alloy phase is present, thus pure metals tend to be more ductile than impure metals and alloys. These cavities join up to form a crack which is roughly normal to the axis of the test piece. This weakens the metal still further as well as concentrating the stress, and fracture finally occurs by simple shear at an angle of 45° to the axis of stress.

Since ductile fracture is preceded by plastic deformation, it is reasonable to assume that the normal mechanisms of dislocation and slip occur in the crystal structure of the metal. Figure 10.19 shows how dislocations 'pile-up' when they are obstructed by a separate alloy phase or by an impurity. Such a 'pile up' nucleates micro-cracks in the metal. These micro-cracks tend to join together (coalesce) and spread until the material is weakened sufficiently for fracture to occur.

Ductile failure can also occur in compression, although this is not so common since the applied axial force tends to close any internal discontinuities. However, as the metal starts to bulge, axial surface cracks appear around the sides of the component. These cracks reduce the cross-sectional area of the remaining coherent metal and the stress builds up until failure occurs. Such cracks can be seen around the head of a rivet when it has been excessively cold-worked or hot-worked at too low a temperature.

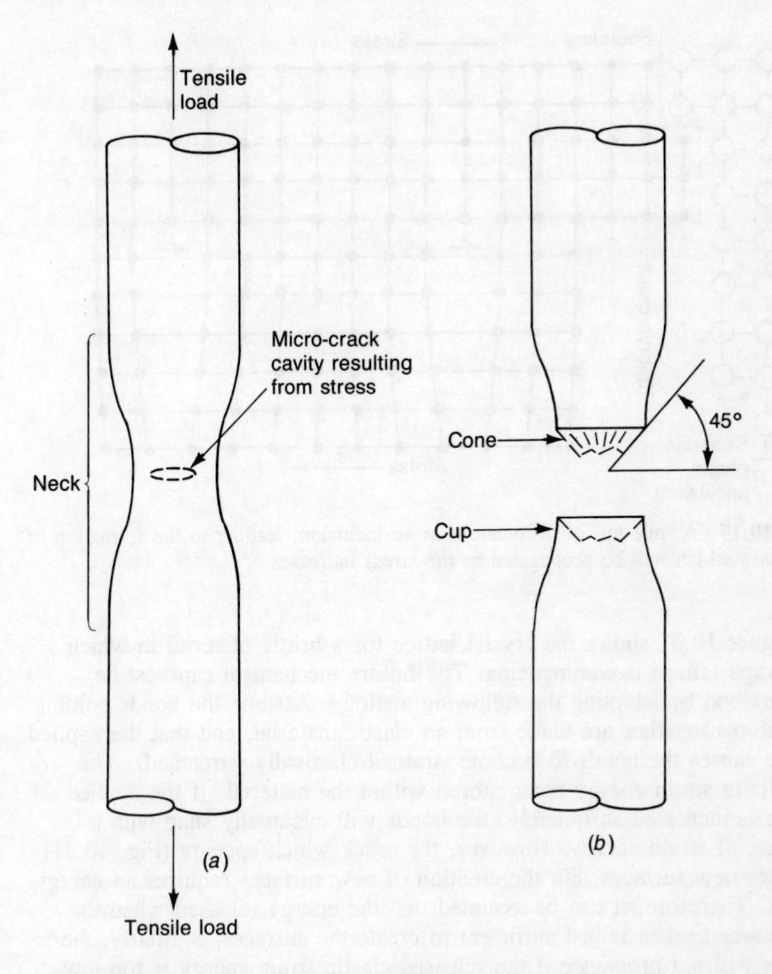

Fig. 10.18 Tensile test specimen (ductile) (*a*) Necking immediately prior to fracture (*b*) Typical ductile fracture showing cup and cone

10.12 Brittle fracture

This type of fracture is associated with non-metals such as glass, concrete and thermosetting plastics. In metals, brittle fracture occurs mainly when body-centred-cubic crystals and close-packed-hexagonal crystals are present, and rarely when face-centred-cubic crystals are present. Specimens fractured in this way show little or no plastic deformation. Figure 10.20 compares the stress/strain curves for a medium carbon steel when it is in the annealed and ductile condition and after it has been quench hardened and is brittle and lacking in ductility.

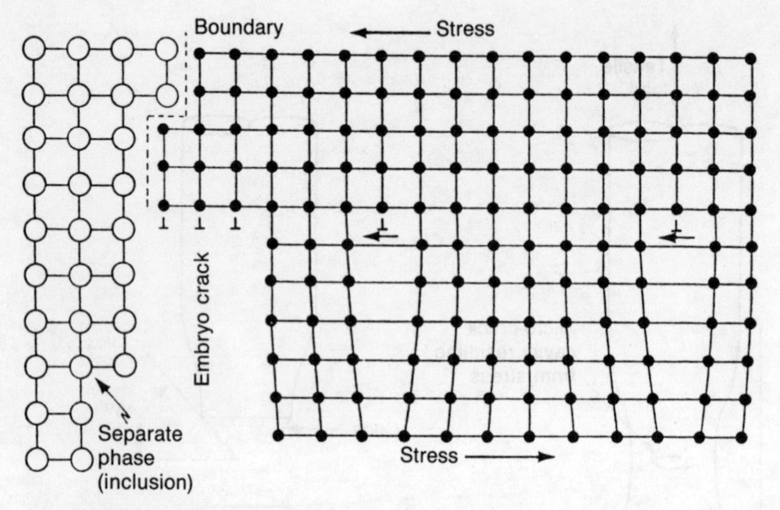

Fig. 10.19 A 'pile-up' of dislocations at an inclusion, leading to the formation of a fissure which will be propagated as the stress increases

Figure 10.21 shows the crystal lattice for a brittle material in which cleavage failure is commencing. The failure mechanism can best be understood by adopting the following analogy. Assume the bonds holding the atoms together are made from an elastic material, and that the applied force causes the bonds to become strained elastically (stretched). This results in strain energy being stored within the material. If the applied force is increased sufficiently, the bonds will eventually snap with a release of strain energy. However, the crack which appears (Fig. 10.21) creates new surfaces, but the creation of new surfaces requires an energy input. Therefore, it can be assumed that the energy released when the bond was broken is just sufficient to create the surfaces. Similarly, the crack will not propagate if the released elastic strain energy is too low.

10.13 Griffith crack theory

According to A.A. Griffith, the energy required to fracture a brittle material is not uniformly distributed over the volume of the material since minute faults and cracks at the surface or within the material create regions of energy concentration. Thus the real strength of the material is substantially lower than the theoretical strength of the material as derived from the inter-atomic forces.

Figure 10.22 shows a small elliptical crack in a rod of brittle non-metallic material. The relationship between the stress required to propagate the crack and the length of the crack can be determined from the expression:

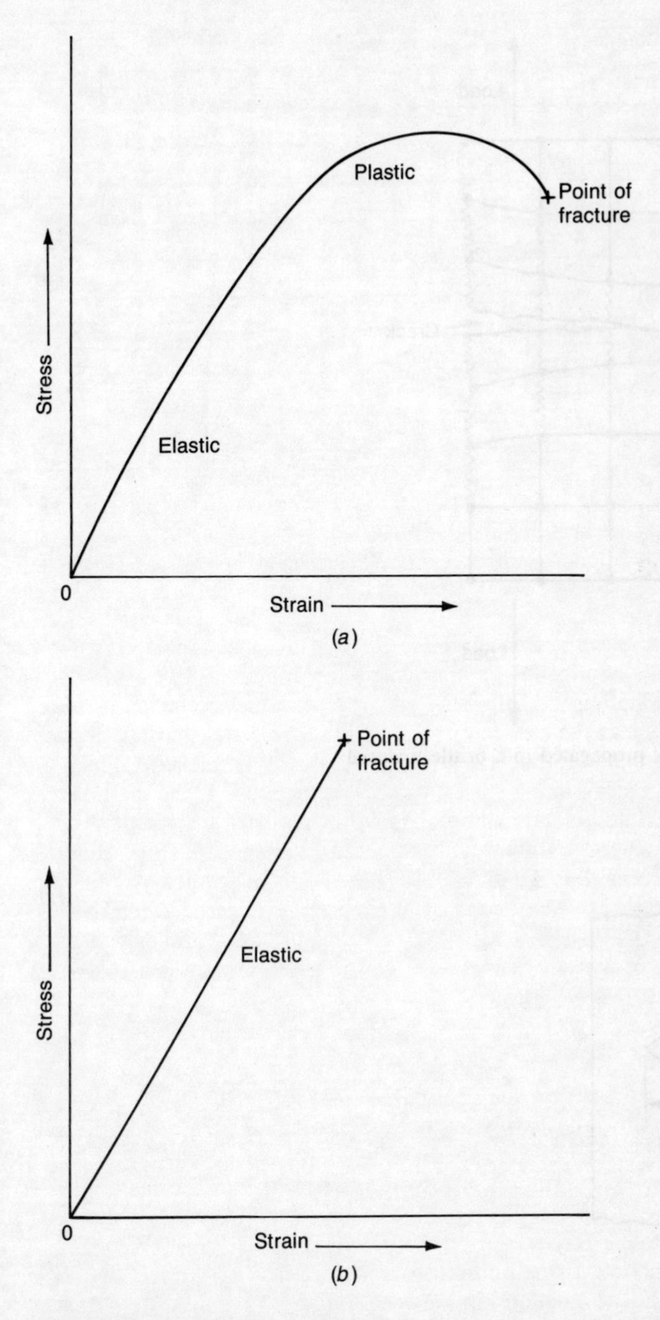

Fig. 10.20 Stress/strain curves for a medium carbon steel (*a*) In the annealed condition (*b*) After quench hardening

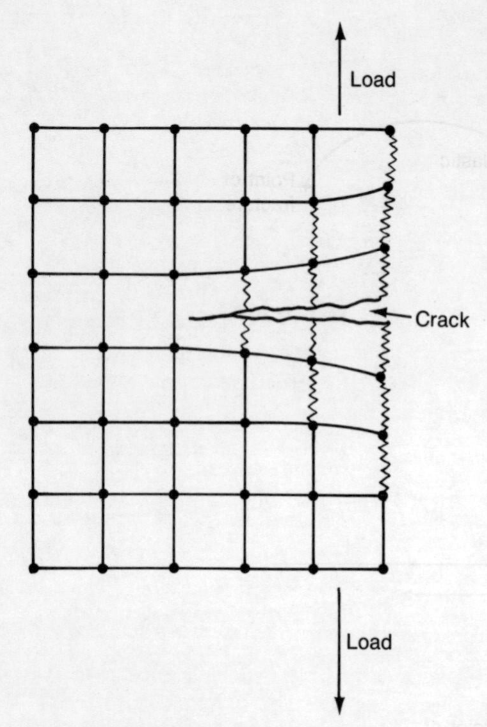

Fig. 10.21 Crack propagated in a brittle material

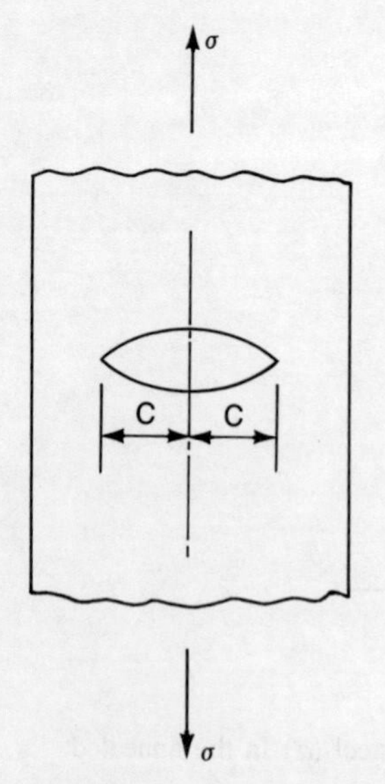

Fig. 10.22 Stress concentration due to micro-crack

$$\sigma = \sqrt{[(2\gamma E)/(\pi c)]}$$

where: σ = crack propagation stress;
γ = surface energy per unit area;
E = tensile modulus;
c = half length for major axis of crack.

Note that the multiplying factor 2 is derived from the fact that the crack which is propagated has two surfaces.

When the applied stress reaches the value σ, the micro-cracks present can begin to propagate and coalesce so that c increases. It can be seen that as c increases the value of σ decreases. Eventually, when fracture occurs and c spreads completely across the section so that the two portions separate, σ becomes zero. The Griffith relationship is sometimes approximated to:

$$\sigma = \sqrt{[(\gamma E)/c]}$$

from which expression, for a given material, the following proportionality can be derived:

$$c \propto 1/\sigma^2$$

Thus the larger the stress, the smaller will be the critical crack length to precipitate catastrophic failure.

The Griffith relationship was derived for brittle materials such as concrete, glass, ceramics and thermosetting plastics. It is not valid for ductile metals nor even for brittle metals where some plastic deformation, however small, always precedes fracture. When applied to metals, the relationship has to be modified to take into account the energy required for plastic deformation, as follows:

$$\sigma = \sqrt{\{[E(2\gamma + \gamma_p)]/(\pi c)\}}$$

where γ_p = the energy required for plastic deformation per unit area of crack.

Note: For most metals γ is smaller than γ_p.

10.14 Factors affecting crack formation

The effects of inclusions on crack nucleation and propagation in a material have already been discussed. However, other factors must also be taken into account. Any sudden change in section which will produce a stress concentration can lead to cracking. Under highly stressed conditions even a deep scratch or a tooling mark can lead to crack propagation. Faults in design or manufacture which cause stress concentrations are referred to as *incipient cracks*.

The amount by which the stress is raised is proportional to the depth of the notch and inversely proportional to the radius at the root of the notch. Cracks in brittle materials have a very small radius, so the stress concentration at the root of the crack is very high. The propagation of the crack can be arrested by drilling a hole at its root to increase its effective radius and thus reduce the stress concentration. Early attempts to produce ships with welded hulls led to disaster when cracks spread unhindered so that the ships broke up. Ships with riveted hulls did not suffer in this manner since any crack stopped at the next adjacent rivet hole. There is an approximate relationship between the applied stress and the stress at the root of the notch given by the expression:

$$\sigma_n = \sigma[1 + 2\sqrt{(L/r)}]$$

where: σ_n = stress at root of notch
σ = applied stress
L = length of notch
r = root radius of notch.

Thus the increase in stress due to the notch is given by the expression:

$$\sigma_i = 2\sqrt{(L/r)}$$

where σ_i = increase in stress.

In ductile materials, the stress concentration at the root of the notch is less likely to lead to crack propagation and failure since much of the available energy is dissipated in plastic flow and insufficient energy may be available for the creation of new crack surfaces. Thus cracks in ductile metals are less likely to result in fracture than cracks in brittle metals.

The speed of loading can also affect the behaviour of the metal, and this was introduced when discussing material testing in volume 1. This applied to tensile testing and, particularly, to Izod and Charpy impact testing where a notched test piece is loaded by a sudden blow. If the loading is applied too rapidly to an otherwise ductile metal, there may not be time for the dislocations associated with plastic deformation to take place and the metal will behave as though it were brittle.

Temperature can also have a profound effect on the behaviour of metals. As the temperature decreases, the movement of dislocations becomes more sluggish so that the internal stresses may exceed the shear strength for the metal. Therefore, brittle fracture is more likely to occur at low temperatures and may even occur in otherwise ductile metals if the temperature is sufficiently low. This becomes an important design consideration when selecting materials for equipment operating under arctic conditions and for space vehicles. Metals with body-centred-cubic crystals and close-packed-hexagonal crystals such as beryllium, chromium, molybdenum, plain carbon steels (particularly the ferrite content), tungsten and zinc all suffer from low temperature brittleness.

Figure 10.23 shows the effect of temperature on a low carbon steel when subjected to impact loading. At low temperatures the metal has a BCC ferrite structure and, as stated above, this exhibits 100 per cent brittle fracture at low temperatures. As the temperature rises the dislocations become less sluggish so that some ductile fracture can occur

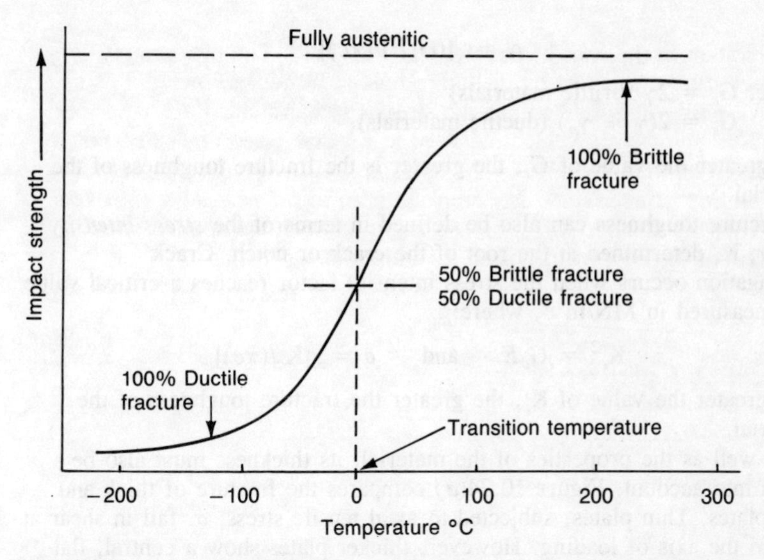

Fig. 10.23 Effect of temperature on the fracture of a low carbon steel

even in BCC ferrite. At the *transition temperature* the failed surface will show 50 per cent brittle fracture and 50 per cent ductile fracture. As the temperature is raised still further, the failed surface will show 100 per cent ductile fracture. The presence of carbon, phosphorus and nitrogen raises the transition temperature for mild steel and for this reason such impurities must be kept to a minimum when the steel has to operate at low temperatures. However, alloying elements such as nickel and manganese depress the transition temperature. Steel plates suitable for the welded hulls of ships, operating in the cold North Atlantic, will contain only 0.14 per cent carbon but as much as 1.3 per cent manganese.

10.15 Fracture toughness

The factors affecting fracture have just been considered in Section 10.14, including the effect of temperature on impact toughness. Toughness is defined as the ability of a material to resist transverse impact loading without failure. In standard tests such as the Izod test and the Charpy test, which were described in volume 1, a notched specimen is struck by a pendulum and the energy absorbed in bending and/or breaking the specimen is stated as a measure of the fracture toughness of the material.

It has been stated earlier that for a crack to propagate the critical elastic strain energy released at the root of the crack or notch must equal the energy to create new surfaces. Thus *fracture toughness* can be defined in terms of the critical elastic strain release rate, G_c, measured in kJ/m^2. To determine G_c the Griffith relationship discussed earlier can be modified as follows.

$$\alpha = \sqrt{[(G_c E)/(\pi c)]}$$

where: $G_c = 2\gamma$ (brittle materials)
$G_c = 2(\gamma + \gamma_p)$ (ductile materials).

The greater the value of G_c, the greater is the fracture toughness of the material.

Fracture toughness can also be defined in terms of the *stress intensity factor*, K, determined at the root of the crack or notch. Crack propagation occurs when the stress intensity factor reaches a critical value K_c, measured in $MN/m^{1.5}$, where:

$$K_c^2 = G_c E \quad \text{and} \quad \sigma = \sqrt{[K_c/(\pi c)]}$$

The greater the value of K_c, the greater the fracture toughness of the material.

As well as the properties of the material, its thickness must also be taken into account. Figure 10.24(*a*) compares the fracture of thick and thin plates. Thin plates, subjected to axial tensile stress, σ, fail in shear at 45° to the axis of loading. However, thicker plates show a central, flat area where fracture is normal to the axis of loading. The thicker the plate the greater is this normal area and the greater becomes its influence on the stress intensity factor. Figure 10.24(*b*) shows how K_c has a high value for thin plate but that the value of K_c becomes less for thick plate, until the curve almost levels out at some value K_{Ic}. This lower limiting value of the critical stress intensity factor is referred to as the *plane strain fracture toughness factor*. In addition to plate thickness, fracture toughness is also affected by such factors as:

(*a*) *Composition*. The presence of alloying elements and/or impurities can have a significant effect upon the plane strain toughness factor. Impurities such as phosphorus and sulphur, even in small amounts, seriously reduce the fracture toughness of ferrous metals.

(*b*) *Heat treatment*. Quench hardening plain carbon and alloy steels substantially reduces their fracture toughness. However, subsequent tempering progressively restores the fracture toughness, as the tempering temperature is raised, more rapidly than the hardness is reduced. The fracture toughness of heat-treatable aluminium alloys is also affected by solution treatment and precipitation hardening. The fracture toughness of other non-ferrous metals and alloys is affected by work-hardening during processing and by stress relief annealing.

(*c*) *Service conditions*. The fracture toughness value can also be affected by such factors as temperature, environment (corrosion) and cyclical loading (fatigue).

10.16 Failure of metals (fatigue)

Since it is estimated that more than 75 per cent of failures in engineering components can be attributed to fatigue failure, and as the reliability and

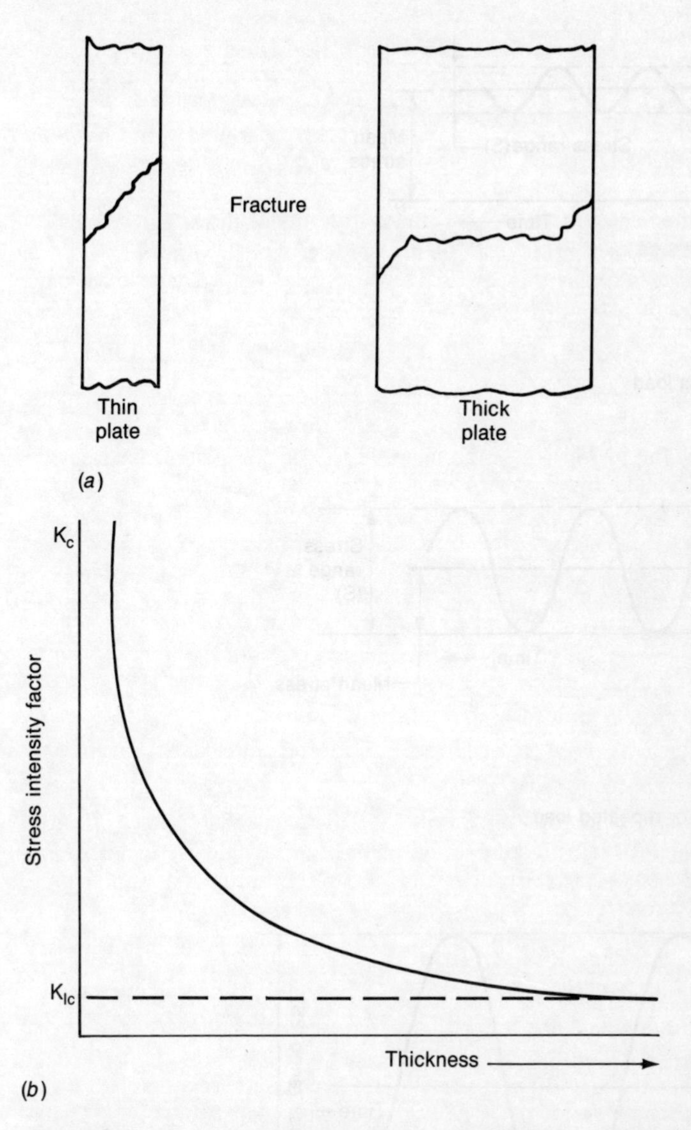

Fig. 10.24 Effect of a plate thickness on crack propagation (*a*) Effect of plate thickness on crack profile (*b*) Effect of plate thickness on stress intensity factor

performance expected from engineering products rises, the need to understand this mode of failure becomes increasingly important. Fatigue failure, the factors affecting fatigue and fatigue testing for both metals and polymers were introduced in volume 1. The effect of mean stress will now be considered.

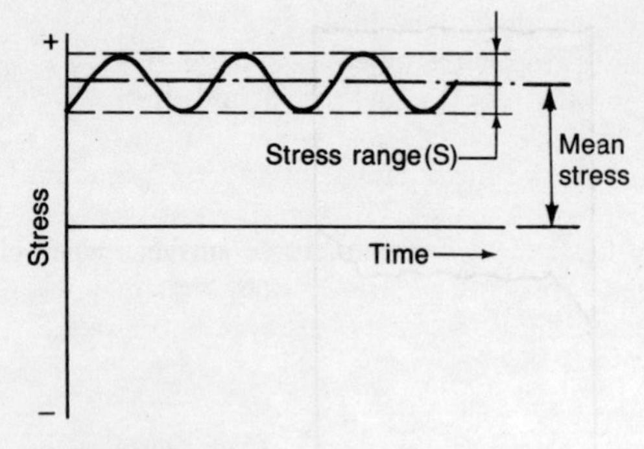

(a) Fluctuating load

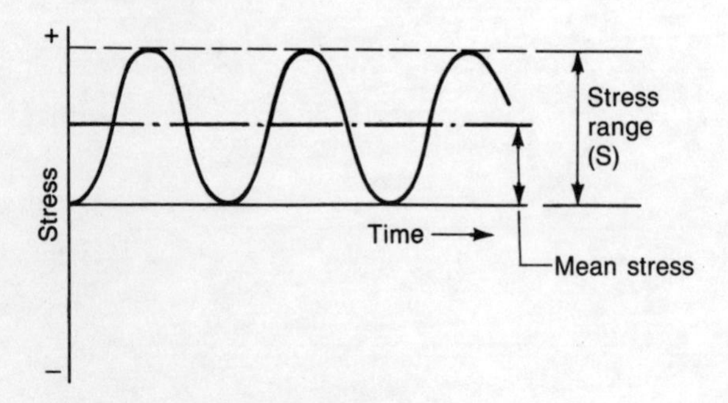

(b) Pulsating or repeated load

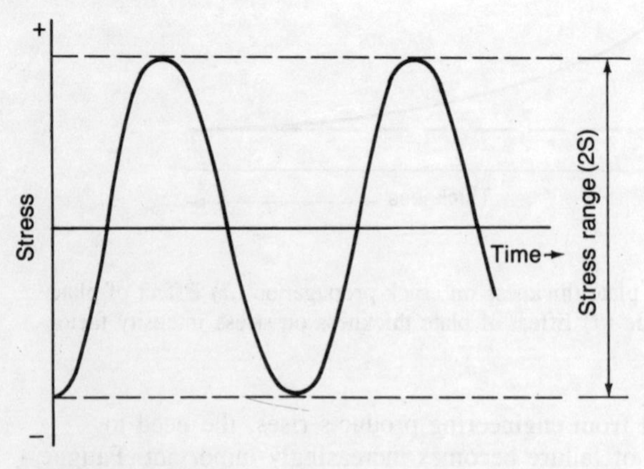

(c) Alternating load

Fig. 10.25 Conditions of fatigue loading

Figure 10.25 shows the conditions for fatigue loading. With a fluctuating load the mean stress is greater than the stress range; with a pulsating or repeated load the mean stress is equal to half the stress range, and with an alternating load, the mean stress is zero. In each instance the stress amplitude is half the stress range. Figure 10.26 shows typical S/N diagrams, as considered in volume 1, where the test piece was subjected to an alternating load. However, this can be misleading if there is a steady state component stress (mean stress). Both Goodman and Soderberg investigated the relationship between stress amplitude, mean stress and fatigue limit.

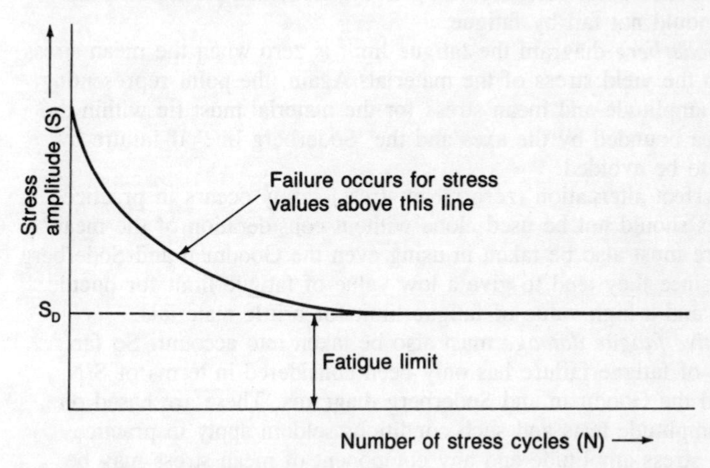

(a) S-N diagram for a typical steel

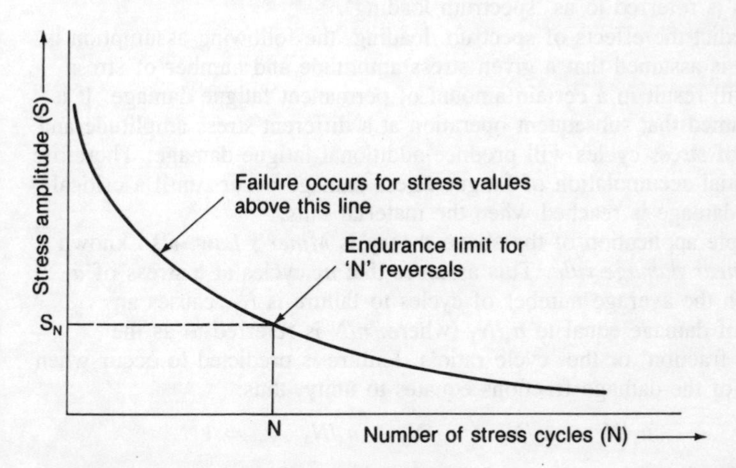

(b) S-N diagram for a typical non-ferrous alloy

Fig. 10.26 Stress reversal curves (S-N diagrams)

Goodman and Soderberg diagrams are shown in Fig. 10.27. When the mean stress is zero (perfect alternation), the fatigue limit is at a maximum value before failure occurs. However, if a steady state stress is superimposed upon the cyclical stress then this must also be taken into account. This steady state stress is the *mean stress* in Fig. 10.25.

In the *Goodman* diagram the fatigue limit is zero when the mean stress is equal to the tensile strength of the material, since the material will fail at this value before any cyclical loading can commence. Therefore, if the point representing the stress amplitude and mean stress for any given set of conditions lies within the area bounded by the axes and the 'Goodman line', the shaded area, then according to the Goodman relationship the material should not fail by fatigue.

In the *Soderberg* diagram the fatigue limit is zero when the mean stress is equal to the yield stress of the material. Again, the point representing the stress amplitude and mean stress for the material must lie within the shaded area bounded by the axes and the 'Soderberg line' if failure by fatigue is to be avoided.

Since perfect alternation (zero mean stress) rarely occurs in practice, S/N curves should not be used alone without consideration of the mean stress. Care must also be taken in using even the Goodman and Soderberg diagrams since they tend to give a low value of fatigue limit for ductile materials, and a high value of fatigue limit for brittle materials.

Cumulative fatigue damage must also be taken into account. So far, prediction of fatigue failure has only been considered in terms of S/N curves and the Goodman and Soderberg diagrams. These are based on constant amplitude tests and such conditions seldom apply in practice where the stress amplitude and any component of mean stress may be constantly varying in an unpredictable manner. Such a spread of stress variables is referred to as 'spectrum loading'.

To predict the effects of spectrum loading, the following assumption is made. It is assumed that a given stress amplitude and number of stress cycles will result in a certain amount of permanent fatigue damage. It is also assumed that subsequent operation at a different stress amplitude and number of stress cycles will produce additional fatigue damage. Therefore a sequential accumulation of fatigue stress damage occurs until a critical level of damage is reached when the material fails.

A simple application of the above theory is *Miner's Law*, also known as the *linear damage rule*. This assumes that n_1 cycles at a stress of σ_1, for which the average number of cycles to failure is N_1, causes an amount of damage equal to n_1/N_1 (where: n/N is referred to as the 'damage fraction' or the 'cycle ratio'). Failure is predicted to occur when the sum of the damage fractions equates to unity, thus:

$$n_1/N_1 + n_2/N_2 + n_3/N_3 + n_4/N_4 \ldots = 1$$

Despite its simplicity, Miner's Law is easy to apply and gives a reasonable level of prediction reliability. However, the ratios n/N are not

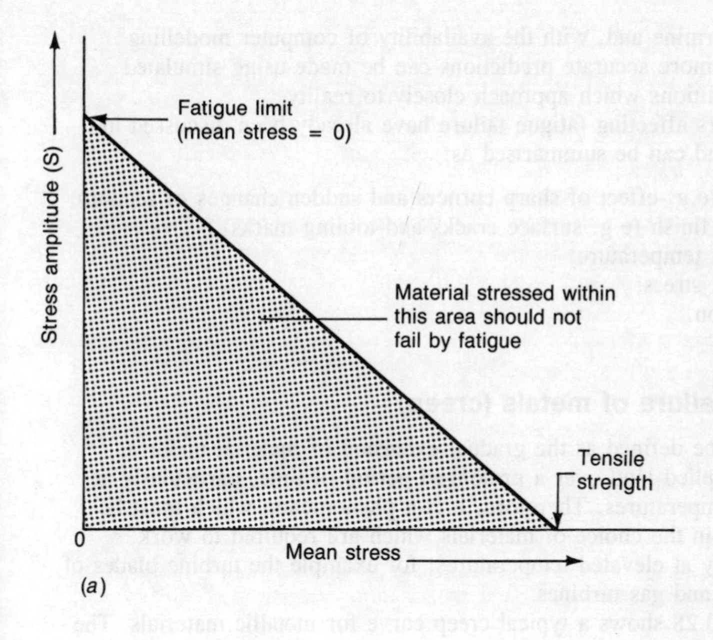

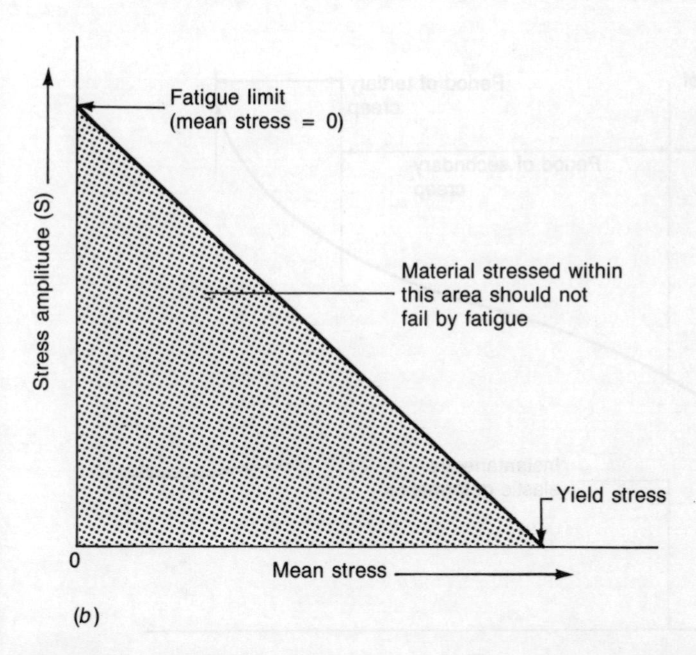

Fig. 10.27 The effect of mean stress (*a*) Goodman diagram (*b*) Soderberg diagram

easy to determine and, with the availability of computer modelling techniques, more accurate predictions can be made using simulated service conditions which approach closely to reality.

The factors affecting fatigue failure have already been discussed in volume 1 and can be summarised as:

(*a*) design (e.g. effect of sharp corners and sudden changes of section);
(*b*) surface finish (e.g. surface cracks and tooling marks);
(*c*) ambient temperature;
(*d*) residual stress;
(*e*) corrosion.

10.17 Failure of metals (creep)

Creep can be defined as the gradual extension of material under a constant applied load over a prolonged period of time, particularly at elevated temperatures. Therefore, it is a phenomenon which must be considered in the choice of materials which are required to work continuously at elevated temperatures, for example the turbine blades of jet engines and gas turbines.

Figure 10.28 shows a typical creep curve for metallic materials. The

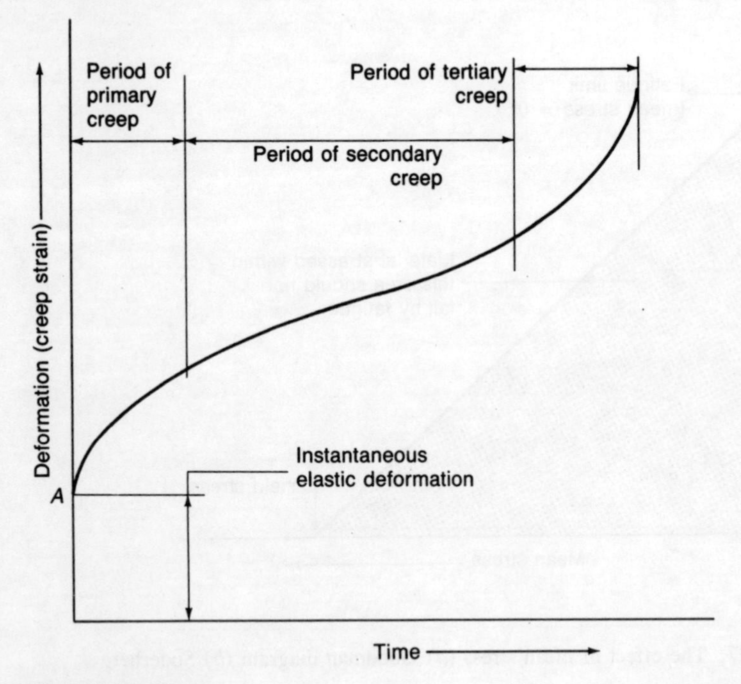

Fig. 10.28 Creep

three periods of creep have already been defined in volume 1. The creep curve for any given metal will change with any variation in applied stress and/or temperature and this is shown in Fig. 10.29.

The variation in the curves is best understood by examining the mechanisms of creep. Creep is due to:

(*a*) plastic deformation caused by normal dislocation along slip planes in crystalline materials (metals);
(*b*) plastic deformation caused by viscous flow at the grain boundaries where misorientation occurs.

Consider curve P in Fig. 10.29. At first, plastic deformation resulting in primary creep is rapid as dislocation can take place relatively unhindered. This is because the thermal agitation, due to the elevated temperature of the metal, enables such barriers as solute atoms and dislocation pile-ups at the grain boundaries to be overcome. However, whilst the temperature of the metal and therefore the thermal agitation remains constant, the barriers to dislocation build up and the rate of creep decreases.

During secondary creep, plastic deformation continues more slowly as the work-hardening effect of the deformation increases but is offset, to some extent, by the recovery processes associated with the elevated temperature of the metal. This balance gives rise to a constant creep rate. Deformation by viscous flow as well as by dislocation occurs during

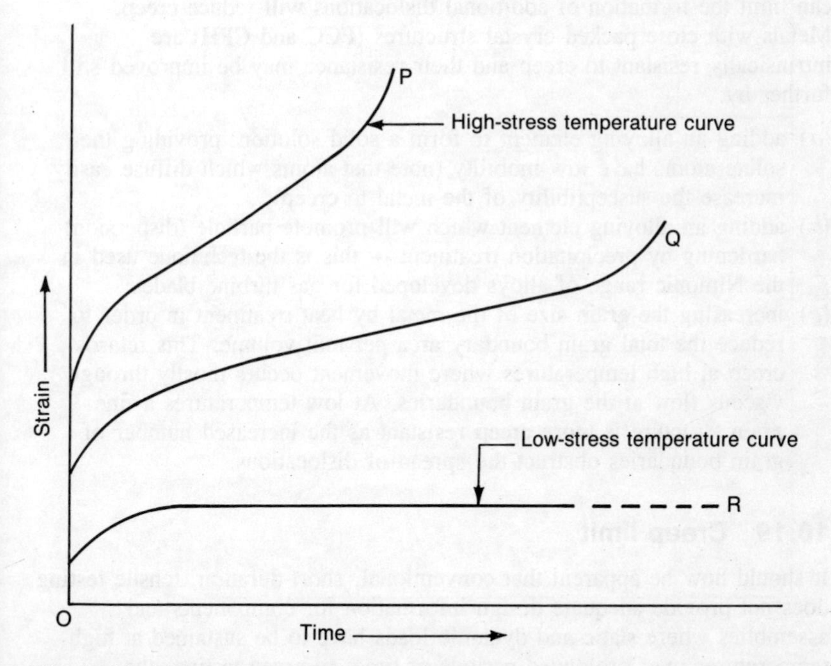

Fig. 10.29 Effect of stress and/or temperature on creep rate

secondary creep. A high level of stress has the same effect on the creep curve as a high temperature since a high level of stress can overcome the barriers to dislocation for a given level of thermal agitation and recovery.

During tertiary creep micro-cracks begin to appear at the grain boundaries as the barriers to dislocation movement become too great for the thermal agitation and the applied stress to overcome. These micro-cracks result in a rapid reduction in cross-section (necking) leading to a rapid increase in creep rate and fracture (creep failure).

Consider curve Q in Fig. 10.29. The creep rate is obviously lower resulting from a lower applied stress and/or lower temperature. Not only does the lower stress reduce the ability of dislocations to overcome such barriers as solute atoms and dislocation pile-ups at the boundaries, but the movement of the dislocations themselves will be more sluggish because of the lower temperature.

Consider curve R in Fig. 10.29. Here the temperature is too low for recovery to occur and the work-hardening which occurred during primary creep results in the stiffness of the metal becoming too great to be overcome by the limited applied stress and creep becomes negligible.

10.18 Creep resistance (metals)

It has already been shown that creep is dependent upon the movement of dislocations, therefore anything which can limit this movement and which can limit the formation of additional dislocations will reduce creep. Metals with close packed crystal structures (FCC and CPH) are intrinsically resistant to creep and their resistance may be improved still further by:

(*a*) adding an alloying element to form a solid solution, providing the solute atoms have low mobility (note that atoms which diffuse easily increase the susceptibility of the metal to creep);
(*b*) adding an alloying element which will promote particle (dispersion) hardening by precipitation treatment — this is the technique used in the Nimonic range of alloys developed for gas turbine blades;
(*c*) increasing the grain size of the metal by heat treatment in order to reduce the total grain boundary area per unit volume. This retards creep at high temperatures where movement occurs mostly through viscous flow at the grain boundaries. At low temperatures a fine grain structure is more creep resistant as the increased number of grain boundaries obstruct the spread of dislocations.

10.19 Creep limit

It should now be apparent that conventional, short-duration, tensile testing does not provide adequate design information for components and assemblies where static and dynamic loads have to be sustained at high temperatures over prolonged periods of time. In creep testing, the specimen is subjected to a constant tensile load whilst it is maintained at a constant high temperature. Usually, several specimens of the same metal are tested at the same temperature but with different applied loads. The maximum stress which can be applied with no measurable creep is called the *creep limit* or *creep stress*.

This is not a particularly satisfactory method of determining the creep resistance of a metal since the test may take several months to complete and the results are dependent upon the sensitivity and accuracy of the extensometers used as the extensions can be extremely small. A more useful measure of creep resistance is the *rupture stress* for a specified time and temperature. For example, a low carbon steel whose rupture stress for a life of 10 k hours is 220 MPa at 400°C, but at 500°C the rupture stress is limited to only 60 MPa.

Alternatively, the *Lawson-Miller parameter* may be used to estimate rupture time for a given stress, or the rupture stress for any given time. A curve, similar to that shown in Fig. 10.30, has to be prepared for each material using experimental data in conjunction with the relationship:

$$P = \theta(C + \log_e t)$$

where: P = *the Lawson-Miller parameter*
θ = the service temperature (K)
C = constant (usually 20)
t = the time to rupture (hours)

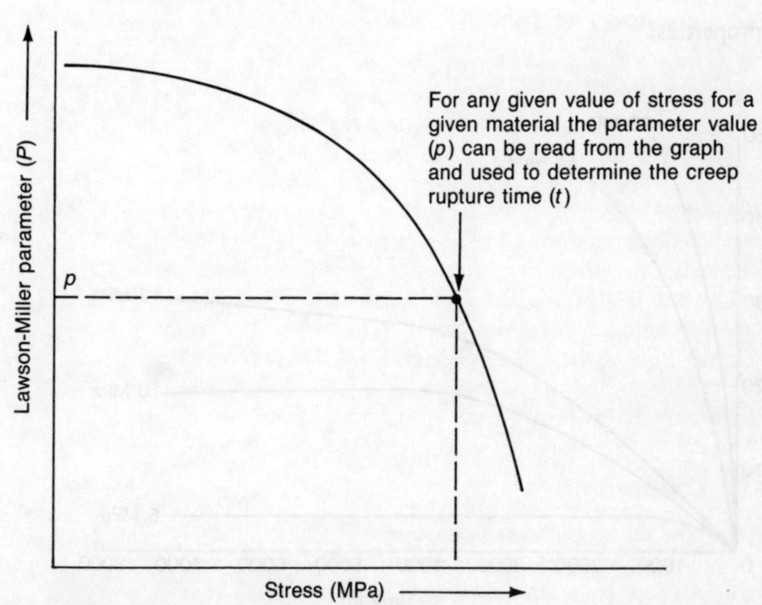

Fig. 10.30 Larson-Miller parameter curve

To use this expression to estimate the time to rupture for a given stress, it is rearranged to make 't' the subject, thus:

$$\log_e t = (P/\theta) - C.$$

10.20 Failure of polymers (creep)

Creep in metal usually occurs at high temperatures, whereas polymeric materials are subject to creep effects even under normal ambient conditions. Figure 10.31 shows typical creep curves for cellulose acetate at a temperature of 25°C. It can be seen that the time to rupture is very much shorter than for a metal and that the degree of creep is substantially greater. These curves are typical for most thermoplastics.

As for metals, the creep properties of plastics are also largely dependent upon service temperature. Below the glass transition temperature, T_g, the material will be rigid and the creep rate will be low. As the temperature rises, the creep rate and the elongation will increase. Above the glass transition temperature, T_g, viscoelastic deformation occurs resulting in greater elongation, but at a lower creep rate, for a given applied stress.

10.21 Material specification

The specification of a material for any given engineering application involves the consideration of three sets of parameters; these are:

(*a*) properties;

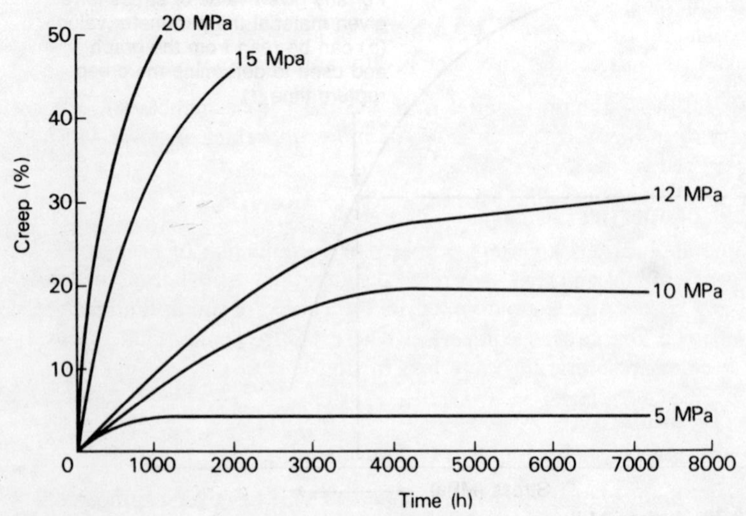

Fig. 10.31 Typical creep values (cellulose acetate)

(*b*) processing;
(*c*) commercial viability.

Inevitably, the final choice will be a compromise between all the above.

Properties

Throughout this book, and in volume 1, the general properties of the main groups of materials have been considered, together with an understanding of how those properties are related to the physical and chemical structure of the materials; also, how the properties may be modified by processing (e.g. plastic deformation, heat treatment, anti-corrosion finishing). From this general knowledge, and from a knowledge of the design requirements and the service requirements, it is possible to refer to British and International Standards and manufacturers' data sheets in order to draw up a 'short list' of potentially suitable materials for further consideration.

Processing

Unless the production run warrants investment in new plant, the materials chosen must be suitable for processing on the existing plant. Alternatively, some of the processing may have to be sub-contracted to firms who have suitable facilities.

Commercial viability

This involves a number of compromises between:

(*a*) cost, including unit material cost, processing cost, after sales servicing costs and warranty claims;
(*b*) availability, use of standard materials and standard sizes wherever possible;
(*c*) repeatability of quality and reliability in service;
(*d*) customer satisfaction, customer loyalty and customer safety.

To select the optimum material from the short list a number of techniques can be used and some of the more important of these will now be considered.

Critical properties

For some applications a certain property or combination of properties may be paramount and may override considerations of material cost and processing costs. An example would be the choice of the optimum material for a key aircraft component where failure could result in a major accident resulting in heavy loss of life.

Property limits

There are three property limits to be considered:

(*a*) *Upper limit*, above which the component will be 'over-engineered'. This is not only wasteful of resources, and often increases the cost of processing unnecessarily, it can result in an excessively high unit material cost. Materials with properties above this upper limit can be excluded from consideration.

(*b*) *Lower limit*, below which the component will not function correctly, will not have an adequate service life, and may prove to be unsafe. Materials with properties below the lower limit must on no account be selected or even considered, however low in cost.

(*c*) *Target value* is, as its name implies, a property value or set of values which the chosen material should achieve within close limits set by the designer.

Cost per unit property

Materials are compared with a particular critical property or set of properties on a 'cost per unit property' basis. Thus, if the requirement of a material for a particular application in an aircraft or motor vehicle is its fatigue strength/density ratio, then a group of suitable materials with the required properties would be evaluated on a relative cost per unit property basis.

Merit rating

This is a property 'handicapping' exercise. Each property for each material is given a relative merit score. This score is then multiplied by a predetermined 'weighting factor' depending upon the relative importance of the property for the application under consideration. The merit rating is the sum of the individual weighted scores, and the material which is finally chosen is the one with the highest merit rating.

This procedure does not take into account the relative costs of the materials considered. A variation on the merit rating procedure is called the *cost-modified merit rating*. Comparable cost-modified merit rating factors can be determined from the expression:

$$\frac{-(\text{material unit cost})}{(\text{maximum allowable cost})} \times (\text{merit rating})$$

Any selection made on the basis of any of the above procedures can only, at best, be a *draft specification*. An initial batch of components should be made from the chosen material and rigorously tested by simulation and field trials. Such procedures are widely used when developing new aircraft and vehicles, and the results of such tests may influence the final material specification used in production. More problematical are major structures, such as bridges, which are manufactured on a 'one off' basis. Two procedures may be adopted when selecting materials for such projects.

(*a*) One is to examine the designs of previous, similar structures and follow accepted practice. Case studies of failures should be examined in order to avoid repetition of the faults in design, manufacture and material selection which caused such failures.

(*b*) Where new and attractive materials have been developed for which there are no precedents, the designer must proceed with caution. Large structures are usually made up of a number of smaller components, units or sub-assemblies, many of which may be used repetitively in relatively large quantities. In this case prototypes of these smaller components, units and sub-assemblies should be given simulation tests and field tests to destruction. Having evaluated the suitability of the new material in this manner, the results may be extrapolated using sophisticated computer modelling techniques.

All successful companies will have an ongoing programme of testing and evaluation of the new materials which are being continually developed, and the assessment of the suitability of these new materials for their products.

11 Exercises

(Questions 1 to 10 inclusive are based mainly upon Chapter 1.)

1. Discuss the relative advantages and limitations of alloy steels compared with plain carbon steels.
2. Select FIVE of the following alloying elements and, in each instance, explain the effect of adding the alloying element to a carbon steel.
   Chromium, cobalt, lead, manganese, molybdenum, nickel, phosphorous, silicon, and tungsten.
3. Explain why nickel and chromium are frequently used together in alloy steels.
4. Explain what is meant by the term *temper brittleness* when related to alloy steels, what causes this phenomenon, and how it can be avoided.
5. (*a*) In terms of the effects on the properties of a steel, explain the differences between an alloying element which forms a solid solution and an alloying element which forms an intermetallic compound with the parent metal.
   (*b*) With the aid of sketches explain the difference between a substitutional solid solution and an interstitial solid solution.
6. (*a*) Compare the properties, composition and applications of the following groups of stainless steels:
   Autensitic;
   Ferritic;
   Martensitic.
   (*b*) Describe the precautions which must be taken when heat treating and welding stainless steels to avoid *temper brittleness* and *weld decay*.

7. Compare the essential differences between a stainless steel and a heat resisting steel in terms of composition, and properties.
8. Compare the essential differences between a maraging steel and a conventional alloy steel in terms of composition, properties, and heat-treatment.
9. Referring to manufacturers' literature, compare the properties, composition and applications of typical alloy steels from the following groups:
   nickel-chromium steels;
   nickel-chromium-molybdenum steels;
   nickel-chromium-vanadium steels.
10. Explain how elements such as cobalt, molybdenum and tungsten improve the hardness of alloy steels when operating at elevated temperatures.

(Questions 11 to 20 inclusive are based mainly upon Chapter 2.)

11. Explain why the iron-carbon phase equilibrium diagram is unsuitable for predicting the changes which take place during the quench hardening of a plain carbon steel.
12. (*a*) Describe the essential difference between *hardness* and *hardenability*.
    (*b*) Explain what is meant by the term *ruling section*.
13. Explain the essential differences between:
    (*a*) a 1 per cent plain carbon steel, and
    (*b*) a 0.3 per cent carbon 3.0 per cent nickel alloy,
    in terms of hardness and hardenability.
14. Sketch a time-temperature-transformation curve for a high-carbon steel and explain how it can be used to determine the critical cooling rate for that steel and how it can be used to predict the changes which will take place during cooling.
15. With reference to a time-temperature-transformation curve for a plain carbon steel of eutectoid composition, explain the significance of the M_s and M_f temperatures.
16. Explain why quench-hardened plain carbon and alloy steels need to be tempered before use.
17. Compare and contrast the following processes for hardening plain carbon and alloy steels in terms of their relative advantages and limitations.
    (*a*) Austempering.
    (*b*) Martempering.
18. Compare the advantages and limitations of either Austempering or Martempering with conventional quench hardening followed by tempering.
19. Compare and contrast the heat treatment of maraging steels with the heat treatment of conventional alloy steels.

20. Compare and contrast the hardening of a *high-speed steel* with the hardening and tempering of a plain carbon steel. Pay particular attention to the difference between secondary hardening and tempering, and the precautions which must be taken to avoid temper brittleness, grain growth, and cracking.

(Questions 21 to 32 inclusive are based mainly upon Chapter 3.)

21. List the basic properties of ceramic materials and explain how these properties make them important engineering materials.
22. In each instance, state the name, properties and a typical application for a ceramic material which is representative of each of the main groups: *crystalline ceramics, amorphous ceramics*, and *bonded ceramics*.
23. Explain the essential differences in properties and behaviour under load between metallic crystals and ionically-bonded ceramic crystals.
24. With the aid of sketches, explain the essential differences between the following ceramic structures: *chain structure, sheet structure, framework structure*, and describe the properties of ceramic materials possessing such structures.
25. For the following groups of ceramic materials, name a typical application and a typical material suitable for that application, giving reasons for your choice: *clay refractories, common refractories*, and *high grade refractories*.
26. In each instance, describe TWO typical engineering applications for metallic: *oxides; borides; nitrides* and *carbides*, giving reasons for your choice based upon the properties, cost and availability of these materials.
27. Describe THREE processes by which ceramic materials may be shaped prior to firing or sintering.
28. (*a*) Select TWO types of glass and describe their composition, properties and typical applications.
    (*b*) Describe the heat treatment processes available for making glass less susceptible to fracture.
29. Compare the properties of 'E'-glass, 'S'-glass, and 'M'-glass as used in producing reinforcement fibres and give a typical application appropriate to each.
30. Compare and contrast the advantages and limitations of the following systems of reinforcing concrete.
    (*a*) Simple reinforcement.
    (*b*) Prestressed reinforcement.
    (*c*) Post-tensioned reinforcement.
31. Explain what is meant by a *particle-hardened* composite and a *dispersion-hardened* composite, giving an example of such a material and a typical application in each case.

32. Describe the general properties and describe suitable applications for: (*a*) carbon fibre, and (*b*) 'whiskers'.

(Questions 33 to 40 inclusive are based mainly upon Chapter 4.)

33. Discuss the essential requirements of tool and die steels and explain how these requirements are met by the addition of suitable alloying elements.
34. Explain, with the aid of examples, what is meant by *intrinsic* properties, and what is meant by *conferred* properties as applied to cutting tool materials.
35. Compare and contrast the properties and applications of the following tool materials: *high carbon steel, high-speed steel, high-carbon-chrome die steel*, and *tungsten carbide*.
36. State the composition, properties and TWO typical applications of 'Stellite'.
37. Explain the essential differences between the following cutting tool materials and state a typical application for each material giving reasons for your choice.
    (*a*) Tungsten carbide.
    (*b*) Mixed carbides.
    (*c*) Titanium carbide.
    (*d*) Coated carbides.
38. Describe typical applications of *cubic boron nitride* and *diamond* as cutting tool materials and explain the precautions which must be taken in their use.
39. List the common abrasives used for grinding and honing operations, giving typical applications for each material with reasons for your choice.
40. (*a*) Describe the main factors which must be considered when selecting a cutting tool material for a given application.
    (*b*) For a cutting or forming operation of your choice present a reasoned argument for the selection of a suitable tool material.

(Questions 41 to 55 inclusive are based mainly upon Chapter 5.)

41. Compare and contrast the differences between welding and brazing processes and show, with the aid of sketches, the essential difference in the design of joints for these processes.
42. Describe the essential differences between *reducing, oxidising* and *neutral* welding flames, stating under what circumstances they would be used.
43. Explain what is meant by the *heat-affected zone* of a welded joint and describe how it affects the properties of the metals being joined and the strength of the joint.
44. Describe the chemical reactions which take place during oxy-acetylene welding and how these reactions can affect the joint.

45. Explain the need for, and essential requirements of, a flux, and describe under what circumstances such a flux would be needed when oxy-acetylene welding.
46. Describe the composition, need for, and function of the flux coating of an arc-welding electrode.
47. Describe the principles of MIG and TIG welding and compare the advantages and limitations of these processes.
48. Describe the principles of *spot* and *butt* welding processes.
49. The principal factors to be assessed when judging the quality of a weld are:
    (*a*) shape of the profile;
    (*b*) uniformity of surface and freedom from surface defects;
    (*c*) degree of undercutting;
    (*d*) smoothness of join where weld is recommenced;
    (*e*) penetration of bead and degree of root penetration;
    (*f*) degree of fusion;
    (*g*) non-metallic inclusions and cavities.
    Describe the effect of FIVE of these factors on the strength of the joint, suggest possible causes of defects in the factors chosen, and suggest suitable remedies in technique to prevent the defects occurring in the first place.
50. Discuss the essential differences between silver-soldering and brazing, and with the aid of examples, explain where each process could be used to the greatest advantage.
51. Compare the effects of welding and brazing on the properties of the materials being joined.
52. List the advantages and limitations of hard-soldering compared with welding.
53. Compare the advantages and limitations of welding with riveting as a means of making strong, permanent joints between metal components.
54. Explain what is meant by *weld-decay* in alloy steels and describe its effect on the strength of the joint and explain how this effect can be avoided.
55. Explain how maraging steels may be welded and describe any precautions which must be taken to ensure a sound joint.

(Quesitons 56 to 72 inclusive are based mainly upon Chapter 6.)

56. With the aid of sketches show how component design can influence the onset of corrosion.
57. Describe the essential reactions of *dry corrosion*, the causes of such reactions, and how they may be resisted.
58. Describe the essential reactions of *galvanic* or *bi-metallic* corrosion, the causes of such reactions, and how they may be resisted.
59. With the aid of sketches, compare the mechanisms of *uniform corrosion* with *preferential corrosion*, and give examples where these are likely to occur.

60. With the aid of sketches describe the mechanism of crevice corrosion, give an example of the circumstances under which it is liable to occur and how it can be prevented.
61. Describe the causes and methods of preventing:
    (*a*) pitting;
    (b) intergranular corrosion.
62. Discuss the effect of intergranular corrosion on welded joints, and describe how such corrosion can be avoided.
63. Describe the mechanism of *leaching*, together with TWO examples of this form of corrosion and suggest ways to prevent it.
64. With the aid of sketches, show what is meant by erosion corrosion and suggest ways in which it may be avoided.
65. Discuss the effect of corrosion fatigue on the properties of metals, quoting case-studies of structural failures resulting from this form of metal fatigue.
66. With the aid of examples, explain what is meant by:
    (*a*) hydrogen damage;
    (*b*) biological corrosion.
67. Discuss the factors affecting aqueous corrosion. Give examples of how this form of corrosion is affected by good and bad design and the choice of materials.
68. Describe, with examples, how corrosion can be prevented by the use of:
    (*a*) sacrificial anodes;
    (*b*) anodic passivation;
    (*c*) cathodic protection.
69. (*a*) Explain why the successful application of a corrosion-resistant coating is dependent upon the correct preparatory treatment of the substrate surface.
    (*b*) Discuss the effects of finishing processes on material properties.
70. Discuss how the following surfaces could be prepared ready for the application of a paint system.
    (*a*) Bright low carbon steel pressings protected by an oil film.
    (*b*) New structural steelwork.
    (*c*) Rusted structural steelwork which has been previously painted and is in need of refurbishment.
    (*d*) Aluminium alloy components.
71. With the aid of typical examples, discuss the use and relative merits of bitumastic, plastic and elastomer protective coatings.
72. With the aid of typical examples, discuss the use and relative merits of paint systems as decorative and protective coatings.

(Questions 73 to 84 inclusive are based mainly upon Chapter 7.)

73. With the aid of diagrams explain what is meant by:
    (*a*) Bohr model;
    (*b*) quantum shell;

(*c*) valency shell;
(*d*) Pauli's exclusion principle.

74. Explain how materials may be classified as conductors, insulators or semiconductors in terms of their 'energy levels', 'valency bands' and 'conduction bands'.
75. Explain how an electric current flows through a conductor material and distinguish between conventional current flow and electron current flow.
76. (*a*) Describe the effects of temperature change, composition, impurities and structure on the electrical properties of conductor and insulator materials.
    (*b*) Explain what is meant by *superconductivity*.
77. Discuss the essential differences between *intrinsic* and *extrinsic* semiconductor materials.
78. Explain how an intrinsic semiconductor material conducts electricity in terms of its n-type charge carriers and its p-type charge carriers.
79. Explain how extrinsic semiconductor materials are given their n-type or p-type characteristics by the addition of dopants.
80. (*a*) Explain the basic principle of operation of a junction diode and suggest TWO typical applications of this device, giving reasons for your choice.
    (*b*) Explain the basic principle of operation of a bipolar junction transistor and show how it may be used as:
    (i) a direct current amplifier;
    (ii) an alternating current (AF or RF) amplifier.
81. Explain what is meant by the terms:
    (*a*) atomic magnetic moment;
    (*b*) magnetic domain;
    (*c*) ferromagnetism;
    (*d*) ferrimagnetism;
    (*e*) paramagnetism;
    (*f*) diamagnetism.
    In the cases of (*c*) to (*f*) above give examples of typical materials with these characteristics.
82. With reference to their hysteresis diagrams, discuss the essential differences between 'hard' and 'soft' magnetic materials and, in each instance, suggest TWO typical applications, giving reasons for your choice.

(Questions 83 to 92 inclusive are based mainly upon Chapter 8.)

83. Discuss the relative merits of germanium and silicon as semiconductor materials.
84. Describe how high-purity polycrystalline silicon is produced.
85. Describe how polycrystalline silicon is converted into monocrystalline silicon by the *Czochralski (CZ)* process.

86. Describe how polycrystalline silicon is converted into monocrystalline silicon by the *float-zone (FZ)* process.
87. Explain why monocrystalline silicon, free from impurities and of high structural integrity, is required for the manufacture of solid-state electronic devices.
88. Explain how the Dash technique is used to prevent dislocations occurring during the production of monocrystalline silicon.
89. Explain at what stages during the manufacture of wafers dopants may be added, and how they may be added. Name the dopants in general use and describe the electrical characteristics they impose upon the silicon.
90. With the aid of diagrams, outline the production of 'wafers' from monocrystalline silicon and indicate the dimensional and electrical characteristics which may be expected from such wafers when the monocrystalline material has been produced by the CZ process.
91. With the aid of diagrams, explain:
    (*a*) how junction diodes are produced;
    (*b*) what is meant by an *epitaxial layer*;
    (*c*) how a bipolar junction transistor is produced by planar fabrication processes.
92. With the aid of diagrams, explain the basic principles of the manufacture of field effect transistors (FET) and simple integrated circuits, using metal-oxide-semiconductor (MOS) technology.

(Questions 93 to 102 inclusive are based mainly upon Chapter 9.)

93. Discuss the essential differences between gums, glues, and synthetic adhesives, and suggest typical examples where each would be used giving reasons for your choice.
94. (*a*) Explain what is meant by the terms *adherend* and *adhesive*.
    (*b*) Explain what is meant by *adhesion* and *cohesion*.
    (*c*) Explain what is meant by a *mechanical bond* and a *specific bond*.
95. With the aid of diagrams show how bonded joints may fail in service.
96. With the aid of diagrams, show how a welded assembly of your choice should be redesigned to make use of adhesive bonding.
97. Describe how the substrate joint surface should be prepared for adhesive bonding.
98. Name an example of each of the following types of thermoplastic adhesive and, in each instance, suggest a typical application giving reasons for your choice.
    (*a*) Heat activated adhesives.
    (*b*) Solvent activated adhesives.
    (*c*) Impact adhesives.
99. Explain the essential differences in processing and properties

between thermosetting adhesives and thermoplastic adhesives. Suggest typical applications where thermosetting adhesives should be used, naming the type of adhesive chosen and giving reasons for your choice.

100. Discuss, in detail, the factors which affect adhesion.
101. For TWO examples of your choice, explain the factors influencing your selection of a suitable adhesive and state the adhesive type chosen.
102. Discuss the health and safety hazards, and the precautions which must be taken when using:
    (*a*) paint systems on an industrial scale;
    (*b*) synthetic adhesives on an industrial scale.

(Questions 103 to 119 inclusive are based mainly upon Chapter 10.)

103. With the aid of diagrams show how crystal orientation notation is derived from the Miller indices for the crystal.
104. With the aid of diagrams, describe how plastic flow in crystalline materials occurs by:
    (*a*) block slip;
    (*b*) dislocation;
    (*c*) twinning.
105. (*a*) Explain how interaction occurs between dislocations.
    (*b*) Explain how the generation of dislocations occur in terms of:
        (i) The Frank-Read source;
        (ii) misorientation at the grain boundaries.
106. Discuss the inter-relationship between dislocation, work hardening, and dispertion hardening.
107. (*a*) Discuss stress-relief and recrystallisation in terms of dislocation theory.
    (*b*) Explain the significance of positive and negative 'climb'.
108. Discuss the essential differences between the deformation of metallic materials and polymeric materials in terms of their relative structures.
109. Explain the essential differences between *ductile* and *brittle* fracture.
110. With reference to the *Griffith Crack Theory*, explain why the real strength of a material is substantially lower than the theoretical strength based upon inter-atomic forces.
111. Explain how the *Griffith Crack Theory* can be adapted so that it can be applied to metallic materials.
112. Discuss the main factors affecting the nucleation and propagation of cracks and, with the aid of diagrams, show how component design can minimise the cracking.
113. Explain what is meant by the term *fracture toughness* and the factors, other than its inherent properties, which affect the fracture toughness of a material.

114. Explain why the S/N curve for a material can be misleading if there is a steady state component force in addition to the alternating load.
115. Explain how *Goodman diagrams* and *Soderberg diagrams* are prepared and how these diagrams are used to determine the operational parameters for a material if failure by fatigue is to be avoided.
116. Explain what is meant by *cumulative fatigue damage*, and how *Miner's law* is applied when predicting the fatigue life of a component under 'real life' service conditions.
117. Discuss the factors affecting the creep resistance of metals and describe what is meant by the terms:
     (*a*) creep limit (creep stress);
     (*b*) rupture stress;
     (*c*) Larson-Miller parameter.
118. Choose a component with which you are familiar, and select a suitable material for that component taking into account the following factors:
     (*a*) Properties.
     (*b*) Processing.
     (*c*) Commercial viability.
119. With reference to suitable examples of your own choice, explain the meaning of:
     (*a*) critical properties;
     (*b*) property limits;
     (*c*) cost per unit property;
     (*d*) merit rating;
     (*e*) cost-modified merit rating.

Index